Advanced Luminescent
Materials and Devices

Advanced Luminescent Materials and Devices

Editor

Dirk Poelman

Basel • Beijing • Wuhan • Barcelona • Belgrade • Novi Sad • Cluj • Manchester

Dirk Poelman
Department of Solid State Sciences
Ghent University
Ghent
Belgium

Editorial Office
MDPI AG
Grosspeteranlage 5
4052 Basel, Switzerland

This is a reprint of the Special Issue, published open access by the journal *Materials* (ISSN 1996-1944), freely accessible at: www.mdpi.com/journal/materials/special_issues/Luminescent_Materials_Devices.

For citation purposes, cite each article independently as indicated on the article page online and using the guide below:

Lastname, A.A.; Lastname, B.B. Article Title. *Journal Name* **Year**, *Volume Number*, Page Range.

ISBN 978-3-7258-1536-4 (Hbk)
ISBN 978-3-7258-1535-7 (PDF)
https://doi.org/10.3390/books978-3-7258-1535-7

© 2024 by the authors. Articles in this book are Open Access and distributed under the Creative Commons Attribution (CC BY) license. The book as a whole is distributed by MDPI under the terms and conditions of the Creative Commons Attribution-NonCommercial-NoDerivs (CC BY-NC-ND) license (https://creativecommons.org/licenses/by-nc-nd/4.0/).

Contents

About the Editor vii

Preface ix

Lily Bossin, Igor Plokhikh, Jeppe Brage Christensen, Dariusz Jakub Gawryluk, Yuuki Kitagawa and Paul Leblans et al.
Addressing Current Challenges in OSL Dosimetry Using MgB_4O_7:Ce,Li: State of the Art, Limitations and Avenues of Research
Reprinted from: *Materials* **2023**, *16*, 3051, doi:10.3390/ma16083051 1

George Kitis and Vasilis Pagonis
On the Need for Deconvolution Analysis of Experimental and Simulated Thermoluminescence Glow Curves
Reprinted from: *Materials* **2023**, *16*, 871, doi:10.3390/ma16020871 18

Georgios S. Polymeris
Thermally Assisted Optically Stimulated Luminescence (TA-OSL) from Commercial BeO Dosimeters
Reprinted from: *Materials* **2023**, *16*, 1494, doi:10.3390/ma16041494 34

Magdalena Biernacka, Alida Timar-Gabor, Zuzanna Kabacińska, Piotr Palczewski and Alicja Chruścińska
Trap Parameters for the Fast OSL Signal Component Obtained through Analytical Separation for Various Quartz Samples
Reprinted from: *Materials* **2022**, *15*, 8682, doi:10.3390/ma15238682 52

Alexander Vokhmintsev, Ahmed Henaish, Taher Sharshar, Osama Hemeda and Ilya Weinstein
Study of Spectrally Resolved Thermoluminescence in Tsarev and Chelyabinsk Chondrites with a Versatile High-Sensitive Setup
Reprinted from: *Materials* **2021**, *14*, 6518, doi:10.3390/ma14216518 67

Valentyn Laguta, Lubomir Havlak, Vladimir Babin, Jan Barta, Jan Pejchal and Martin Nikl
Charge Transfer and Charge Trapping Processes in Ca- or Al-Co-doped Lu_2SiO_5 and $Lu_2Si_2O_7$ Scintillators Activated by Pr^{3+} or Ce^{3+} Ions
Reprinted from: *Materials* **2023**, *16*, 4488, doi:10.3390/ma16124488 74

Adriana Popa, Maria Stefan, Sergiu Macavei, Laura Elena Muresan, Cristian Leostean and Cornelia Veronica Floare-Avram et al.
Photoluminescence and Photocatalytic Properties of MWNTs Decorated with Fe-Doped ZnO Nanoparticles
Reprinted from: *Materials* **2023**, *16*, 2858, doi:10.3390/ma16072858 88

Maxime Delaey, Seppe Van Bogaert, Ewoud Cosaert, Wout Mommen and Dirk Poelman
Neodymium-Doped Gadolinium Compounds as Infrared Emitters for Multimodal Imaging
Reprinted from: *Materials* **2023**, *16*, 6471, doi:10.3390/ma16196471 105

Guanyu Cai, Teresa Delgado, Cyrille Richard and Bruno Viana
ZGSO Spinel Nanoparticles with Dual Emission of NIR Persistent Luminescence for Anti-Counterfeiting Applications
Reprinted from: *Materials* **2023**, *16*, 1132, doi:10.3390/ma16031132 120

Victor Lisitsyn, Dossymkhan Mussakhanov, Aida Tulegenova, Ekaterina Kaneva, Liudmila Lisitsyna and Mikhail Golkovski et al.
The Optimization of Radiation Synthesis Modes for YAG:Ce Ceramics
Reprinted from: *Materials* **2023**, *16*, 3158, doi:10.3390/ma16083158 135

Victor Lisitsyn, Aida Tulegenova, Ekaterina Kaneva, Dossymkhan Mussakhanov and Boris Gritsenko
Express Synthesis of YAG:Ce Ceramics in the High-Energy Electrons Flow Field
Reprinted from: *Materials* **2023**, *16*, 1057, doi:10.3390/ma16031057 150

Andreas Herrmann, Mohamed Zekri, Ramzi Maalej and Christian Rüssel
The Effect of Glass Structure on the Luminescence Spectra of Sm^{3+}-Doped Aluminosilicate Glasses
Reprinted from: *Materials* **2023**, *16*, 564, doi:10.3390/ma16020564 161

Anastasiia Molchanova, Kirill Boldyrev, Nikolai Kuzmin, Alexey Veligzhanin, Kirill Khaydukov and Evgeniy Khaydukov et al.
Manganese Luminescent Centers of Different Valence in Yttrium Aluminum Borate Crystals
Reprinted from: *Materials* **2023**, *16*, 537, doi:10.3390/ma16020537 173

Suchinder K. Sharma, Jinu James, Shailendra Kumar Gupta and Shamima Hussain
UV-A,B,C Emitting Persistent Luminescent Materials
Reprinted from: *Materials* **2022**, *16*, 236, doi:10.3390/ma16010236 187

Hao Shen, Huabao Shang, Yuhan Gao, Deren Yang and Dongsheng Li
Efficient Sensitized Photoluminescence from Erbium Chloride Silicate via Interparticle Energy Transfer
Reprinted from: *Materials* **2022**, *15*, 1093, doi:10.3390/ma15031093 203

Radosław Lisiecki, Jarosław Komar, Bogusław Macalik, Michał Głowacki, Marek Berkowski and Witold Ryba-Romanowski
Exploring the Impact of Structure-Sensitivity Factors on Thermographic Properties of Dy^{3+}-Doped Oxide Crystals
Reprinted from: *Materials* **2021**, *14*, 2370, doi:10.3390/ma14092370 213

José Rosa, Mikko J. Heikkilä, Mika Sirkiä and Saoussen Merdes
Red Y_2O_3:Eu-Based Electroluminescent Device Prepared by Atomic Layer Deposition for Transparent Display Applications
Reprinted from: *Materials* **2021**, *14*, 1505, doi:10.3390/ma14061505 232

Oleksandr Bezvikonnyi, Ronit Sebastine Bernard, Viktorija Andruleviciene, Dmytro Volyniuk, Rasa Keruckiene and Kamile Vaiciulaityte et al.
Derivatives of Imidazole and Carbazole as Bifunctional Materials for Organic Light-Emitting Diodes
Reprinted from: *Materials* **2022**, *15*, 8495, doi:10.3390/ma15238495 242

About the Editor

Dirk Poelman

Prof. Dirk Poelman is a tenured senior professor leading the research group LumiLab, Department of Solid State Sciences at Ghent University, Belgium. He has experience in different fields of solid state physics research, including thin film deposition and optical characterization; photo-, electro- and cathodoluminescent materials; structural and electrical defects in semiconductors; x-ray analytical techniques and human vision. His current research is focused on phosphor materials for various applications, from glow-in-the-dark compounds, near-infrared emitters for bio-imaging to phosphors for thermometry and dosimetry. He also has an active research line on photocatalytic air purification.

Preface

Luminescent materials or phosphors have been studied for decades and were previously optimized for use in cathode ray tubes and fluorescent lamps. Luminescent materials have obtained a revived interest in lighting and display applications as wavelength converters for blue LEDs. Nowadays, very good materials are available for these applications, and advances are mostly incremental.

However, owing to these developments in materials research, new applications, requiring specific phosphor properties, have emerged. Wavelength ranges have expanded toward UV and near-IR emission; highly performant persistent phosphors have been developed for safety illumination and bio-imaging; phosphors have been proposed for anti-counterfeiting, energy storage, photocatalysis, thermometry, scintillators and dosimetry; and new synthesis and characterization techniques for luminescent materials have been developed. This text presents recent developments in a wide range of applications of luminescent materials. It compiles contributions from a number of leading scientists in the field, and should be an inspiration for both established and new researchers in this exciting area of research.

Dirk Poelman
Editor

Article

Addressing Current Challenges in OSL Dosimetry Using MgB$_4$O$_7$:Ce,Li: State of the Art, Limitations and Avenues of Research

Lily Bossin [1,*], Igor Plokhikh [1,2], Jeppe Brage Christensen [1], Dariusz Jakub Gawryluk [2], Yuuki Kitagawa [3], Paul Leblans [4], Setsuhisa Tanabe [5], Dirk Vandenbroucke [4] and Eduardo Gardenali Yukihara [1]

[1] Department of Radiation Safety and Security, Paul Scherrer Institute, Forschungsstrasse 111, 5232 Villigen, Switzerland; eduardo.yukihara@psi.ch (E.G.Y.)
[2] Laboratory for Multiscale Materials Experiments, Paul Scherrer Institute, Forschungsstrasse 111, 5232 Villigen, Switzerland
[3] National Institute of Advanced Industrial Science and Technology (AIST), Ikeda, Osaka 563-8577, Japan
[4] Radiology Division, Agfa NV, 2640 Mortsel, Belgium
[5] Graduate School of Human and Environmental Studies, Kyoto University, Kyoto 606-8501, Japan
* Correspondence: lily.bossin@psi.ch; Tel.: +41-56-310-54-30

Citation: Bossin, L.; Plokhikh, I.; Christensen, J.B.; Gawryluk, D.J.; Kitagawa, Y.; Leblans, P.; Tanabe, S.; Vandenbroucke, D.; Yukihara, E.G. Addressing Current Challenges in OSL Dosimetry Using MgB$_4$O$_7$:Ce,Li: State of the Art, Limitations and Avenues of Research. *Materials* 2023, 16, 3051. https://doi.org/10.3390/ma16083051

Academic Editor: Dirk Poelman

Received: 28 February 2023
Revised: 29 March 2023
Accepted: 3 April 2023
Published: 12 April 2023

Copyright: © 2023 by the authors. Licensee MDPI, Basel, Switzerland. This article is an open access article distributed under the terms and conditions of the Creative Commons Attribution (CC BY) license (https://creativecommons.org/licenses/by/4.0/).

Abstract: The objective of this work is to review and assess the potential of MgB$_4$O$_7$:Ce,Li to fill in the gaps where the need for a new material for optically stimulated luminescence (OSL) dosimetry has been identified. We offer a critical assessment of the operational properties of MgB$_4$O$_7$:Ce,Li for OSL dosimetry, as reviewed in the literature and complemented by measurements of thermoluminescence spectroscopy, sensitivity, thermal stability, lifetime of the luminescence emission, dose response at high doses (>1000 Gy), fading and bleachability. Overall, compared with Al$_2$O$_3$:C, for example, MgB$_4$O$_7$:Ce,Li shows a comparable OSL signal intensity following exposure to ionizing radiation, a higher saturation limit (ca 7000 Gy) and a shorter luminescence lifetime (31.5 ns). MgB$_4$O$_7$:Ce,Li is, however, not yet an optimum material for OSL dosimetry, as it exhibits anomalous fading and shallow traps. Further optimization is therefore needed, and possible avenues of investigation encompass gaining a better understanding of the roles of the synthesis route and dopants and of the nature of defects.

Keywords: MgB$_4$O$_7$; OSL; luminescence dosimetry; phosphors

1. Introduction

Despite a wealth of materials reported in the literature as potential candidates for optically stimulated luminescence (OSL) dosimetry, only two OSL materials, BeO and Al$_2$O$_3$:C, are routinely used for personal dosimetry [1]. This stems from the difficulty in satisfying the many requirements in dosimetry; see Yukihara et al. [2] for common pitfalls.

There are, however, areas in which those two materials have shown limitations and where the need for a new material was identified. Laser-scanning spatially-resolved dosimetry using OSL, for example, requires a material with a fast luminescence lifetime (of the order of microseconds, depending on the readout resolution and speed needed) to prevent pixel bleeding during fast laser scanning readout [3]. The rapid increase in particle therapy centers worldwide [4] requires detectors with a response independent of the linear energy transfer (LET) [5]. Al$_2$O$_3$:C quenches at high LET values and has an effective atomic number Z_{eff} = 11.3; therefore, it is not tissue equivalent. BeO is not produced specifically for dosimetry, and it has dosimetric properties that are not controlled during production. Furthermore, it is acutely toxic when inhaled, and, because of this, it is not suitable to be developed in versatile mediums where particles could come loose.

To address these issues, MgB$_4$O$_7$:Ce,Li has been put forward as a suitable candidate in response to these challenges, as it exhibits a luminescence lifetime of 31.5 ns and a reduced quenching at high LET values; in addition, its matrix is tissue equivalent (Z_{eff} = 8.5), it has not shown effects of aging, can be reused, and it is not toxic [6–8]. Additionally, De Souza et al. [9] tested the OSL dose response of MgB$_4$O$_7$:Ce,Li irradiated in photon beams (6 MV and 10 MV), in the dose range (0.1–100) Gy, and found a response proportional to the absorbed dose at those energies. Despite its novelty, there is now a substantial amount of information in the literature from different laboratories reporting on the synthesis of this material, its properties and its potential.

This contribution focuses on painting a comprehensive picture of the properties of MgB$_4$O$_7$:Ce,Li, combining results from the literature with new experimental results to fill in the gaps. The dosimetric performance of MgB$_4$O$_7$:Ce,Li will be weighted against those of Al$_2$O$_3$:C. Finally, where properties and understanding of the material were not satisfactory, future research directions are outlined. This helps us to assess whether MgB$_4$O$_7$:Ce,Li can respond the needs in areas where a new OSL material is needed. By doing so, we wil try to answer to the question: *"What is the added value of MgB$_4$O$_7$:Ce,Li to luminescence dosimetry, compared with existing solutions?"*.

2. Reported MgB$_4$O$_7$ Materials Developed for Luminescence Dosimetry

MgB$_4$O$_7$:Dy,Tm was first introduced as a new material for thermoluminescence (TL) dosimetry in 1980 [10]. With a sensitivity seven times higher than that of TLD-100, near tissue equivalence and a fading of less than 10% over 60 days, it exhibited a clear potential for dosimetry. From there on, the MgB$_4$O$_7$ matrix was found capable of hosting a number of dopants, mostly from the lanthanide series. The luminescence properties for a range of doped and co-doped MgB$_4$O$_7$ samples have been reported (see Table 1).

A systematic search for the best dopants in MgB$_4$O$_7$ was carried out by Yukihara et al. [11]. In their work, the luminescence properties of lanthanide-doped MgB$_4$O$_7$ samples were juxtaposed with theoretical models, predicting the energy levels created by their introduction in the matrix. From this work, MgB$_4$O$_7$:Ce,Li was singled out for its bright signal featuring an intense TL peak with few shallow traps. The TL signal of MgB$_4$O$_7$ compound with various dopants has been proposed for applications in personal dosimetry [12] and temperature sensing [13,14].

Several materials based on an MgB$_4$O$_7$ matrix have subsequently been developed for optically stimulated luminescence (OSL) dosimetry (see Table 1). For a comparison of the luminescence properties of MgB$_4$O$_7$ materials with different dopants, we refer to Yukihara et al. [11]. However, amongst those, MgB$_4$O$_7$:Ce,Li showed favorable characteristics in terms of OSL properties. Adding up to 10 % Li enhanced the brightness of the OSL signal by a factor of 10 [15]. Besides its signal brightness, the short lifetime (31.5 ns) of its luminescence signal originating from Ce^{3+} meant it was a suitable candidate for spatially resolved dosimetry [6,15]. Finally, recent work showed less ionization quenching in ion beams compared with other dosimeters [8].

Table 1. List of materials using an MgB$_4$O$_7$ matrix tested for their TL or OSL signal, as reported in the literature.

Dopants	Author	Measurement Method Tested
Ce,Li	Gustafson et al. [6], De Souza et al. [9], Yukihara et al. [11,15], Souza et al. [16], Kitagawa et al. [17]	TL, OSL
Dy,Li	Yukihara et al. [11], Souza et al. [16]	TL, OSL
Pr,Li	Yukihara et al. [11]	TL
Nd,Li	Yukihara et al. [11]	TL
Sm,Li	Yukihara et al. [11]	TL

Table 1. *Cont.*

Dopants	Author	Measurement Method Tested
Eu,Li	Yukihara et al. [11]	TL
Tb,Li	Yukihara et al. [11]	TL
Ho,Li	Yukihara et al. [11]	TL
Er,Li	Yukihara et al. [11]	TL
Tm,Li	Yukihara et al. [11]	TL
Yb,Li	Yukihara et al. [11]	TL
Gd,Li	Yukihara et al. [11], Annalakshmi et al. [18]	TL
Dy,Na	Karali et al. [19], Bahl et al. [20], Kitis et al. [21], De Oliveira et al. [22]	TL
Dy,Tm	Prokić [10], Souza et al. [13], Karali et al. [19]	TL
Ce,Gd	Altunal et al. [23]	TL, OSL
Ce,Na	Ozdemir et al. [24]	TL, OSL
Pr,Dy	Ozdemir et al. [25]	TL
Mn,Tb	Sahare et al. [26]	TL
Dy,Tb	Karali et al. [19]	TL
Nd,Dy	Souza et al. [13]	TL
Tm,Ag	González et al. [27]	TL
Dy,Mn	Zhijian et al. [28]	TL
Ag	Palan et al. [29]	TL, OSL
Ce	Dogan and Yazici [30]	TL
Mn	Prokic [31]	TL
Tb	Kawashima et al. [32]	TL
Dy	Barbina et al. [33], Campos and Fernandes Filho [34], Lochab et al. [35], Legorreta-Alba et al. [36], De Souza et al. [37], İflazoğlu et al. [38]	TL
Tm	Porwal et al. [39]	TL
undoped glass	Bakhsh et al. [40]	TL

3. Prospective Applications

MgB_4O_7:Ce,Li has been proposed for the following applications, where its properties could complement the existing materials for OSL dosimetry.

3.1. Spatially-Resolved Dosimetry

Spatially-resolved dosimetry based on laser scanning requires an OSL material with a fast luminescence lifetime ($<\mu s$) for faster scanning [7]. The luminescence emission of Al_2O_3:C exhibits a longer lifetime component (35 ms) that persists after the laser has moved away from the respective part of the film. In spatially-resolved dosimetry, for example, where the laser is scanned across a sheet of materials at a reasonable speed (ca 2 μs, see Crijns et al. [41]), this needs to be accounted for, subtracted and corrected using pixel bleeding algorithms [3]. Although BeO exhibits a lifetime of 27 μs [42], its toxicity means that it is confined to a ceramic form, and producing films at low cost can be challenging [43].

In contrast, the luminescence emission of MgB_4O_7:Ce,Li is dominated by a fast component (31.5 ns). This not only means that MgB_4O_7:Ce,Li requires fewer corrections when used for laser-scanning spatially resolved dosimetry [7], but also that the proportion of signal usable for spatially-resolved dosimetry is de facto greater for MgB_4O_7:Ce,Li compared with Al_2O_3:C.

3.2. Dosimetry in the kGy Range

Doses in the kGy range may have to be measured for medical sterilization [44–46], food processing [47,48] or tomography imaging of biological samples [49–51]. In those contexts, luminescence dosimetry offers many advantages compared with other dosimetry methods. It is dose-rate independent [52–54] and detectors can be produced as small as 1 mm^2 to be fitted in narrow fields or at the sample's position. However, the two materials currently used for OSL dosimetry, Al_2O_3:C and BeO, each exhibit a saturation limit of the order of 100s Gy [2], hindering their use in the high dose range.

3.3. Dosimetry for Ion Beam Therapy

Although commercially available OSL detectors present many advantages for ion beam dosimetry (e.g., practicality, spatially-resolved dosimetry, insensitivity to dose-rate and magnetic fields), they are not exempt from the reduced relative response with increasing linear energy transfer (LET) exhibited by solid state detectors, the so-called ionization quenching [5,55–59]. To circumvent this, the possibility of measuring the LET of the incident particle using Al_2O_3:C and the ratio of its two emission bands has been used to apply LET correction factors [57,60]. As already hinted by its high saturation limit [15], MgB_4O_7:Ce,Li exhibits a reduced quenching to high-LET compared with Al_2O_3:C and provides a negligible LET-correction for dosimetry in clinically relevant proton beams [8].

4. Material and Methods

The original measurements presented here were obtained using MgB_4O_7:Ce (0.3%), Li (10%) samples synthesized through solution synthesis, solid state synthesis or glass synthesis, where the dopant concentration is the nominal concentration in mol% with respect to the Mg concentration added to the initial reagents. The samples were produced from a mixture of H_3BO_3 (Alfa Aesar, 5N5), $Mg(NO_3)_2 \times 6H_2O$ (Alfa Aesar, 5N), $LiNO_3$ (ROTH, 2N5) and $Ce(NO_3)_3 \times 6H_2O$ (Thermo scientific 2N5) in 1:6:0.1:0.003 molar ratio. The starting materials were thoroughly ground in agate mortar, and the resulting mixture was split into three parts and used for glass synthesis (according to Kitagawa et al. [17]), solution synthesis (according to Gustafson et al. [6], except that the urea was omitted, as it was not found to significantly enhance the brightness of the signal), and solid state reaction. The samples were hence produced from the same mixture of starting reagents, reducing the variation in the properties due to contamination of the reagents.

The luminescence measurements were carried out using a Risø TL/OSL reader TL/OSL-DA-20 (DTU Nutech, Denmark). The TL and OSL signals were detected using a bi-alkali photomultiplier tube (PMT) (model 9235QB, Electron Tubes Ltd., Uxbridge, UK). TL emission spectra were collected using an Andor iXon Ultra 888 EMCCD attached to Andor's Kymera 193i spectrometer (grating 150 lines/mm with a center wavelength at 500 nm, CCD pre-cooled to $-60\,^\circ$C). A spectral correction obtained using a calibration lamp (Bentham, CL2 irradiance standard) was applied. The system was also equipped with a photon timer (Photon Timer PicoQuant TimeHarp 260, 0.25 ns base resolution, deadtime <2 ns) for time-stamped time-resolved OSL measurements. Hoya U-340 filters (7.5 mm thickness, Hoya Corp. Tokyo, Japan), Edmund Optics UV/VIS neutral density filters (ND OD 2.0, 3.0 mm, EO 47-210) or a silica window were placed in front of the PMT during OSL or TL readouts. The detection setup specific to each measurement is indicated in the caption of each figure. Continuous-wave OSL measurements (CW-OSL) were conducted at a stimulation power of 90% (LEDs: 470 nm 2xSMBB470-1100-TINA-RS GG420 maximal power: 90 mW/cm^2) over 300 s, linear-modulated OSL (LM-OSL) measurements were conducted by linearly ramping up the power of the diodes from 0% to 90% over 1000 s of stimulation, and the TL measurements were recorded at a rate of 1 $^\circ$C/s. UV (365 nm, 11 W) and green (525 nm, 40 mW/cm^2) LEDs were also used to bleach the samples. Irradiations were performed using a 1.48 GBq ^{90}Sr/^{90}Y beta source integrated in the Risø reader. The source was calibrated in air kerma for MgB_4O_7:Ce,Li films relative to a ^{137}Cs reference irradiation.

The MgB_4O_7:Ce,Li film samples used to assess the dose response were prepared by Agfa (Belgium) by mixing samples prepared by the solution method with a binder and spreading onto a plastic film, in a fashion similar to that described by Shrestha et al. [7]. The 22 µm median particle size phosphor was dispersed in Kraton FG1901 in a toluene solution. The lacquer was coated on white PET with a bar coater and dried to obtain phosphor coatings with thicknesses between 50 and 100 µm. The OSL signal of MgB_4O_7:Ce,Li was compared with that of similar Al_2O_3:C films, as described by Ahmed et al. [3].

The mass energy absorption coefficient of MgB_4O_7:Ce,Li was calculated for different nominal concentrations of cerium by adding up the mass energy absorption coefficient of each one of the elements present, weighted by their respective atomic weight fraction. This was divided by the mass energy coefficient of water, obtained in a similar way, to get a response respective to that of water. The mass absorption energy coefficients were obtained from the NIST database [61].

5. State-of-the-Art
5.1. Luminescence Properties
5.1.1. Luminescence Spectroscopy

MgB_4O_7:Ce,Li radioluminescence (RL) emission spectra reported by Gustafson et al. [6] showed a peak at 350 nm for samples synthesized through solution combustion. They attributed this to the Ce^{3+} emission. This emission band was also found in OSL, TL and photoluminescence spectra and ascribed to the 5d-4f transition of Ce^{3+}. This is consistent with the RL spectra obtained by Kitagawa et al. [17] for glass synthesis samples (340 nm peak) but differs from those reported by Souza et al. [16] for samples synthesized using solid stated synthesis, where a 412 nm emission was predominant. As the only apparent difference between the three samples listed above appeared to be the synthesis route, we gathered TL spectra for samples synthesized for the present study using solution synthesis, solid state synthesis and glass synthesis.

Figure 1 shows the emission spectra of the main TL glow peak (integrated over the temperature range 225–275 °C) for MgB_4O_7:Ce,Li synthesized by the three different routes (solution synthesis, solid state synthesis and glass synthesis), as described in Section 4. The emission is similar in all three samples, with a main peak at 360 nm across all three samples. These data are consistent with those of Gustafson et al. [6] and Kitagawa et al. [17]. Furthermore, we did not observe a shift in the UV emission as a function of the synthesis route, indicating the emission was not synthesis-dependent and that Ce^{3+} was incorporated in the matrix.

5.1.2. Signal Intensity

Figure 2 compares the intensity of the luminescence signal per milligram of powder sample under blue stimulation, green stimulation and thermal stimulation with that of Al_2O_3:C. Regardless of the light stimulation source, the MgB_4O_7:Ce,Li brightness is comparable to that of Al_2O_3:C. However, if a broad BG-39 filter is used for TL measurements, Al_2O_3:C's signal is brighter than that of MgB_4O_7:Ce,Li, because the Hoya U-340 filter blocks most of the main emission of Al_2O_3:C, which is a broad band centered at 420 nm. The OSL and TL signals of BeO and MgB_4O_7:Ce,Li also show similar brightness under blue light stimulation (Supplementary Materials Figure S1). The prototype MgB_4O_7:Ce,Li films produced by Agfa NV yielded a signal 25% weaker than Al_2O_3:C films under blue light stimulation and 40% weaker under green light stimulation, if the signal is taken as the integral of the first 100 s of illumination. Although the exact amount of powder for both of those films is not known and differs from the two films, the results are intended to demonstrate that comparable performances can be achieved.

5.1.3. Step-Annealing

Step-annealing measurements were conducted by irradiating MgB_4O_7:Ce,Li and Al_2O_3:C samples with a β dose of 35 mGy, preheating the sample to a temperature in the range 30–350 °C and recording the subsequent OSL signal.

Figure 3 shows the results, normalized by the OSL intensity measured at room temperature (25 °C). The sharp drop in OSL intensity at 200 °C for MgB_4O_7:Ce,Li indicates that the OSL signal originates mostly from traps associated to the main TL peak and not from shallower traps (Figure 3a). This is similar to the behavior observed in Al_2O_3:C, where the OSL signal also appears to be associated with traps linked to a 200 °C temperature region, corresponding to its main TL peak (Figure 3b).

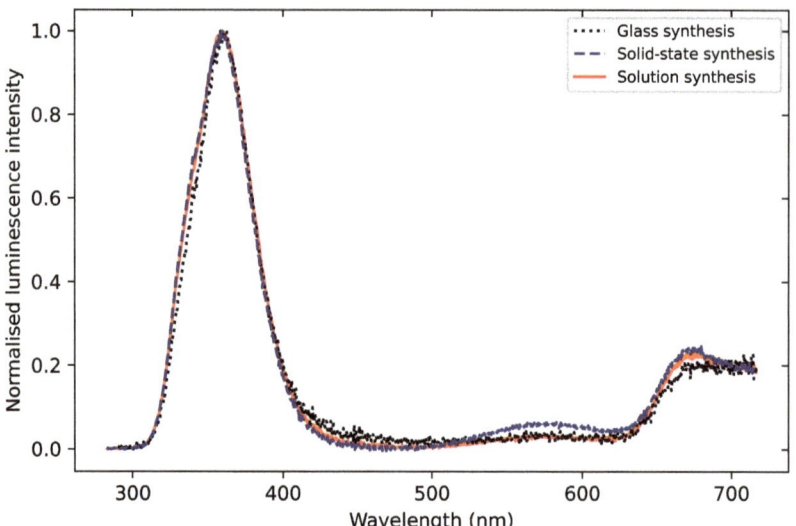

Figure 1. TL spectra of MgB$_4$O$_7$:Ce,Li samples prepared through solution synthesis, solid state synthesis and glass synthesis, following β irradiation 35 Gy. The spectra were normalized by their respective maximum intensity and were obtained through integration over the temperature range 225–275 °C, corresponding to the TL peak maximum. The 360 nm emission corresponds to the Ce^{3+} emission. The spectra were corrected for instrument response. Detection unit: Andor spectrometer; silica window; heating rate: 1 °C/s.

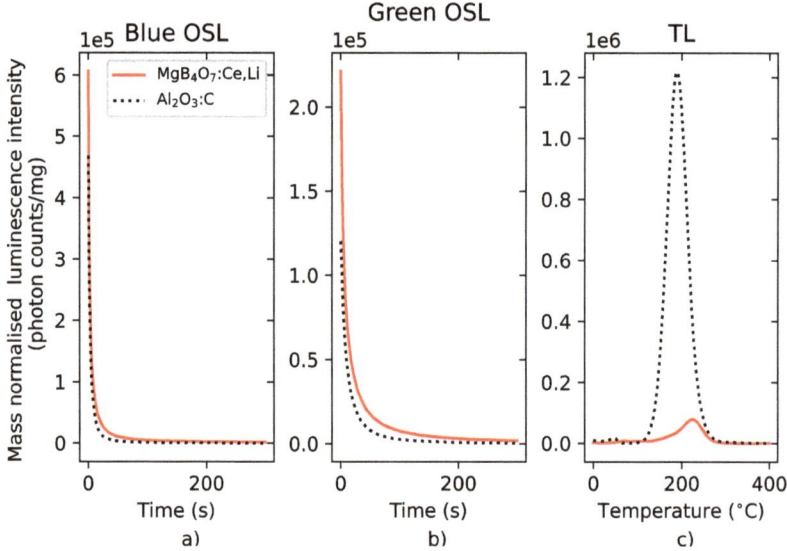

Figure 2. Intensity comparison of the mass normalized signals of MgB$_4$O$_7$:Ce,Li and Al$_2$O$_3$:C under continuous-wave blue stimulation (**a**), green stimulation (**b**) and thermal stimulation (**c**) following a 350 mGy β irradiation. Detection unit: PMT 9235QB; (**a**,**b**) Hoya U-340 filter, (**c**) Schott BG-39 filter.

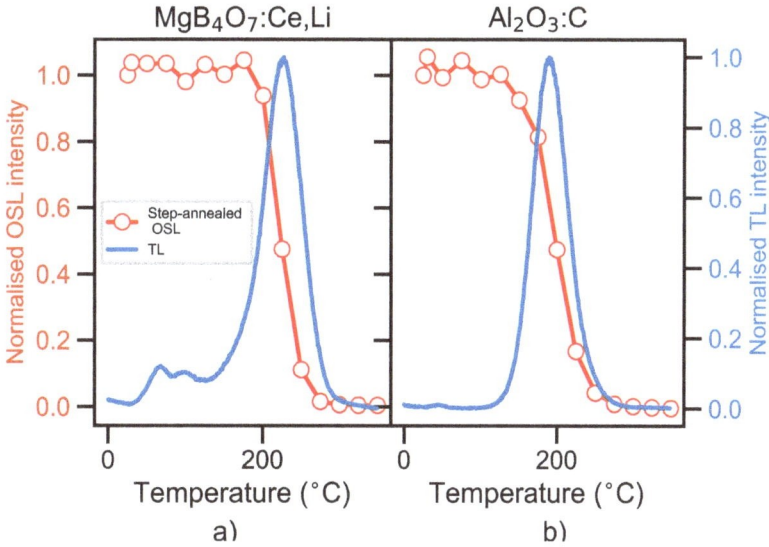

Figure 3. OSL signal of MgB$_4$O$_7$:Ce,Li (**a**) and Al$_2$O$_3$:C (**b**) following 35 mGy β irradiation and preheat to 30–350 °C (open red circles). The OSL signal, taken as the integral of the OSL decay curve minus a background subtraction, was normalized by the signal at room temperature (25 °C, no preheat). These data are compared with the TL glow curve of each material (blue continuous line). The TL was recorded following a 35 mGy β irradiation, at a heating rate of 1 °C/s, and each curve was normalized by its respective maximal intensity. Detection unit: PMT 9235QB; Hoya U-340 filter.

5.1.4. Time-Resolved Luminescence

Whereas Al$_2$O$_3$:C is characterized by a fast and a slow component (7 ns and 35 ms respectively), MgB$_4$O$_7$:Ce,Li's luminescence signal exhibits one dominating component of the order of ns, associated with the Ce^{3+}'s emission. This is illustrated in Figure 4, where the photon arrival time distribution curves are shown for both materials under stimulation pulses of 10 µs. The timing of the pulses and stimulation period does not allow for a full decay of the slow component of Al$_2$O$_3$:C, and this results in a higher background during the off time. In contrast, MgB$_4$O$_7$:Ce,Li's signal decays within the time resolution of the LEDs following the end of the pulse, which produces a lower background in comparison to Al$_2$O$_3$:C. The luminescence lifetime could not be extracted from this data, as the LED rise and fall time was too slow compared with the Ce^{3+} lifetime. The portion of the signal that would be accounted for in spatially resolved dosimetry is displayed as the difference in intensity between the signal as the end of the stimulation pulse and the background and is around five times greater for MgB$_4$O$_7$:Ce,Li than for Al$_2$O$_3$:C. Similarly, the OSL signal of BeO exhibits a much longer lifetime than MgB$_4$O$_7$:Ce,Li, which would extend the duration of measurements and reduce the portion of useful signal (Supplementary Materials Figure S2).

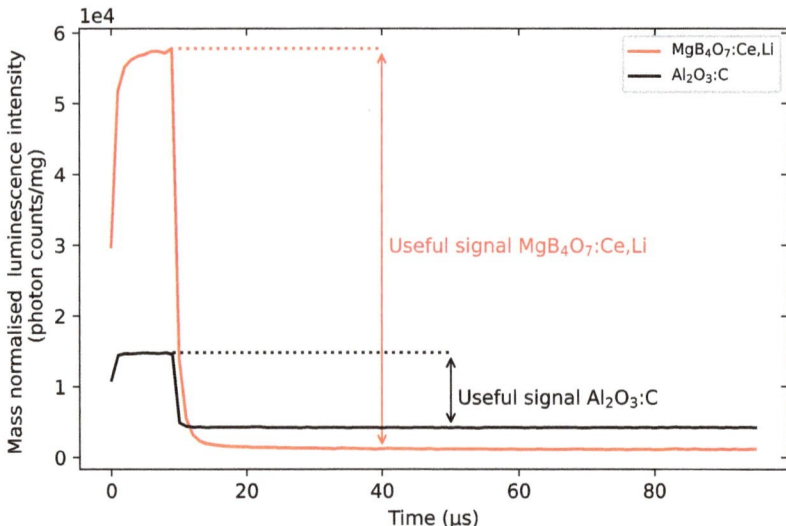

Figure 4. Photon arrival time distribution curves of MgB_4O_7:Ce,Li and Al_2O_3:C following blue stimulation pulse 10 µs, following a 350 mGy β irradiation. Detection unit: PMT 9235QB; Hoya U-340 filter.

5.1.5. Dose Response

Contrasting data exists regarding the dose-response behavior of MgB_4O_7:Ce,Li. Where the results from Souza et al. [16] indicate saturation around 100 Gy, those of Yukihara et al. [15] indicate supralinearity above 100 Gy and no saturation up to 1000 Gy. Samples produced using the solution method were selected to test the dose response, as they exhibited the highest intensity. They were tested in the framework of this study under linear-modulated stimulation (LM-OSL) to prevent the saturation of the PMT.

The dose-response behavior is shown in Figure 5. As indicated by the dotted line representing the linearity region, the dose response is linear up to 100 Gy before exhibiting a supra-linear behavior above 100 Gy and saturating around 7000 Gy. This is in agreement with the data reported by Gustafson et al. [6] for samples prepared using solution combustion.

5.1.6. Fading

Differing fading behaviors have been reported in the literature, and it is unclear whether they result from different synthesis routes. Yukihara et al. [15] observed an OSL signal that decays 10–15% in the first 72 h following irradiation, stabilizing afterwards. Since the OSL is associated with the 200 °C TL peak (see Section 5.1.3), this would indicate anomalous fading. Souza et al. [16] reported less than 1% fading of the OSL signal following 40 days of storage in the dark. TL fading has so far only been presented by Kitagawa et al. [17] for samples synthesized through glass synthesis, for which anomalous fading, in the temperature region >200 °C of the TL glow curve, was not detected.

Preliminary results were obtained for the TL fading of the main peak (200–300 °C) of samples synthesized through solid state, solution and glass synthesis. The fading data of the integral of the main TL peak (200–300 °C), shown in Figure 6a, points towards a difference in loss of signal of 3% after 24 h between the sample with the least fading (solid state synthesis) and the one with the most (glass synthesis), and longer fading measurements would be needed to better constrain this difference. Thermal fading for TL peaks in the 200–300 °C region is not expected for samples stored at room temperature in the dark. Therefore, this therefore points towards the occurrence of anomalous fading in all three samples.

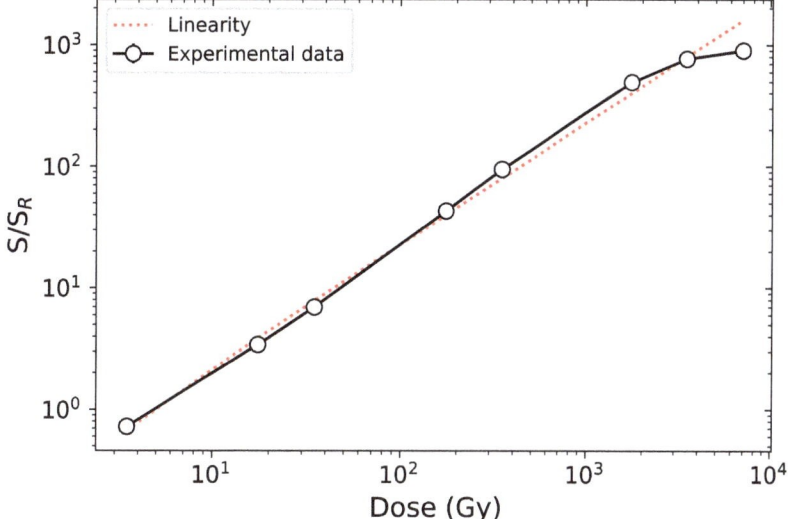

Figure 5. Blue LM-OSL dose response of MgB$_4$O$_7$:Ce,Li synthesized through solution method following β irradiation with doses in the 3.5–7000 Gy range. Each datapoint represents the average response of three samples (open circles), the continuous lines, the interpolation between experimental datapoints. The response to dose was calculated as the integrated LM-OSL signal over the first 200 s of stimulation (S), normalized to the sample-specific LM-OSL response to a test dose of 3.5 Gy (S_R). The test dose response was measured before the administration and measurement of the nominal β dose, to minimize sensitivity changes induced by a hard-to-bleach component. The samples were readout one week following irradiation, to avoid signal originating from the shallower traps. The dotted line indicates linearity. Detection unit: PMT 9235QB; Hoya U-340 and ND OD 2 filters. The uncertainties were evaluated as the standard deviation of the signal from three samples, but are too small to be visible on the graph.

Furthermore, the comparison of the TL and OSL fading of a sample synthesized through solution synthesis corroborates an OSL signal originating from the 175–250 °C region of the TL glow curve. In the data displayed in Figure 6b, the OSL signal was calculated as the integral over the entire OSL curve, and its fading behavior is plotted alongside the TL fading in the TL region 175–250 °C, as pulse-anneal measurements had indicated traps in this TL region as also responsible for the OSL signal (Section 5.1.3). The fading behaviors of the OSL and TL signals in the corresponding regions are identical, within uncertainties. Further measurements of multiple aliquots over extended periods of time will be conducted to confirm those results.

5.1.7. Bleachability

Figure 7 illustrates the behavior of the TL signal of MgB$_4$O$_7$:Ce,Li and Al$_2$O$_3$:C following bleaching with either blue, green or UV light for various durations (0–120 s). From these graphs, it is clear that, regardless of the wavelength of the bleaching light, the main MgB$_4$O$_7$:Ce,Li trap has a lower optical cross-section than Al$_2$O$_3$:C, which results in a harder-to-bleach signal. Whereas 50 s of blue illumination was sufficient to bleach the main dosimetric trap of Al$_2$O$_3$:C, it only resulted in a loss of signal of 55% in MgB$_4$O$_7$:Ce,Li. Furthermore, exposure to light resulted in a photo-transfer process in MgB$_4$O$_7$:Ce,Li, giving rise to a peak at 75 °C. The reduced bleachability indicates an OSL process of reduced efficiency.

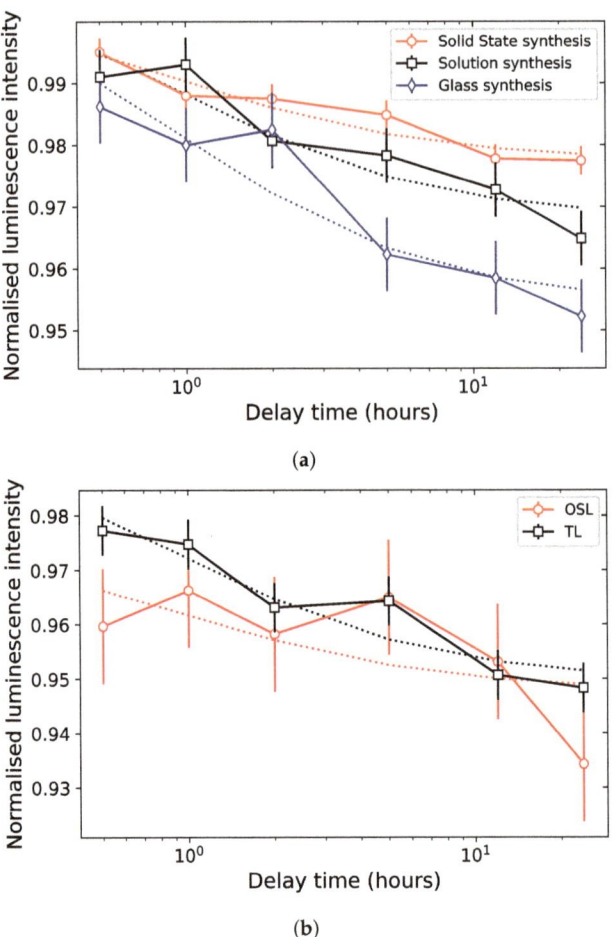

Figure 6. (a) Normalized loss of signal of the main TL peak of MgB$_4$O$_7$:Ce,Li samples synthesized through solid state synthesis (red circles), solution synthesis (black squares) and glass synthesis (blue diamonds). The signal was calculated as the integral in the temperature region 200–300 °C. (b) Comparison of the loss of OSL signal (red circles) of a sample synthesized through solution synthesis, as calculated integrating over the entire OSL decay curve and the loss of TL signal (black squares) of the same sample, calculated as the integral in the region 175–250 °C, to which the OSL signal was associated by pulse-annealed data (see Section 5.1.3). The TL and OSL signals were obtained following a 0.35 Gy β irradiation and storage in the dark (0.5–24 h) and normalized to the signals following a similar irradiation and readout within 5 min. The dotted lines indicates a function $y(t) = a + \frac{b}{1+t}$ fitted to the experimental datapoints. The uncertainties were evaluated as the fits' residuals standard deviation. Detection unit: PMT 9235QB; Hoya U-340 filter.

In practice, this means that, for applications in personal dosimetry, for example, MgB$_4$O$_7$:Ce,Li dosimeters will have to be subjected to a more aggressive (e.g., longer and/or at shorter wavelengths) bleaching to be zeroed and re-used. It also indicates that the same amount of stimulation energy is capable of releasing less trapped charges in MgB$_4$O$_7$ than in Al$_2$O$_3$:C, which can be a disadvantage for laser-scanning readouts. On the other hand, neighboring pixels may be bleached less by scattered light when scanning neighboring regions of the film.

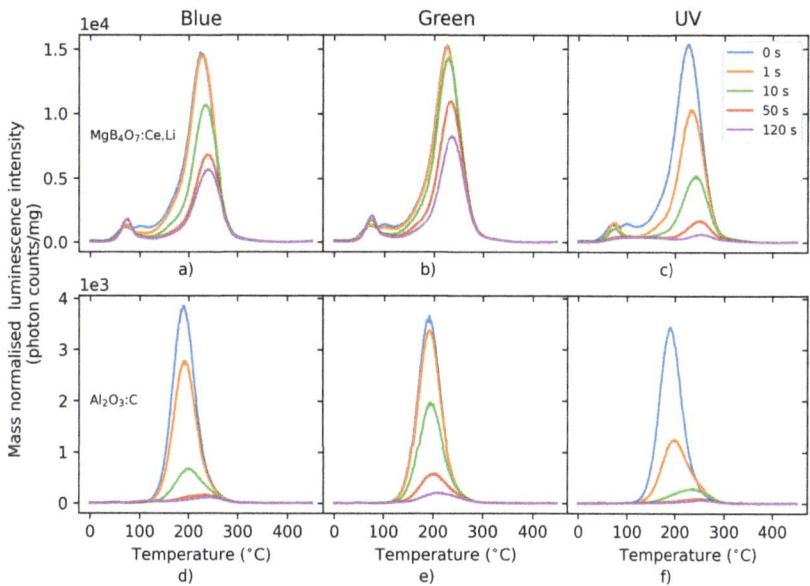

Figure 7. TL signal of MgB$_4$O$_7$:Ce,Li (**a–c**) and Al$_2$O$_3$:C (**d–f**) following bleaching using blue (**a,d**), green (**b,e**) or UV (**c,f**) LEDs for 0–120 s. Detection unit: PMT 9235QB; Hoya U-340 filter.

5.1.8. Photon Energy Response

The photon energy response reported by De Souza et al. [9] for MgB$_4$O$_7$:Ce,Li pellets synthesized by solid-state synthesis shows an over-response of a maximum of 20% below 83 keV and close to unity above this threshold (83–1250 keV). Barbina et al. [33] and Prokić [10] found an energy over-response for MgB$_4$O$_7$:Dy of ca 1.5 and 2 at 50 keV, respectively.

We calculated mass energy absorption coefficients of MgB$_4$O$_7$:Ce,Li for various contents of cerium (0–1%). Normalized by the water mass energy absorption coefficients, they can be used as a predictor of the absorbed dose energy response (see Ch. 3, p. 123 in Yukihara and McKeever [62]), which is illustrated in Figure 8. The peak around 50–60 keV is caused by the presence of cerium—an element with a higher atomic number, and whose K-edge absorption occurs at 40 keV.

Although these results remain to be experimentally confirmed, they show that, even in small proportion, the presence of cerium causes an over-response with a maximum at 60 keV. For a cerium content of 0.3%, this is calculated to be of a factor of 1.5. Moreover, a relatively small increase in the concentration of cerium significantly impacts the energy response. For example, for a cerium concentration of 1%, our calculations predict an over-response by a factor of 2.5, compared with 1.5 for 0.3% cerium.

It is therefore recommended to keep the cerium concentration as low as possible, while guaranteeing the desired luminescent and dosimetric properties.

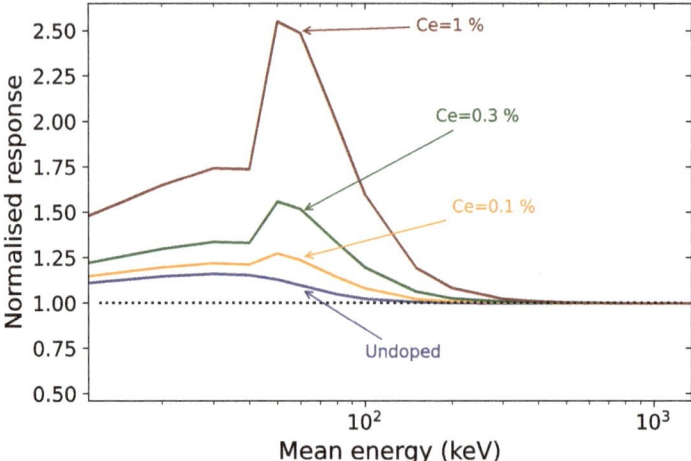

Figure 8. Calculated mass energy absorption coefficients of MgB$_4$O$_7$:Ce,Li for various contents of cerium (0–1%) normalized to those of water, $\frac{\mu_{en}/\rho}{\mu_{en}/\rho(Water)}$.

6. Current Challenges and Avenues of Research

6.1. Current Challenges

6.1.1. Eliminating Shallow Traps

The TL signal of MgB$_4$O$_7$:Ce,Li is composed of a main TL peak centered around 220 °C and shallower traps between 50–150 °C. Although step-annealing data shows that the latter traps do not directly contribute to the OSL signal (Figure 3), they could delay the luminescence emission by re-trapping charges or acting as competing centers. This is supported by the bleaching of the TL glow curve (Figure 7), where a photo-transfer peak is produced upon blue and green light exposure, in the shallow trap region.

At the present stage, the exact nature of the trapping centers in MgB$_4$O$_7$:Ce,Li is unknown, and therefore the species introducing those traps remain unidentified. It is possible that they originate from amorphous regions, or grain boundaries, as the samples produced so far are not single crystals.

6.1.2. Reducing the Fading

The data presented in Figure 6 indicates the presence of anomalous fading for samples synthesized through three different routes, whether it is for the main TL peak or OSL. Longer-term fading and isothermal decay experiments will be carried out to better constrain the differences in fading between samples, and linking this with a deeper material characterization could pinpoint why this fading rate differs from one sample to another.

One can correct for the loss of the signal when the delay time between irradiation and measurement is known. Alternatively, one could delay each measurement for approximately two days following the administration of a dose for the stable component to dominate the signal, although this would increase the overall delay to final dose results.

6.1.3. Improving the Bleachability

MgB$_4$O$_7$:Ce,Li samples exhibit poorer bleachability compared with Al$_2$O$_3$:C (Figure 7). In terms of operationability, this means that the material will have to be subjected to prolonged bleaching treatment or the use of a shorter wavelength in order to be reused, and that a given amount of stimulation energy releases less signal. This may be an intrinsic property of MgB$_4$O$_7$:Ce,Li, resulting from a lower optical cross-section of the trapping centers. Further measurements will aim at testing the bleachability of the OSL signal under UV light.

6.2. Future Research Directions

6.2.1. Understanding the Material

The results presented here are purely empirical, and a complete comprehensive theoretical framework is yet to be built. This would help not only in further optimizing the material but would also guide the design of future OSL materials.

In general, there seems to be a lack of understanding from the material properties' side. For example, this includes the nature of the defects giving rise to the trapping centers, as Ce^{3+} only acts as a recombination center. X-ray absorption data could help to constrain the local environment of the cerium ions and evaluate the degree of crystallinity in the samples. The nature of the role of lithium also remains obscure; the 10% added as dopant is unlikely to enter the lattice in its entirety.

It has been reported that an excess of boric acid is essential to obtain the pure phase [11]. Evaporation has been advanced as an explanation for high-temperature glass synthesis [17], but the same observation has been made for synthesis requiring temperature below the evaporation threshold of boric acid.

6.2.2. Investigating the Role of the Synthesis Route

Although the data in Figure 1 has shown that the synthesis route does not influence the TL emission spectra, more research is needed to understand the influence of the synthesis on the luminescent and dosimetric properties of MgB_4O_7:Ce,Li.

The preliminary results of the TL spectra for three samples synthesized through solid state synthesis, solution synthesis and glass synthesis did not seem to indicate an influence of the synthesis route on the emission spectra (Figure 1). However, the mass normalized TL glow curves for glass, solid-state and solution syntheses shown in Figure 9 exhibit differences, which may be related to the synthesis route. Glass synthesis, for example, appeared to produce a relative larger amount of shallower traps that resulted in a weaker 200 °C peak, compared with solution synthesis or solid-state synthesis.

Differences in TL curves for MgB_4O_7:Dy samples synthesized either through solid-state or precipitation methods have already been reported by De Souza et al. [37]. They found significant differences in the TL glow curve, with an additional higher temperature peak being created through precipitation synthesis.

Finally, samples synthesized through solid state synthesis appeared to exhibit less fading than samples synthesized through glass synthesis (Figure 6). Further work will focus on building a more complete picture of the influence of the synthesis route on the luminescence properties but also on understanding the structural differences causing possible variation in luminescence properties.

6.2.3. Photon Energy Response

If the calculations shown in Figure 8 are experimentally confirmed, they will indicate that, despite the MgB_4O_7 matrix being tissue equivalent and the cerium concentration being low, it can result in an over-response for low photon energies. This should be investigated for samples with different cerium concentrations to determine the optimum cerium concentration that increases the photon energy response as little as possible, while preserving the luminescent and dosimetric properties.

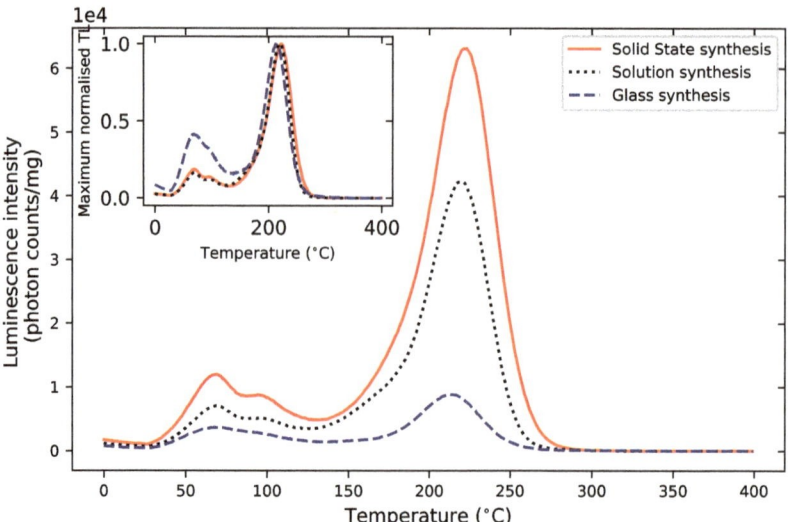

Figure 9. Mass normalized TL intensity of MgB$_4$O$_7$:Ce,Li samples prepared through solution synthesis, solid state synthesis and glass synthesis, following a 350 mGy β irradiation. The inset shows the TL glow curves normalized by their maximum. Detection unit: PMT 9235QB; Hoya U-340 filter.

7. Conclusions

In this work, we summarized the dosimetric properties of MgB$_4$O$_7$:Ce,Li, outlining and comparing literature results and complementing them by original measurements, using Al$_2$O$_3$:C as a reference material. The comparison of literature results highlighted some discrepancies, for example, in terms of fading or emission spectroscopy. We hypothesized that this could be caused by differences in synthesis routes. A side-by-side comparison of samples through solid-state, solution or glass did not show differences in terms of emission spectra, but a slight variability in terms of fading behavior. Overall, whereas the OSL signal of MgB$_4$O$_7$:Ce,Li is comparable to that of Al$_2$O$_3$:C in terms of brightness, it underperforms in terms of bleachability and fading. Although the OSL signal of MgB$_4$O$_7$:Ce,Li appears to originate from traps associated with a similar temperature region of the TL glow curve, anomalous fading seems to cause a loss of signal of 2–6% within a day. However, MgB$_4$O$_7$:Ce,Li also exhibits a saturation limit in the kGy range and a higher useful signal in time-resolved measurement, confirming its potential for dosimetry in the kGy range and spatially-resolved dosimetry.

Future work will focus on better understanding whether the luminescence properties differ from one synthesis route to another and experimentally assessing the influence of the cerium content on the energy response. Finally, material characterization tools will be applied to gain a better understanding of the structural properties.

Supplementary Materials: The following supporting information can be downloaded at: https://www.mdpi.com/article/10.3390/ma16083051/s1, Figure S1: OSL and TL intensity—comparison with BeO; Figure S2: Time-resolved luminescence—comparison with BeO.

Author Contributions: Conceptualization, L.B. and E.G.Y.; methodology, L.B., I.P., D.J.G., Y.K., S.T. and E.G.Y.; formal analysis, L.B. and I.P.; investigation, L.B., I.P. and D.J.G.; resources, L.B., I.P., D.J.G., P.L., D.V. and E.G.Y.; data curation, L.B. and I.P.; writing—original draft preparation, L.B.; writing—review and editing, L.B., I.P., J.B.C., D.J.G., Y.K., P.L., S.T., D.V. and E.G.Y.; supervision, L.B., D.J.G. and E.G.Y.; project administration, D.J.G.; funding acquisition, L.B. and D.J.G. All authors have read and agreed to the published version of the manuscript.

Funding: This work was funded by a Paul Scherrer Institute research grant (no. 2021 01346). The Risø TL/OSL-DA-20365 reader (DTU Nutech, Denmark) was acquired with partial support from the Swiss National Science Foundation (R'Equip project 206021 177028).

Institutional Review Board Statement: Not applicable.

Informed Consent Statement: Not applicable.

Data Availability Statement: The data that support the findings of this study are available from the corresponding author upon reasonable request.

Conflicts of Interest: The authors declare no conflict of interest.

References

1. Yukihara, E.G.; McKeever, S.W.; Andersen, C.E.; Bos, A.J.; Bailiff, I.K.; Yoshimura, E.M.; Sawakuchi, G.O.; Bossin, L.; Christensen, J.B. Luminescence dosimetry. *Nat. Rev. Methods Prim.* **2022**, *2*, 1–21. [CrossRef]
2. Yukihara, E.G.; Bos, A.J.; Bilski, P.; McKeever, S.W. The quest for new thermoluminescence and optically stimulated luminescence materials: Needs, strategies and pitfalls. *Radiat. Meas.* **2022**, *158*, 106846. [CrossRef]
3. Ahmed, M.; Schnell, E.; Ahmad, S.; Yukihara, E. Image reconstruction algorithm for optically stimulated luminescence 2D dosimetry using laser-scanned Al_2O_3:C and Al_2O_3:C,Mg films. *Phys. Med. Biol.* **2016**, *61*, 7484. [CrossRef] [PubMed]
4. Grau, C.; Durante, M.; Georg, D.; Langendijk, J.A.; Weber, D.C. Particle therapy in Europe. *Mol. Oncol.* **2020**, *14*, 1492–1499. [CrossRef]
5. Yukihara, E.G.; Doull, B.A.; Ahmed, M.; Brons, S.; Tessonnier, T.; Jäkel, O.; Greilich, S. Time-resolved optically stimulated luminescence of Al_2O_3: C for ion beam therapy dosimetry. *Phys. Med. Biol.* **2015**, *60*, 6613. [CrossRef]
6. Gustafson, T.D.; Milliken, E.; Jacobsohn, L.; Yukihara, E. Progress and challenges towards the development of a new optically stimulated luminescence (OSL) material based on MgB_4O_7:Ce,Li. *J. Lumin.* **2019**, *212*, 242–249. [CrossRef]
7. Shrestha, N.; Vandenbroucke, D.; Leblans, P.; Yukihara, E. Feasibility studies on the use of MgB_4O_7:Ce,Li-based films in 2D optically stimulated luminescence dosimetry. *Phys. Open* **2020**, *5*, 100037. [CrossRef]
8. Yukihara, E.; Christensen, J.; Togno, M. Demonstration of an optically stimulated luminescence (OSL) material with reduced quenching for proton therapy dosimetry: MgB_4O_7:Ce,Li. *Radiat. Meas.* **2022**, *152*, 106721. [CrossRef]
9. De Souza, L.F.; Souza, D.N.; Rivera, G.B.; Vidal, R.M.; Caldas, L.V. Dosimetric characterization of MgB_4O_7:Ce,Li as an optically stimulated dosimeter for photon beam radiotherapy. *Perspect. Sci.* **2019**, *12*, 100397. [CrossRef]
10. Prokić, M. Development of highly sensitive $CaSO_4$: Dy/Tm and MgB_4O_7: Dy/Tm sintered thermoluminescent dosimeters. *Nucl. Instrum. Methods* **1980**, *175*, 83–86. [CrossRef]
11. Yukihara, E.; Milliken, E.; Doull, B. Thermally stimulated and recombination processes in MgB_4O_7 investigated by systematic lanthanide doping. *J. Lumin.* **2014**, *154*, 251–259. [CrossRef]
12. Prokić, M. Individual monitoring based on magnesium borate. *Radiat. Prot. Dosim.* **2007**, *125*, 247–250. [CrossRef]
13. Souza, L.F.; Antonio, P.L.; Caldas, L.V.; Souza, D.N. Neodymium as a magnesium tetraborate matrix dopant and its applicability in dosimetry and as a temperature sensor. *Nucl. Instrum. Methods Phys. Res. Sect. A Acceler. Spectrom. Detect. Assoc. Equipm.* **2015**, *784*, 9–13. [CrossRef]
14. Yukihara, E.; Coleman, A.; Bastani, S.; Gustafson, T.; Talghader, J.; Daniels, A.; Stamatis, D.; Lightstone, J.; Milby, C.; Svingala, F. Particle temperature measurements in closed chamber detonations using thermoluminescence from $Li_2B_4O_7$:Ag,Cu,MgB_4O_7: Dy,Li and $CaSO_4$: Ce, Tb. *J. Lumin.* **2015**, *165*, 145–152. [CrossRef]
15. Yukihara, E.; Doull, B.; Gustafson, T.; Oliveira, L.; Kurt, K.; Milliken, E. Optically stimulated luminescence of MgB_4O_7:Ce,Li for gamma and neutron dosimetry. *J. Lumin.* **2017**, *183*, 525–532. [CrossRef]
16. Souza, L.; Silva, A.; Antonio, P.; Caldas, L.; Souza, S.; d'Errico, F.; Souza, D. Dosimetric properties of MgB_4O_7:Dy,Li and MgB_4O_7:Ce,Li for optically stimulated luminescence applications. *Radiat. Meas.* **2017**, *106*, 196–199. [CrossRef]
17. Kitagawa, Y.; Yukihara, E.G.; Tanabe, S. Development of Ce^{3+} and Li^+ co-doped magnesium borate glass ceramics for optically stimulated luminescence dosimetry. *J. Lumin.* **2021**, *232*, 117845. [CrossRef]
18. Annalakshmi, O.; Jose, M.; Madhusoodanan, U.; Venkatraman, B.; Amarendra, G. Synthesis and thermoluminescence characterization of MgB_4O_7. *Radiat. Meas.* **2013**, *59*, 15–22. [CrossRef]
19. Karali, T.P.; Rowlands, A.; Prokic, M.D.; Townsend, P.; Halmagean, E. Thermoluminescent spectra of rare earth doped MgB_4O_7 dosemeters. *Radiat. Protect. Dosim.* **2002**, *100*, 333–336. [CrossRef]
20. Bahl, S.; Pandey, A.; Lochab, S.; Aleynikov, V.; Molokanov, A.; Kumar, P. Synthesis and thermoluminescence characteristics of gamma and proton irradiated nanocrystalline MgB_4O_7:Dy,Na. *J. Lumin.* **2013**, *134*, 691–698. [CrossRef]
21. Kitis, G.; Polymeris, G.S.; Sfampa, I.K.; Prokic, M.; Meric, N.; Pagonis, V. Prompt isothermal decay of thermoluminescence in MgB_4O_7: Dy, Na and LiB_4O_7: Cu, In dosimeters. *Radiat. Meas.* **2016**, *84*, 15–25. [CrossRef]
22. De Oliveira, L.S.S.; De Souza, L.F.; Donald, G.G.; D'Emidio, M.F.S.; Novais, A.d.L.F.; Souza, D. Challenges in personal and clinical dosimetry using $Li_2B_4O_7$ and MgB_4O_7 as TLD and OSLD. *Braz. J. Radiat. Sci.* **2022**, *10*, 2A. [CrossRef]
23. Altunal, V.; Abusaid, W.; Guckan, V.; Ozdemir, A.; Yegingil, Z. Luminescence characterization of Ce and Gd doped MgB_4O_7 phosphors. *J. Lumin.* **2022**, *246*, 118815. [CrossRef]

24. Ozdemir, A.; Altunal, V.; Guckan, V.; Kurt, K.; Yegingil, Z. Luminescence characteristics of newly-developed $MgB_4O_7:Ce^{3+},Na^+$ phosphor as an OSL dosimeter. *J. Alloys Comp.* **2021**, *865*, 158498. [CrossRef]
25. Ozdemir, A.; Guckan, V.; Altunal, V.; Kurt, K.; Yegingil, Z. Thermoluminescence in MgB_4O_7:Pr,Dy dosimetry powder synthesized by solution combustion synthesis method. *J. Lumin.* **2021**, *230*, 117761. [CrossRef]
26. Sahare, P.; Singh, M.; Kumar, P. Synthesis and TL characteristics of MgB_4O_7:Mn,Tb phosphor. *J. Lumin.* **2015**, *160*, 158–164. [CrossRef]
27. González, P.; Ávila, O.; Mendoza-Anaya, D.; Escobar-Alarcón, L. Effect of sintering temperature on sensitivity of MgB_4O_7:Tm,Ag obtained by the solution combustion method. *Appl. Radiat. Isotopes* **2021**, *167*, 109459. [CrossRef]
28. Tan, Z.; Lan, T.; Gao, L.; Zhang, C.; Liu, X.; Luo, D.; Tang, Q. Comparison of thermoluminescence spectra of MgB_4O_7 doped with dysprosium and manganese. *J. Rare Earths* **2013**, *31*, 1039–1042. [CrossRef]
29. Palan, C.; Chauhan, A.; Sawala, N.; Bajaj, N.; Omanwar, S. Thermoluminescence and optically stimulated luminescence properties of MgB_4O_7:Ag phosphor. *Int. J. Lumin. Appl.* **2015**, *5*, 408–410.
30. Dogan, M.; Yazici, A. Thermoluminescence properties of Ce-doped MgB_4O_7 phosphor. *J. Optoelectron. Adv. Mater.* **2009**, *11*, 1783.
31. Prokic, M. MgB_4O_7: Mn as a new TL dosemeter. *Radiat. Protect. Dosim.* **1993**, *47*, 191–193. [CrossRef]
32. Kawashima, Y.S.; Gugliotti, C.F.; Yee, M.; Tatumi, S.H.; Mittani, J.C.R. Thermoluminescence features of MgB_4O_7:Tb phosphor. *Radiat. Phys. Chem.* **2014**, *95*, 91–93. [CrossRef]
33. Barbina, V.; Contento, G.; Furetta, C.; Malisan, M.; Padovani, R. Preliminary results on dosimetric properties op MgB_4O_7:Dy. *Radiat. Effects* **1982**, *67*, 55–62. [CrossRef]
34. Campos, L.; Fernandes Filho, O. Thermoluminescent characterisation of MgB_4O_7:Dy sintered pellets. *Radiat. Protect. Dosim.* **1990**, *33*, 111–113. [CrossRef]
35. Lochab, S.; Pandey, A.; Sahare, P.; Chauhan, R.; Salah, N.; Ranjan, R. Nanocrystalline MgB_4O_7:Dy for high dose measurement of gamma radiation. *Phys. Status Solidi A* **2007**, *204*, 2416–2425. [CrossRef]
36. Legorreta-Alba, O.; Cruz-Zaragoza, E.; Díaz, D.; Marcazzó, J. Synthesis of MgB_4O_7:Dy^{3+} and thermoluminescent characteristics at low doses of beta radiation. *J. Nucl. Phys. Mater. Sci. Radiat. Appl.* **2018**, *6*, 71–76. [CrossRef]
37. De Souza, L.F.; Caldas, L.V.; Junot, D.O.; Silva, A.M.; Souza, D.N. Thermal and structural properties of magnesium tetraborate produced by solid state synthesis and precipitation for use in thermoluminescent dosimetry. *Radiat. Phys. Chem.* **2019**, *164*, 108382. [CrossRef]
38. İflazoğlu, S.; Yılmaz, A.; Kafadar, V.E.; Topaksu, M.; Yazıcı, A. Neutron+ Gamma response of undoped and Dy doped MgB_4O_7:Dy thermoluminescence dosimeter. *Appl. Radiat. Isotopes* **2019**, *147*, 91–98. [CrossRef]
39. Porwal, N.; Kadam, R.; Seshagiri, T.; Natarajan, V.; Dhobale, A.; Page, A. EPR and TSL studies on MgB_4O_7 doped with Tm: Role of BO_3^{2-} in TSL glow peak at 470 K. *Radiat. Meas.* **2005**, *40*, 69–75. [CrossRef]
40. Bakhsh, M.; Wan Abdullah, W.S.; Mustafa, I.S.; Al Musawi, M.S.A.; Razali, N.A.N. Synthesis, characterisation and dosimetric evaluation of MgB_4O_7 glass as thermoluminescent dosimeter. *Radiat. Effects Defects Solids* **2018**, *173*, 446–460. [CrossRef]
41. Crijns, W.; Dirk, V.; Paul, L.; Tom, D. A reusable OSL-film for 2D radiotherapy dosimetry. *Phys. Med. Biol.* **2017**, *62*, 8441.
42. Yukihara, E.G. Luminescence properties of BeO optically stimulated luminescence (OSL) detectors. *Radiat. Meas.* **2011**, *46*, 580–587. [CrossRef]
43. Schalch, D.; Scharmann, A.; Weiß, A. Characterization of reactively sputtered BeO films. *Thin Solid Films* **1985**, *124*, 351–358. [CrossRef]
44. Sadat, T.; Morisseau, M.; Ross, M. Electron beam sterilisation of heterogeneous medical devices. *Radiat. Phys. Chem.* **1993**, *42*, 491–494. [CrossRef]
45. Eagle, M.; Rooney, P.; Lomas, R.; Kearney, J. Validation of radiation dose received by frozen unprocessed and processed bone during terminal sterilisation. *Cell Tissue Bank.* **2005**, *6*, 221–230. [CrossRef]
46. Alariqi, S.A.; Kumar, A.P.; Rao, B.; Singh, R. Biodegradation of γ-sterilised biomedical polyolefins under composting and fungal culture environments. *Polym. Degrad. Stab.* **2006**, *91*, 1105–1116. [CrossRef]
47. Kumar, S.; Saxena, S.; Verma, J.; Gautam, S. Development of ambient storable meal for calamity victims and other targets employing radiation processing and evaluation of its nutritional, organoleptic, and safety parameters. *LWT Food Sci. Technol.* **2016**, *69*, 409–416. [CrossRef]
48. Marathe, S.; Deshpande, R.; Khamesra, A.; Ibrahim, G.; Jamdar, S.N. Effect of radiation processing on nutritional, functional, sensory and antioxidant properties of red kidney beans. *Radiat. Phys. Chem.* **2016**, *125*, 1–8. [CrossRef]
49. Le Cann, S.; Tudisco, E.; Turunen, M.J.; Patera, A.; Mokso, R.; Tägil, M.; Belfrage, O.; Hall, S.A.; Isaksson, H. Investigating the mechanical characteristics of bone-metal implant interface using in situ synchrotron tomographic imaging. *Front. Bioeng. Biotechnol.* **2019**, *6*, 208. [CrossRef]
50. Yan, L.; Cinar, A.; Ma, S.; Abel, R.; Hansen, U.; Marrow, T.J. A method for fracture toughness measurement in trabecular bone using computed tomography, image correlation and finite element methods. *J. Mech. Behav. Biomed. Mater.* **2020**, *109*, 103838. [CrossRef]
51. Dejea, H.; Schlepütz, C.M.; Méndez-Carmona, N.; Arnold, M.; Garcia-Canadilla, P.; Longnus, S.L.; Stampanoni, M.; Bijnens, B.; Bonnin, A. A tomographic microscopy-compatible Langendorff system for the dynamic structural characterization of the cardiac cycle. *Front. Cardiovasc. Med.* **2022**, *9*, 3682. [CrossRef]

52. Karsch, L.; Beyreuther, E.; Burris-Mog, T.; Kraft, S.; Richter, C.; Zeil, K.; Pawelke, J. Dose rate dependence for different dosimeters and detectors: TLD, OSL, EBT films, and diamond detectors. *Med. Phys.* **2012**, *39*, 2447–2455. [CrossRef]
53. Christensen, J.B.; Togno, M.; Nesteruk, K.P.; Psoroulas, S.; Meer, D.; Weber, D.C.; Lomax, T.; Yukihara, E.G.; Safai, S. Al_2O_3:C optically stimulated luminescence dosimeters (OSLDs) for ultra-high dose rate proton dosimetry. *Phys. Med. Biol.* **2021**, *66*, 085003. [CrossRef]
54. Motta, S.; Christensen, J.B.; Togno, M.; Schäfer, R.M.; Safai, S.; Lomax, A.J.; Yukihara, E.G. Characterization of LiF:Mg,Ti thermoluminescence detectors in low-LET proton beams at ultra-high dose rates. *Phys. Med. Biol.* **2023**. [CrossRef]
55. Yasuda, H.; Kobayashi, I. Optically stimulated luminescence from Al_2O_3:C irradiated with relativistic heavy ions. *Radiat. Prot. Dosim.* **2001**, *95*, 339–343. [CrossRef]
56. Edmund, J.M.; Andersen, C.E.; Greilich, S.; Sawakuchi, G.; Yukihara, E.; Jain, M.; Hajdas, W.; Mattsson, S. Optically stimulated luminescence from Al_2O_3:C irradiated with 10–60 MeV protons. *Nucl. Instrum. Methods Phys. Res. Sect. A Acceler. Spectrom. Detect. Assoc. Equipm.* **2007**, *580*, 210–213. [CrossRef]
57. Sawakuchi, G.O.; Yukihara, E.; McKeever, S.; Benton, E.; Gaza, R.; Uchihori, Y.; Yasuda, N.; Kitamura, H. Relative optically stimulated luminescence and thermoluminescence efficiencies of Al_2O_3: C dosimeters to heavy charged particles with energies relevant to space and radiotherapy dosimetry. *J. Appl. Phys.* **2008**, *104*, 124903. [CrossRef]
58. Kerns, J.R.; Kry, S.F.; Sahoo, N. Characteristics of optically stimulated luminescence dosimeters in the spread-out Bragg peak region of clinical proton beams. *Med. Phys.* **2012**, *39*, 1854–1863. [CrossRef]
59. Teichmann, T.; Torres, M.G.; van Goethem, M.; van der Graaf, E.; Henniger, J.; Jahn, A.; Kiewiet, H.; Sommer, M.; Ullrich, W.; Weinhold, C.; et al. Dose and dose rate measurements in proton beams using the luminescence of beryllium oxide. *J. Instrument.* **2018**, *13*, P10015. [CrossRef]
60. Christensen, J.B.; Togno, M.; Bossin, L.; Pakari, O.V.; Safai, S.; Yukihara, E.G. Improved simultaneous LET and dose measurements in proton therapy. *Sci. Rep.* **2022**, *12*, 1–10. [CrossRef]
61. Hubbell, J.; Seltzer, S. *NIST Standard Reference Database 126*; National Institute of Standards and Technology: Gaithersburg, MD, USA, 1996.
62. Yukihara, E.G.; McKeever, S.W. *Optically Stimulated Luminescence: Fundamentals and Applications*; John Wiley & Sons: Hoboken, NJ, USA, 2011.

Disclaimer/Publisher's Note: The statements, opinions and data contained in all publications are solely those of the individual author(s) and contributor(s) and not of MDPI and/or the editor(s). MDPI and/or the editor(s) disclaim responsibility for any injury to people or property resulting from any ideas, methods, instructions or products referred to in the content.

Article

On the Need for Deconvolution Analysis of Experimental and Simulated Thermoluminescence Glow Curves

George Kitis [1,*] and Vasilis Pagonis [2]

[1] Nuclear Physics and Elementary Particles Physics Section, Physics Department, Aristotle University of Thessaloniki, 54124 Thessaloniki, Greece
[2] Physics Department, McDaniel College, Westminster, MD 21157, USA
* Correspondence: gkitis@auth.gr

Abstract: Simulation studies of thermoluminescence (TL) and other stimulated luminescence phenomena are a rapidly growing area of research. The presence of competition effects between luminescence pathways leads to the complex nature of luminescence signals, and therefore, it is necessary to investigate and validate the various methods of signal analysis by using simulations. The present study shows that in simulations of luminescence signals originating from multilevel phenomenological models, it is not possible to extract mathematically the individual information for each peak in the signal. It is further shown that computerized curve deconvolution analysis is the only reliable tool for extracting the various kinetic parameters. Simulation studies aim to explain experimental results, and therefore, it is necessary to validate simulation results by comparing with experiments. In this paper, testing of simulation results is performed using two methods. In the first method, the influence of competition effects is tested by comparing the input model parameters with the output values from the deconvolution analysis. In the second method, the agreement with experimental results is tested using the properties of well-known glow peaks with very high repeatability among TL laboratories, such as the 110 °C glow peak of quartz.

Keywords: thermoluminescence; stimulated luminescence; kinetic parameters; superposition principle; competition between levels; computerized glow curve deconvolution

Citation: Kitis, G.; Pagonis, V. On the Need for Deconvolution Analysis of Experimental and Simulated Thermoluminescence Glow Curves. *Materials* **2023**, *16*, 871. https://doi.org/10.3390/ma16020871

Academic Editors: Wiesław Stręk and Efrat Lifshitz

Received: 30 November 2022
Revised: 6 January 2023
Accepted: 12 January 2023
Published: 16 January 2023

Copyright: © 2023 by the authors. Licensee MDPI, Basel, Switzerland. This article is an open access article distributed under the terms and conditions of the Creative Commons Attribution (CC BY) license (https://creativecommons.org/licenses/by/4.0/).

1. Introduction

Simulation studies of thermoluminescence (TL) and other stimulated luminescence phenomena are a rapidly growing area of research [1,2]. Simulations of phenomenological models consisting of many energy levels responsible for TL peaks result in complex TL glow curves, which are very similar to experimental glow curves. In both experimental and simulated glow curves, it is important to extract the information regarding the TL intensity used for dosimetric applications, as well as the kinetic parameters of each peak, which can then be used to evaluate the signal lifetimes for specific dosimetric applications [3].

In traditional TL literature, the TL intensity from experimental glow curves is evaluated (Figure 1) by selecting the intensity at some point of the glow curves (termed the peak height), or by integrating the signal between two temperatures [3,4].

During the last decade, the technique of computerized glow curve deconvolution (CGCD) has been applied extensively [5–8], although some textbooks have raised objections to its wide application [9–11]. The skepticism shown by some researchers is based on the fulfillment or not of the superposition principle (SP), which postulates that the energy levels responsible for each individual TL peak do not depend on each other.

In this work, the best techniques for analyzing the complex glow curves resulting from a simulation will be investigated, attempting to answer the question: is it possible to analyze simulated TL signals using the same methods as for experimental TL signals?

The aims of the present work are: (i) To simulate complex TL glow curves and investigate whether it is possible to extract the quantitative characteristics of each component in the TL signal; (ii) To study the influence of competition effects between traps on the parameters extracted from the simulated TL signals; (iii) To investigate the relation between the input kinetic parameters in the model, and the output values obtained by analyzing the TL signals using CGCD methods; (iv) To establish criteria for the validity of the results obtained in TL simulations; (v) To examine how experimental TL glow curves widely available in the literature can be used for testing the results from simulations.

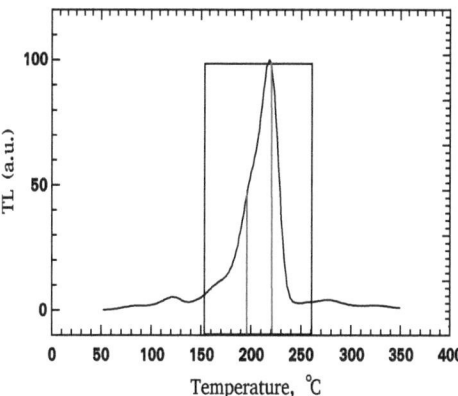

Figure 1. Traditional methods to extract the TL intensity from experimental glow curves either by selecting the peak height at some point of the glow curves (perpendicular lines) or by integrating the signal between two temperatures (region within the box).

2. Materials and Methods
2.1. The Phenomenological Model

The simulation model used in this paper consists of $i = 1, \ldots, 6$ electron traps and one hole trap [3]. The electron traps which can both trap and release electrons are termed active traps, whereas electron traps which can trap only electrons without releasing them at the temperature region of the active traps are termed thermally disconnected deep traps (TDDT). The hole trap in the model acts also as the recombination center. Figure 2 show schematically the energy levels for this model. The simulation consist of three stages termed the irradiation stage, relaxation stage and heating stage. The differential equations governing the traffic of electrons and holes in this model are:

$$\frac{dT}{dt} = \beta \tag{1}$$

$$\frac{dR}{dt} = Drate \tag{2}$$

$$\frac{dn_i}{dt} = -\sum n_i s_i e^{-\frac{E}{kT}} + A_i (N_i - n_i) n_c \tag{3}$$

$$\frac{dn_d}{dt} = A_d (N_d - n_d) n_c, \tag{4}$$

$$\frac{dm}{dt} = A_h (M - m) n_v - A_m m n_c, \tag{5}$$

$$\frac{dn_v}{dt} = Drate - A_h (M - m) n_v, \tag{6}$$

$$\frac{dn_c}{dt} = Drate - \sum \frac{dn_i}{dt} - \frac{dn_d}{dt} - A_m m n_c, \tag{7}$$

where the index $i = 1, \ldots, 5$ stands for the active electron traps, E_i (eV) is the activation energy, s_i (s^{-1}) the frequency factor, N_i (cm^{-3}) is the concentration of available electron traps, n_i (cm^{-3}) the concentration of trapped electrons, M (cm^{-3}) is the concentration of available luminescence centers, m_i (cm^{-3}) concentration of trapped holes. N_d, n_d (cm^{-3}) are the concentrations of available and occupied traps in a thermally disconnected deep trap (TDDT), n_c (cm^{-3}) and n_v (cm^{-3}) are the concentration of electrons in the conduction and holes in the valence band, A_i (cm^3 s^{-1}) are the trapping coefficients in electron traps n_i, A_m (cm^3 s^{-1}) is the recombination coefficient, A_h (cm^3 s^{-1}) is the trapping coefficient for holes in luminescence centers, A_d (cm^3 s^{-1}) trapping coefficient in TDDT, β (K/s) is the heating rate and Drate is the rate of production of ion pairs (i.p) per second (i.p/s) which is proportional to the dose rate.

Irradiation stage

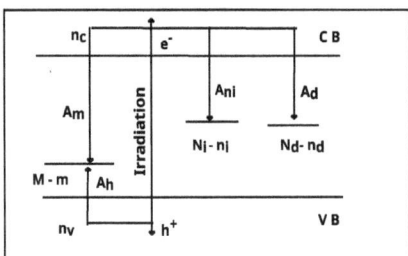

Relaxation stage

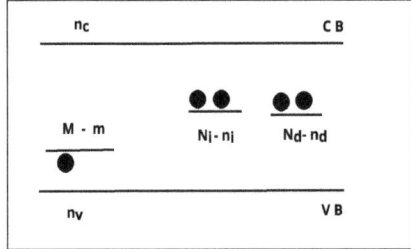

Heating stage

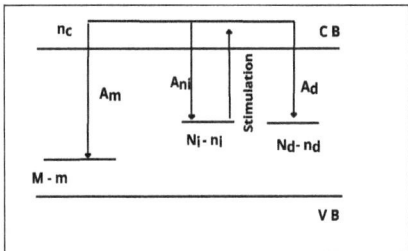

Figure 2. Energy band model used for simulation. (**Upper**): Irradiation stage. (**Middle**): Relaxation stage. (**Down**): Heating stage.

Equation (1) evaluates the temperature as a function of time, whereas Equation (2) evaluates the generated (i.p/s) used in Equations (6) and (7). Equation (3) describes the trapping and thermal release of electrons by the active traps, while Equation (4) refers to the competitor TDDT.

In the above model, all traps with the index $i = 1 \cdots 5$ can trap and release electrons by thermal excitation, and they are referred to as active traps. On the other hand, the TDDT with index d can only trap electrons while thermal stimulation of electrons is not allowed.

In the present work, the TL signals are simulated by solving the system of ordinary differential equations (ODE) described by Equations (1)–(7). The system of ODE is solved using the standard Scientific Python (SciPy) package in the Python programming language.

We specifically simulate four reference TL glow peaks (REFERENCE-01 to REFERENCE-04), with the trap parameters listed below. The four reference glow curves were chosen so that the amount of competition between traps is highest for REFERENCE-01, and it decreases progressively up to REFERENCE-04:

- $N_0 = 2.5 \times e^{10}$;
- $N_1, N_2, N_3, N_4, N_5, N_6 = 0.1 \times 10^{10}, 0.2 \times 10^{10}, 0.2 \times 10^{10}, 0.5 \times 10^{10}, 0.5 \times 10^{10}, 1.0 \times 10^{10}$, in cm^{-3};
- $E_1, E_2, E_3, E_4, E_5 = 1.0, 1.38, 1.48, 1.6, 2.01$, in eV;
- $s_1, s_2, s_3, s_4, s_5 = 1 \times e^{13}, 3.9 \times 10^{16}, 2 \times 10^{16}, 2 \times 10^{16}, 4 \times 10^{19}$ in s^{-1};
- Drate = 2.5×10^9 i.p/s;
- REFERENCE-01-($A_{1,2,3,4} = 10^{-9}$, $A_5 = 4 \times 10^{-9}$, $A_m = A_d = 10^{-7}$), in cm^3 s^{-1};
- REFERENCE-02-($A_{1,2,3,4} = 5 \times 10^{-8}$, $A_5 = 9 \times 10^{-8}$, $A_m = A_d = 10^{-7}$), in cm^3 s^{-1};
- REFERENCE-03-($A_{1,2,3,4} = 10^{-9}$, $A_5 = 4 \times 10^{-9}$, $A_m = 10^{-7}$, $A_d = 10^{-10}$), in cm^3 s^{-1};
- REFERENCE-04-($A_{1,2,3,4} = 5 \times 10^{-8}$, $A_5 = 9 \times 10^{-8}$, $A_m = 10^{-7}$, $A_d = 10^{-10}$), in cm^3 s^{-1}.

2.2. CGCD Analysis Method

The complex TL glow curves resulting from simulation, as well as a large number of experimental glow curves of natural quartz, were analyzed using the CGCD analysis technique. Specifically, the CGCD equation used here is based on the one trap one recombination center (OTOR) model. The solution of the OTOR model is based on Lambert $W(Z)$ function [12,13] and was derived by Kitis and Vlachos [14]. Later, Singh and Gartia [15] developed a similar analytical equation based on the Wright function. The analytical equation was used in the following form, which was also previously used [3,4,16]:

$$I = I_m \exp\left(\frac{E(T - T_m)}{k T T_m}\right) \cdot \frac{W[e^{z_{1m}}] + W[e^{z_{1m}}]^2}{W[e^{z_1}] + W[e^{z_1}]^2}. \tag{8}$$

with

$$z_1 = \frac{R}{1 - R} - \ln\left(\frac{1-R}{R}\right) + \frac{E e^{\frac{E}{kT_m}}}{k T_m^2} \frac{F(T, E)}{1 - 1.05 R^{1.26}}. \tag{9}$$

and

$$F(T, E) = T \cdot \exp(-E/kT) + \frac{E}{k} \cdot Ei(-E/kT) \tag{10}$$

where T_m the temperature at the maximum peak intensity I_m, $R = A_n/A_m$ is the ratio or re-trapping over recombination coefficients and z_{1m} is the value of z_1 at the peak maximum temperature T_m.

The function F(T,E) is the well-known integral appearing in TL theory, which is used here in terms of the built-in exponential integral function $Ei(x)$, instead of the asymptotic series approximation used previously in the literature [1,9].

In modern software packages, the Lambert function $W(z)$ is a built-in function, similar to any other transcendental function like sine, cosine, etc. The Lambert function is termed ProductLog[(0,1),z] in Mathematica, Lambert w_0 and w_1 in MATLAB and EXCEL, lambertw in Python, gsl-sf-lambert-$w_0(z)$, gsl-sf-lambert-$w_1(z)$ in GNU GSL. w_0 and w_1 stands for the first and second real branch of the Lambert function, respectively.

As $z \to \infty$, the numerical value of e^z in the above equations overflows, and the Lambert function $W(Z)$ in Python does not return a value. In such cases, the Lambert W can be approximated by

$$W(e^z) = z - ln(z) \quad (11)$$

Most software packages have implemented the Wright $\omega(Z)$ function, and the above overflow issue is overcome by using the Wright $\omega(Z)$ function [15], instead of the Lambert function $W(Z)$. The two functions are related according to:

$$W(e^z) = \omega(z). \quad (12)$$

2.3. CGCD Analysis Software

The CGCD analysis was applied using the ROOT data Analysis Framework [17]. All TL glow curve fittings were performed using the MINUIT program [18] released in ROOT, which is a physics analysis tool for function minimization. The Lambert function $W(z)$ and the exponential integral function $Ei[-\frac{E}{kT}]$ are implemented in ROOT through the GNU scientific library (GNU GSL) [19].

The goodness of fit was tested using the figure of merit (FOM) [20], which was initially proposed to test the goodness of fit of gamma ray spectroscopy data. Currently, the FOM is widely used by the TL/OSL community within the CGCD deconvolution analysis [21]. It is defined as

$$FOM = 100\% \times \frac{\sum \left| y_i^0 - y_i^f \right|}{\sum y_i^f} \quad (13)$$

where y_i^0 corresponds to experimental points and y_i^f to fitted points.

2.4. Materials and Experiments

Simulations and experiments must support each other. In the present work, we investigate whether experimental results can be used to validate simulation results. The original quartz samples were large crystals of hydrothermal and metamorphic origin which occur in vein—associated metamorphic rocks, collected from different locations spanning Africa (Nigeria), Europe (Greece) and Asia (Nepal) (more details in [22–24]). The experimental data analyzed here concern the low-temperature TL peak of quartz (the 110 °C TL peak), which was studied for both pre-dosed and natural aliquots.

All the TL measurements on the quartz samples were carried out using a RISØ TL/OSL reader (model TL/OSL–DA–15) equipped with a 0.075 Gy/s $^{90}Sr/^{90}Y$ beta ray source [25]. The reader was fitted with a 9635QA photomultiplier tube. The detection optics consisted of a 2.5 mm Hoya U–340 (kp 340 nm, FWHM 80 nm) filter.

The experimental protocol is as follows:

Step 1: Readout up to 250 °C at β = 1 °C/s).

Step 2: Irradiate with a small test dose (less than 1 Gy).

Step 3: Readout up to 250 °C at β = 1 °C/s).

Step 4: Give the same test dose of Step 2.

Step 5: Readout up to 500 °C at β = 1 °C/s).

3. The First Requirement for a Valid Simulation Test: Using Three Simulation Stages

Before presenting the results of the simulations, we first discuss the optimal method of carrying out TL simulations.

All TL simulations should contain three distinct simulation stages, namely, the irradiation stage, relaxation stage and the heating stage. The parameters in these stages should be set up as follows:

- **Irradiation stage:** Set the initial values of all parameters at time t = 0, i.e., T_0 = 273 K, $n_{i0} = 0$, $m_0 = 0$, $nc_0 = 0$, $nv_0 = 0$. Set also $\beta = 0$ since there is no heating, and set the irradiation dose rate (in the present work $Drate = 10^5$ e-h pairs/s). Store the last values of all concentrations at the end of the irradiation stage, which will be used as the initial concentrations for the next stage.

- **Relaxation stage:** This stage simulates the time interval between the end of irradiation stage and the beginning of the heating stage. Set as initial values the last concentrations of the previous irradiation stage. Set also $\beta = 0$ and $Drate = 0$, since there is no heating and no irradiation. Store the last values of all concentrations at the end of this stage, which will be the initial concentrations for the next stage.

- **Heating stage:** Set as initial values the last values of the previous relaxation stage. Since there is no irradiation and hole trapping, set $Drate = 0$ and $A_h = 0$. Set the heating rate β (in our simulations β = 2 K/s). At the end of this stage, the TL glow peak is evaluated.

In order to show the importance of using three simulation stages, we will describe two cases in the literature, in which the three stage requirement is not followed.

The first example concerns the Randall–Wilkins model [26,27], which consists of only the heating stage. This model produces the very well known analytical equation for first-order kinetics. However, despite its great role in TL research, this model has a restricted physical basis. The reason is the requirement for zero re-trapping during the heating stage, which also means zero trapping during the irradiation stage. This means that the initial condition leading to the Randall–Wilkins model cannot be achieved at all during the irradiation stage. The Randall–Wilkins kinetics is an extreme boundary condition of the OTOR model.

The second example concerns TL simulation studies with extreme pessimistic results concerning the validity of the TL phenomenon itself [28–30]. For many decades, these pessimistic conclusions are, unfortunately, reproduced by several TL textbooks without any further study [9–11].

Sadek and Kitis [31] examined in detail the simulations of Kelly et al. [28,29] and Opanowitz [30] and they found that the common point of both studies was that they contain only the heating stage. Sadek and Kitis [31] investigated these previous results in two ways: first by deriving the TL peaks and then fitting them using analytical expressions for single TL peaks. Several of the TL peaks were fitted excellently, reproducing exactly the values of the activation energy used in the simulations, contrary to the conclusions of Kelly et al. [28,29] and Opanowitz [30]. On the other hand, there were cases of peaks which could be fitted well with the analytical equations but gave erroneous values of the activation energy, and also cases that was impossible to fit, in agreement with the pessimistic conclusions. In this last group of cases, Sadek and Kitis [31] rewrote the models of Kelly et al. [28,29] and of Opanowitz [30] so that they also contained the irradiation and relaxation stages. The final conclusion of Sadek and Kitis [31], after extensive trials, was that the parameters used by Kelly et al. [28,29] and Opanowitz [30] cannot be attained by using appropriate irradiation stages.

These two examples from the literature show that using three-stage simulations is a first necessary requirement for the validity of simulations. An arbitrary selection of parameter values for the heating stage should be avoided since it can easily lead to non-physical results.

4. Is It Possible to Evaluate the Contribution of Each Trap Using the Numerical Solution of the Differential Equations?

In this section, we discuss whether one can isolate the contribution of each trap to the TL signal mathematically by using the numerical solution of the differential equations. The simulated luminescence signal corresponds to the total number of recombination events in the luminescence centers, i.e.,

$$TL(t) = A_m\, m(t)\, n_c(t) \qquad (14)$$

In an attempt to separate the photons originating from each specific electron trap during the heating process, we use the neutrality condition and write Equation (14) in the form:

$$TL(t) = A_m\, (n_1(t) + n_2(t) + n_3(t) + \ldots n_i(t) + n_c(t))\, n_c(t) \qquad (15)$$

Based on simple inspection of this equation, one may suppose that the contribution of each individual trap to the TL signal will be given by an expression of the form:

$$TL_1(t) = A_m\, n_1(t)\, n_c(t); \quad TL_2(t) = A_m\, n_2(t)\, n_c(t); \quad TL_3(t) = A_m\, n_3(t)\, n_c(t); \quad , etc. \qquad (16)$$

However, this is incorrect, since the concentration of electrons n_i from each trap is multiplied by $n_c(t)$, which is due to electrons originating not only from trap n_i, but from all traps. It is a basic property of the delocalized multilevel phenomenological models that once the electrons are released into the conduction band, they have no memory of their origin. As a result of this memory loss, the correspondence between the number of trapped electrons n_i and the number of recombined electrons is lost.

The basic consequence of this situation is that it is not possible to extract mathematically from the numerical solution of the differential equations useful information regarding the intensity of the individual components in the TL signal. Due to the presence of the term n_c, the individual peaks in the TL signal are correlated to each other. This contradicts the superposition principle (SP), which is an underlying assumption during the CGCD analysis of complex TL signals. The validity of the SP requires careful examination, and it is discussed in the next section.

5. Is the Superposition Principle Valid When We Apply CGCD Methods of Analysis?

Unfortunately, the computerized techniques for analysis of TL glow-curves are not universally accepted. A part of the TL community argues that deconvolution is valid only when the superposition principle holds [9–11]. According to these authors, this happens only for the Randall–Wilkins type of first-order kinetics [26,27]. We wish now to discuss and clarify this point, before continuing with the two main objectives of the present work.

The SP states that at a given place and time, the response due to two or more stimuli equals to the sum of the response that would have been caused by each stimulus individually [32]. Mathematically, the SP is expressed as

$$f(x_1 + x_2 + \cdots + x_n) = f(x_1) + f(x_2) + \cdots + f(x_n). \qquad (17)$$

where x is a parameter and $y = f(x)$ is the respective response.

As is obvious from Equations (1)–(7), the shape of complex TL curves is due to the competition among electron traps. The existence of competition makes them correlated, and therefore, the basic requirement of independence assumed by Equation (17) is lost. One would then be tempted to conclude that the SP does not hold and that the CGCD method of analysis cannot be applied.

Sadek and Kitis [32] used a phenomenological model similar to the model used in the present work, and simulated in detail the impact of the non-fulfillment of SP on the analysis of TL glow curves. The study of Sadek and Kitis [32] was based on varying the amount of competition between traps, which is controlled by the parameters A_d and N_d in Equation (4).

In cases of strong competition between traps ($A_m = A_d \gg A_i$ and $N_d > N_i$), the conclusions of Sadek and Kitis [32] can be summarized as follows:

- A strong competition from a TDDT practically removes the competition between the active traps;

- In such strong competition cases, a condition of a pseudo-superposition principle is established, causing the individual active traps to behave independently;
- The glow curve shape shows remarkable stability;
- The simulated complex TL glow curves were fitted excellently with the available analytical CGCD expressions;
- The values of the kinetic parameters evaluated with CGCD analysis were in very good agreement with the values used as input values in the simulation.

In the case of weak competition between traps ($A_d < A_m, A_i, N_d < N_i$), the conclusions of these authors were:

- The weak competition from TDDT transfers the competition effects to the active traps;
- The last peak of a complex TL curve acts like an OTOR peak [33];
- The glow curve shape shows significant changes for different doses;
- However, even in these cases, the simulated complex TL glow curves fit very well with the available CGCD expressions;
- The values of the kinetic parameters evaluated with CGCD analysis were in very good agreement with the values used within the simulation;
- Only in cases where $A_i > A_d, A_m$, the CGCD fails to produce accurate values of the activation energies.

Based on the above study, it is argued that there is no physical contradiction between using the CGCD analysis and the SP. The basic conclusion adopted in the present work is that one can use CGCD analysis on the basis of the pseudo-SP established by Sadek and Kitis [32].

6. Simulation of the TL Signal: Attempting to Separate Individual Peaks in the TL Glow Curve

It was concluded that it is not possible to extract mathematically the TL intensity of each peak from the numerical solution of the system of ODE (in Section 4). However, we will show that the CGCD analysis is an effective tool which can evaluate the contribution of each trap to the TL signal. We compare the mathematical and CGCD methods of analysis by the following simulation procedure:

1. Evaluate the integrated number of trapped electrons n_0 in active traps at the end of the irradiation stage;
2. Use Equation (16) to evaluate the integrated signal due to each trap during the heating stage. This is the mathematical approach, which is based on the solution of the differential equations;
3. Analyze the simulated glow curves using CGCD analysis, as an alternative method to obtain the integrated signal due to each TL peak. The results from this CGCD analysis will be compared with the results from using Equation (16).

As a result of application of Equation (16), curve (1) in Figure 3a corresponds to TL peak 1, while curve (2) corresponds to TL peak 2. It is clear that peak 2 has contributions from electrons released from both n_1 and n_2.

Similarly, curve (3) of Figure 3a has contributions from traps n_1, n_2 and n_3, while curve (4) in Figure 3b has contributions from traps n_1, n_2, n_3 and n_4. Finally, peak 5 in Figure 2c has contributions from all 5 traps. The curve in Figure 3d shows the final complex TL glow curve evaluated using Equation (14).

The above results show clearly the effects of competition, namely that numerical simulations of phenomenological models with many interacting trapping levels fail to evaluate the individual TL glow peaks originating from each trapping level. In fact, it is not possible to even obtain a plot of the individual TL peaks using the ODE, and of course, it is impossible to evaluate their integrated signal.

In order to complete the analysis, we use CGCD analysis to evaluate the components in the TL signal. The CGCD analysis was applied to the four reference glow curves described above, along the lines of the work by Sadek and Kitis [32]. The results are shown in Figure 4, and the results of the CGCD analysis are shown in Table 1. The three columns in Table 1 correspond to the three simulation steps 1–3 described previously in this section. The first column corresponds to the number of electron n_0 trapped in each trapping level, at the end of irradiation stage. The second column shows the integrated signal of the contribution of each term $A_m \, n_i \, n_c$. Finally, the third column corresponds to the integrated signal of each peak obtained from the CGCD analysis.

The first observation is that the values of the second column differ widely from the values of first and third column, while the first and third column are very close to each other. Based on the results of Table 1, it is concluded that:

(1) The numerical simulation of the ODE cannot provide an accurate measure of the integrated TL intensity for each peak;

(2) The CGCD analysis can provide a much more accurate estimate of the simulated TL intensity for each peak in a complex simulated glow curve.

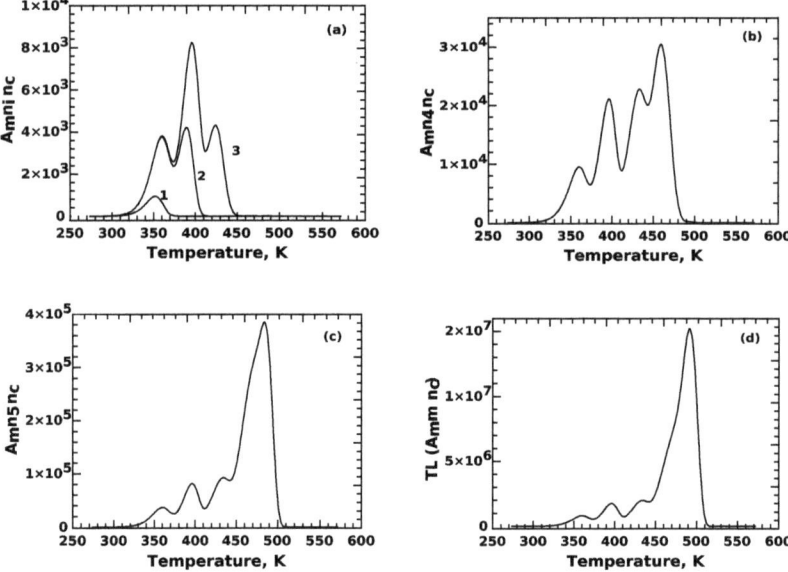

Figure 3. TL peak shape in the case of REFERENCE-01 glow curve, as evaluated from the numerical solutions of the model in Section 2.1. (**a**) Shapes of peaks 1, 2 and 3, (**b**) shape of peak 4, (**c**) shape of peak (5) and (**d**) total glow curve shape. It is obvious that except peak 1, all other peaks, although they must be of single-peak shape, look composite, because the electron distribution within the conduction band contributes to their numerical evaluation.

Table 1. Individual peak integral evaluations. The first column corresponds to the number of electrons n_0 trapped in each trapping level, at the end of irradiation stage. The second column shows the integrated signal of the contribution of each term $A_m \, n_i \, n_c$ and the third column corresponds to the integrated signal of each peak obtained from the CGCD analysis.

Peak	n_0	$A_m \, n_i \, n_c$	CGCD Analysis
REFERENCE-01-($A_{1,2,3,4} = 10^{-9}$, $A_5 = 4 \times 10^{-9}$, $A_m = A_d = 10^{-7}$)			
1	2.8×10^7	2.60×10^4	2.15×10^7
2	4.6×10^7	2.02×10^5	4.02×10^7
3	4.6×10^7	4.13×10^5	4.23×10^7
4	1.2×10^8	2.17×10^6	1.04×10^8
5	2.0×10^9	1.94×10^7	4.06×10^8
REFERENCE-02-($A_{1,2,3,4} = 5 \times 10^{-8}$, $A_5 = 9 \times 10^{-8}$, $A_m = A_d = 10^{-7}$)			
1	4.5×10^8	5.02×10^6	3.05×10^8
2	9.0×10^8	4.00×10^7	6.33×10^8
3	9.0×10^8	8.72×10^7	6.73×10^8
4	2.2×10^9	4.83×10^8	1.82×10^9
5	3.3×10^9	1.42×10^9	2.94×10^9
REFERENCE-03-($A_{1,2,3,4} = 10^{-9}$, $A_5 = 4 \times 10^{-9}$, $A_m = 10^{-7}$, $A_d = 10^{-10}$)			
1	1.2×10^8	2.10×10^6	1.09×10^8
2	2.3×10^8	1.69×10^7	2.22×10^8
3	2.3×10^8	3.61×10^7	2.23×10^8
4	5.9×10^8	2.16×10^8	5.63×10^8
5	2.0×10^9	2.41×10^9	1.98×10^9
REFERENCE-04-($A_{1,2,3,4} = 5 \times 10^{-8}$, $A_5 = 9 \times 10^{-8}$, $A_m = 10^{-7}$, $A_d = 10^{-10}$)			
1	6.6×10^8	1.82×10^7	5.40×10^8
2	1.3×10^9	1.49×10^8	1.13×10^9
3	1.3×10^9	3.49×10^8	1.19×10^9
4	2.3×10^9	2.34×10^9	2.95×10^9
5	4.3×10^9	7.69×10^9	4.97×10^9

7. Testing the Results of Simulations: Comparison of the Output and Input Parameter Values E and s

The values of the kinetic parameters E and s of each electron trap are intrinsic properties of the traps, without any dependence on the traffic of electrons in the conduction band. The big advantage of the CGCD analysis of TL signals is that it evaluates both the integrated intensity of each peak in the glow curve, as well as the kinetic parameters E and s of the individual peaks.

It is clear then that a strong test of the results from simulations is the comparison of the input values of E and s in the model, with the corresponding values obtained by the CGCD analysis.

As mentioned above, an extended initial study comparing the input and output parameters was carried out by Sadek and Kitis [32], and their conclusions were previously summarized in Section 5.

In the present work, we will expand this previous study by applying CGCD analysis to the four simulated reference glow curves (REFERENCE-01 to REFERENCE-02 described above).

The CGCD analysis is shown in Figure 4, and Table 2 shows the comparison between the input and the output values of the kinetic parameters. The output values of the activation energy E are the mean from all reference glow curves. The agreement between the input values of E and s, and the CGCD estimates are excellent.

The peak maximum temperatures (T_m) shown in Table 2 are remarkably stable, although the simulation parameters of the four reference glow curves differ widely. The last column contains the values of the retrapping coefficient $R = A_n/A_m$ (which corresponds to the kinetic order of the TL process). The very low R values for peaks 1–4 indicate first-order kinetics, which is expected due to the strong competition between traps [34].

Also of interest are the values of R in the case of peak 5, which increase as the competition between traps decreases. The values for reference glow curves 1 and 2 correspond to high competition between the traps, and the R values are very small (R = 0.001 and 0.01), indicating first-order kinetics [34].

In the case of reference glow curve 3, the competition is weaker, and the R value is somewhat higher (R = 0.05). Finally, for reference glow curve 4, the competition is the weakest, and the value of R = 0.82. This is in agreement with the results of Sadek and Kitis [32], who found that the last peak of the glow curve (peak 5 in our study) takes the place of the competitor, and that the last peak adopts the properties of the one trap one recombination center (OTOR) model. The value of R = 0.82 means that under weak competition, the TL kinetics tends to second-order kinetics.

These results establish that the stability of the E and s values during the simulation can be used as a unique criterion for simulation testing. We state this criterion as follows:

- *If the input and output values of E and s in a simulation agree with each other, the simulation is valid, and the effects of competition and of the superposition principle have been taken into account successfully;*
- *Any disagreement between input and output values does not necessary mean an invalid simulation, but it could indicate instead the appearance of new physical processes, which need additional study and interpretations.*

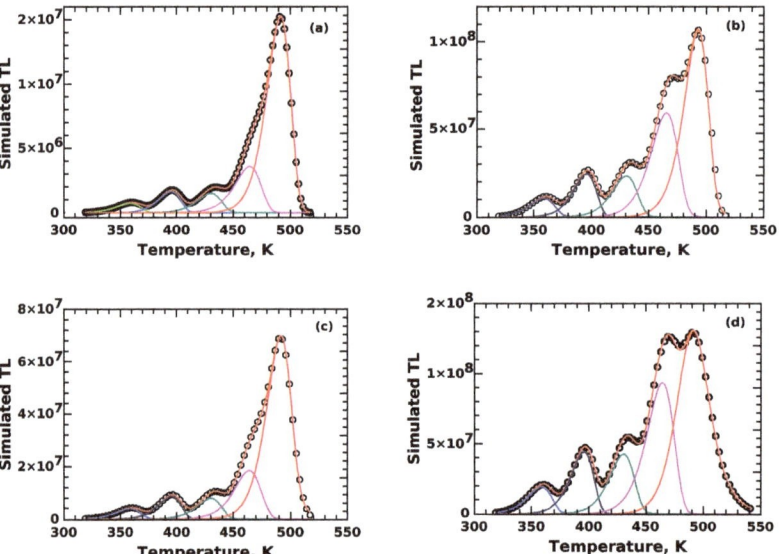

Figure 4. CGCD analysis results of the four REFERENCE glow curves used for simulation. (**a**) REFERENCE-01, (**b**) REFERENCE-02, (**c**) REFERENCE-03, (**d**) REFERENCE-04.

Table 2. Comparison of the kinetic parameters for each energy level in the model, as obtained using CGCD analysis. The input values of the model are compared with the output values from CGCD. The input values for E, s for each peak are common to all reference cases. The units of E are eV, and the units of s are s^{-1}.

	Input Values		Output Values			
Peaks	E	s	E	s	T_m	R
1	1.0	1.0×10^{13}	1.0 ± 0.01	$(1.28 \pm 0.28) \times 10^{13}$	358.9 ± 0.09	0.002
2	1.38	3.9×10^{16}	1.37 ± 0.006	$(3.13 \pm 0.64) \times 10^{16}$	395.8 ± 0.20	0.001
3	1.48	2.0×10^{16}	1.48 ± 0.013	$(1.92 \pm 0.64) \times 10^{13}$	430.6 ± 0.18	0.005
4	1.6	2.0×10^{16}	1.59 ± 0.008	$(1.50 \pm 0.32) \times 10^{13}$	464.9 ± 0.50	0.002
5	2.01	4.0×10^{19}	2.004 ± 0.02	$(3.44 \pm 0.3) \times 10^{13}$	492.3 ± 0.60	0.001, 0.01, 0.05, 0.82

8. Testing the Results of Simulations by Using Experimental Results

A simulation without any experimental control can often lead to unrealistic results. In order to use experimental results for simulation testing, it is necessary to find experimental results which correspond to very specific and stable luminescence signals. As an example, we will consider natural quartz, which is used as a natural dosimeter in archaeological and geological dating.

According to the previous section, the first and most important test is to require the equality between the model input with the output CGCD values of the kinetic parameters of each peak. If a simulation does not reproduce the experimental parameters, then further explanation and interpretation is required. Disagreement between input and output values can provide useful information about the influence of experimental conditions on the values of the parameters.

In a simulation using a multilevel phenomenological model to study a specific material, we can further ask that the simulation reproduce well-known and generally accepted properties of the materials.

In the case of quartz, the low-temperature TL peak, known as the 110 ^{0}C peak in the literature, can act as the ideal tester of the validity of simulated results. The important general properties of this peak are listed below:

- This peak appears in all quartz samples of any origin [35];
- The activation energy of this peak in all types of quartz lies in a narrow range E = 0.75–0.9 eV [35];
- The kinetic order is always of the first order [35];
- The activation energy and peak maximum position T_m remains unchanged, even after very strong external preconditioning of the sample (e.g., high-temperature annealing and the pre-dose effect [36–38]).

Recently, TL and OSL research groups have recognized the importance of the 110 ^{0}C peak, and its properties became the subject of inter-comparison programs between research groups (Schmidt et al. [39,40]).

In the present work, we present new experimental results with very high repeatability. The quartz sources and the experimental procedure were presented in Section 2. The measuring protocols consisted of many thermal heating cycles and temperatures, so that a large number of TL glow curves of different quartz samples was measured. The new data include many pre-conditioned aliquots, which potentially could influence the properties of the TL glow peak at 110 °C. The characteristics studied by CGCD analysis are the activation energy, peak maximum temperature, frequency factor and kinetic order. The results are shown in Figure 5. Figure 5a shows the position of peak maximum temperature T_m = 351.6 ± 2.4 °C, which is extremely stable considering the difference in samples and pre-conditioning of the aliquots. Figure 5b shows the value of the activation energy E = 0.861 ± 0.03 eV. This value is exactly within the limits established in the study by Pagonis et al. [35], and also agrees with the error with the corresponding values obtained in the recent inter-comparison programs ([39,40]). Finally, the values of the logarithm of frequency factor are represented in Figure 5c are also very well concentrated in a narrow region of values. This is a good achievement because the frequency factor is evaluated from the condition for the maximum, so the propagation of the errors of E and T_m in the values of frequency factor has an exponential dependence. For example, an 1% error in E results a 25–30% error in frequency factor [1].

We consider the parameter $R = A_n/A_m$, i.e., the ratio of the retrapping over recombination coefficients. The resulting R values in all cases are less that 0.01, which indicates clearly first-order kinetics [14,16]. Note that the parameter R is more representative of the physical process than the parameter b used in general-order kinetics. The reason is that R is related to both the re-trapping and recombination coefficient, whereas the empirical parameter b is related to the re-trapping process only.

In conclusion, these new results show that the properties of the TL peak 110 °C are an ideal test for simulation studies of quartz. Furthermore, this type of analysis can be used as a general approach while simulating the properties of any stable dosimetric material.

The method can also be used for more complex glow curves. A well-known example is the glow curve of the most commonly used dosimeter LiF:Mg,Ti.

LiF:Mg,Ti has a glow curve which consists of five individual peaks ranging from room temperature up to 250 °C. Its glow curve remains extremely stable from the lowest possible dose up to the onset of saturation after many irradiation–readout cycles. The kinetic parameters of each glow peak, especially of peak 5, are very well known and are generally accepted by dosimetry research groups.

It is suggested that CGCD analysis of any simulation of LiF:Mg,Ti should reproduce the well-known values of its kinetic parameters. These values were extensively tested, showing an excellent inter-laboratory repeatability, in the framework of the GLOCANIN inter-comparison project [5,6].

Additionally, an extensive study of this type was conducted by Kitis et al. [41]. A substantial part of personal dosimetry services is carried out by hot gas TLD readers, under exponential heating function, which gives glow curves shapes very different than that of the linear heating. The LiF:Mg,Ti chips are used in routine monitoring within the Greek Atomic Energy Commission (GAEC). The measuring device is an automatic RADOS reader using nitrogen gas for each readout up to 573 K (300 °C) for 15 s. Prior to their readout, the chips are post-irradiation annealed at 353 K (80 °C) for 1 h. The irradiation was performed at the Secondary Standard Dosimetry Laboratory of GAEC. ^{137}Cs and ^{60}Co were used for the linearity tests. A batch of 10 chips was irradiated at each dose. For the quality control irradiation, a TLD irradiator with ^{90}Sr/^{90}Y source was used.

Using appropriate analytical TL expressions for heating under exponential function [3], Kitis et al. [41] applied the CGCD analysis to a large number (∼100) of glow curves irradiated with doses between 0.1 and 1000 mGy and also to 130 quality assurance glow curves after a dose of 5 mGy. Kitis et al. [41] showed that the glow curves shapes were the same in the whole dose region examined, Furthermore, they were able to analyze all of them using the same values of activation energy, namely 1.24 ± 0.09 eV, 1.45 ± 0.05 eV and 2.28 ± 0.02 eV, for peaks 3, 4 and 5, correspondingly. The CGCD results for the 130 quality assurance glow curves (dose 5 mGy) were 1.16 ± 0.15, 1.46 ± 0.09 and 2.16 ± 0.06 for peaks 3, 4 and 5, correspondingly. The agreement between the above values and those of the GLOCANIN project [5,6] is very good.

It is concluded that the dosimetric LiF:Mg,Ti glow curves can be used reliably for simulation testing, due to their stability and the repeatability of their kinetic parameters. However, care must be taken in two cases where substantial variation of the TL glow curve can take place. Firstly, for aliquots which are pre-conditioned by annealing for many hours between 140 and 160 °C [42–44], and secondly, after very high irradiation doses [45,46].

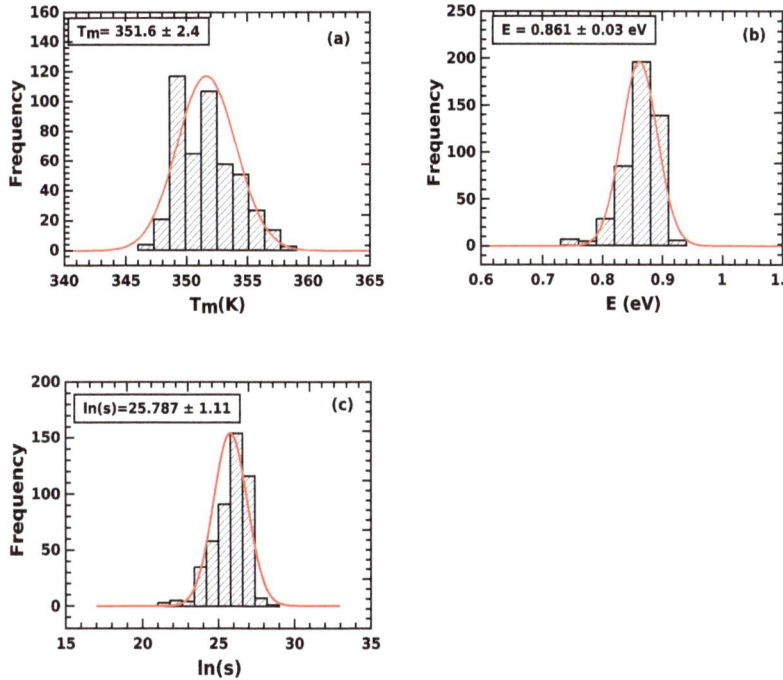

Figure 5. (**a**) Histogram for the position of peak maximum temperature, (**b**) histogram for the values of the activation energy and (**c**) histogram of the logarithm of the frequency factor.

9. Conclusions

It is shown by numerical simulation that there is no reliable mathematical method to extract the information from simulated complex TL glow curves. It is also shown that the CGCD is the only reliable method to extract all information from simulated complex TL glow curves. Furthermore, there is no physical contradiction between the superposition principle and CGCD analysis. The CGCD analysis reproduces all characteristic of a complex glow curve very accurately, especially when the competitor removes the competition between active traps. The CGCD is also accurate for cases of low competition between the traps. The only exception is in cases when the last peak of a complex glow curve takes the role of competitor, due to absence of competition from deep traps. Furthermore, this last peak can become a peak with an OTOR behavior and second-order kinetics. The presence of strongly competing TDDT results in very stable TL glow curve shapes. As the TDDT competition weakens, the shape of the glow curves changes. The agreement between input parameter values and output values evaluated by CGCD analysis is a very powerful criterion for the validity of simulations. Experimental glow curve characteristics with high repeatability can contribute substantially to simulation testing. The most characteristic examples are the experimental properties of the 110 °C glow peak of quartz, as well as the dosimetric glow curves of LiF:Mg,Ti.

Author Contributions: G.K. and V.P. contributed to the whole manuscript equally. All authors have read and agreed to the published version of the manuscript.

Funding: This research received no external funding.

Institutional Review Board Statement: Not applicable.

Informed Consent Statement: Not applicable.

Data Availability Statement: All results can be provided by the authors via E-mail.

Conflicts of Interest: The authors declare no conflict of interest.

References

1. Pagonis, V.; Kitis, G.; Furetta, C. *Numerical and Practical Exercises in Thermoluminescence*, 1st ed.; Springer: Springer New York, NY, USA, 2006.
2. Chen, R.; Pagonis, V. *Thermally and Optically Stimulated Luminescence: A Simulation Approach*, 1st ed.; Wiley: Chichester, UK, 2011.
3. Kitis, G.; Polymeris, G.S.; Pagonis, V. Stimulated luminescence emission: From phenomenological models to master analytical equations. *Appl. Radiat. Isot.* **2019**, *153* 108797. [CrossRef] [PubMed]
4. Peng, J.; Kitis, G.; Sadek, A.M.; Karsu Asal, E.C.; Li, Z. Thermoluminescence glow-curve deconvolution using analytical expressions: A unified presentation. *Appl. Radiat. Isot.* **2021**, *168* 109440. [CrossRef]
5. Bos, A.J.J.; Piters, T.M.; Gomez Ros, J.M.; Delgado, A. An intercomparison of glow curves analysis computer programs: I. Synthetic glow curves. *Radiat. Prot. Dosim.* **1993**, *51*, 257–264. [CrossRef]
6. Bos, A.J.J.; Piters, T.M.; Gomez Ros, J.M.; Delgado, A. An intercomparison of glow curves analysis computer programs: II. Measured glow curves. *Radiat. Prot. Dosim.* **1994**, *47*, 473–477. [CrossRef]
7. Pagonis, V. *Luminescence: Data Analysis and Modeling Using R, Use R!*, 1st ed.; Springer International Publishing: Berlin/Heidelberg, Germany, 2021.
8. Pagonis, V. *Luminescence: Signal Analysis Using Python*, 1st ed.; Springer International Publishing: Berlin/Heidelberg, Germany, 2022.
9. Chen, R.; Chen, R.; McKeever, S.W. *Theory of Thermoluminescence and Related Phenomena*; World Scientific: Singapore, 1997.
10. Bøtter-Jensen, L.; McKeever, S.W.S.; Wintle, A.G. *Optically Stimulated Luminescence Dosimetry*; Elsevier Science B.V.: Amsterdam, The Netherlands, 2003.
11. Bos, A.J.J. Thermoluminescence as a Research Tool to Investigate Luminescence Mechanism. *Materials* **2017**, *10*, 1357. [CrossRef] [PubMed]
12. Corless, R.M.; Gonnet, G.H.; Hare, D.G.E.; Jeffrey, D.J.; Knuth, D.E. On the Lambert W function. *Adv. Comput. Math.* **1996**, *5*, 329–359. [CrossRef]
13. Corless, R.M.; Jeffrey, D.J.; Knuth, D.E. A sequence series for the Lambert W function. In Proceedings of the International Symposium on Symbolic and Algebraic Computation, ISSAC, Maui, HI, USA, 21–23 July 1997; pp. 133–140
14. Kitis, G.; Vlachos, N.D. General semi-analytical expressions for TL,OSL and other luminescence stimulation modes derived from OTOR model using the Lambert W-function. *Radiat. Meas.* **2013**, *482*, 47–54. [CrossRef]
15. Singh, L.L.; Gartia, R.K. Theoretical derivation of a simplified form of the OTOR/GOT differential equation. *Radiat. Meas.* **2013**, *59*, 160–164. [CrossRef]
16. Sadek, A.M.; Eissa, H.M.; Basha, A.M.; Kitis, G. Resolving the limitation of peak fitting and peak shape methods in determinations of the activation energy of thermoluminescence glow peaks. *J. Lumin.* **2014**, *146*, 418–423. [CrossRef]
17. ROOT, A Data Analysis Framework. Available online: https://root.cern.ch (accessed on 29 November 2022).
18. MINUIT, a Physics Analysis Tool for Function Minimization, accessed from Released in ROOT. Available online: https://root.cern.ch (accessed on 29 November 2022).
19. GSL-GNU Scientific Library. Available online: https://www.gnu.org/software/gsl (accessed on 29 November 2022).
20. Balian, H.G.; Eddy, N.W. Figure of Merit(FOM): An improved criterion over the normalized chi-squared test for assessing the goodness-of-fit of gamma ray spectral peaks. *Nucl. Instr. Meth.* **1977**, *145*, 389–395. [CrossRef]
21. Horowitz, Y.S.; Yossian, D. Computerized glow curve deconvolution: Application to thermoluminescence dosimetry. *Radiat. Prot. Dosim.* **1995**, *60*, 1–114.
22. Subedi, B.; Oniya, E.; Polymeris, G.S.; Afouxenidis, D.; Tsirliganis, N.C.; Kitis, G. Thermal quenching of thermoluminescence in quartz samples of various origin. *Nucl. Instruments Methods Phys. Res. Sect. B* **2011**, *269*, 572–581. [CrossRef]
23. Oniya, E.O; Polymeris, G.S.; Tsirliganis, N.C.; Kitis, G. On the pre-dose sensitization of the various components of the LM-OSL signal of annealed quartz; comparison with the case of 110 °C TL peak. *Radiat. Meas.* **2012**, *47*, 864–869. [CrossRef]
24. Polymeris, G.S.; Oniya, E.O.; Jibiri, N.N.; Tsirliganis, N.C.; Kitis, G. In-homogeneity in the pre-dose sensitization of the 110 °C TL peak in various quartz samples: The influence of annealing. *Nucl. Instruments Methods Phys. Res. Sect. B* **2012**, *284*, 105–110 [CrossRef]
25. Bøtter-Jensen, L.; Bulur, E.; Duller, G.A.T.; Murray, A.S. Advances in luminescence instrument systems. *Radiat. Meas.* **2000**, *32*, 523–528. [CrossRef]

26. Randall, J.T.; Wilkins, M.H.F. Phosphorescence and electron traps I. The study of trap distributions. *Proc. R. Soc. Lond.* **1945**, *184*, 366–389.
27. Randall, J.T.; Wilkins, M.H.F. Phosphorescence and electron traps II. The interpretation of long-period phosphorescence. *Proc. R. Soc. Lond.* **1945**, *184*, 390–407.
28. Kelly, P.J.; Braunlich, P. Phenomenological theory of TL. *Phys. Rev. B.* **1970**, *1*, 1587 –1595. [CrossRef]
29. Kelly, P.J.; Laubitz, L.; Braunlich, P. Exact solutions of the kinetic equations governing thermally stimulated luminescence and conductivity. *Phys. Rev.* **1971**, *4*, 1960–1968. [CrossRef]
30. Opanowicz, A. Effect of initial trap occupancy on thermoluminescence characteristics of insulating crystals. *Phys. Stat. Sol. A* **1992**, *130*, 207–217. [CrossRef]
31. Sadek, A.M.; Kitis, G. A critical look at the kinetic parameter values used in simulating the thermoluminescence glow-curve. *J. Lumin.* **2017**, *183*, 533–541. [CrossRef]
32. Sadek, A.M.; Kitis, G. Impact of non-fulfillment of the super position principle on the analysis of Dthermoluminescence glow-curve. *Radiat. Meas.* **2018**, *116*, 14–23. [CrossRef]
33. Chen, R.; Pagonis, V. On the expected order of kinetics in a series of thermoluminescence (TL) and thermally stimulated conductivity (TSC) peaks. *Nucl. Instr. Meth. Phys. Res. B* **2013**, *312*, 60–69. [CrossRef]
34. Pagonis, V.; Kitis, G. Prevalence of first order kineticsG in thermoluminescence materials: An explanation based on multiple competition processes. *Phys. Stat. Sol. B* **2012**, *249*, 1590–1601. [CrossRef]
35. Pagonis, V.; Tatsis, E.; Kitis, G.; Drupieski, C. Search for common characteristics in the glow curves of quartz of various origin. *Radiat. Prot. Dosim.* **2002**, *100*, 373–376. [CrossRef]
36. Chen, R.; Yang, X.H.; McKeever, S.W. The strongly superlinear dose dependence of thermoluminescence in synthetic quartz. *J. Phys. D Appl. Phys.* **1988**, *21*, 1452–1457. [CrossRef]
37. Zimmerman, J. The radiation-induced increase of the 100 °C thermoluminescence sensitivity of fired quartz. *J. Phys. C Sol. St. Phys.* **1971**, *4*, 3265–3291 . [CrossRef]
38. Fleming, S.J.; Thompson, J. Quartz as a heat-resistant dosimeter. *Health Phys.* **1976**, *18*, 567–568. [CrossRef]
39. Schmidt, C.; Chruścińska, A.; Fasoli, M.; Biernacka, M.; Kreutzer, S.; Polymeris, S.G.; Sanderson, D.C.W.; Cresswell, A.; Adamiec, G.; Martini, M. How reproducible are kinetic parameter constraints of quartz luminescence? An interlaboratory comparison for the 110 °C TL peak. *Radiat. Meas.* **2018**, *110*, 14–24. [CrossRef]
40. Schmidt, C.; Chruścińska, A.; Fasoli, M.; Biernacka, M.; Kreutzer, S.; Polymeris, S.G.; Sanderson, D.C.W.; Cresswell, A.; Adamiec, G.; Martini, M. A systematic multi–technique comparison of luminescence characteristics of two reference quartz samples. *J. Lumin.* **2022**, *250*, 119070. [CrossRef]
41. Kitis, G.; Carinou, E.; Askounis, P. Glow–curve de–convolution analysis of TL glow–curve from constant temperature hot gas readers. *Radiat. Meas.* **2012**, *47*, 258–265. [CrossRef]
42. Zimmerman, D.W.; Rhyner, C.R.; Cameron, J.R. Thermal annealing effects on the thermoluminescence of LiF. *Health Phys.* **1966**, *12*, 525–531. [CrossRef] [PubMed]
43. Kitis, G.; Furetta, C. Thermoluminescence characteristics of monocrystaline LiF:Mg,Ti (DTG-4). *Nucl. Instr. Meth. Phys. Res. B* **1994**, *94*, 441–448. [CrossRef]
44. Kitis, G.; Tzima, A.; Cai, G.G.; Furetta, C. Low-temperature (80-C) annealing characteristics of LiF: Mg, Cu, P. *J. Phys. D Appl. Phys.* **1996**, *29*, 1601–1612. [CrossRef]
45. Charalambous, S.; Petridou, C. The thermoluminescence behaviour of LiF(TLD–100) for doses up to 10 MRad. *Numcl. Instr. Meth.* **1976**, *137*, 441–444. [CrossRef]
46. Obryk, B.; Khoury, H.J.; Barros, V.C.; Guzzo, P.L.; Bilski, P. On LiF:Mg,Cu,P and LiF:Mg,Ti phosphors high & ultra-high dose features. *Radiat. Meas.* **2014**, *71*, 25–30.

Disclaimer/Publisher's Note: The statements, opinions and data contained in all publications are solely those of the individual author(s) and contributor(s) and not of MDPI and/or the editor(s). MDPI and/or the editor(s) disclaim responsibility for any injury to people or property resulting from any ideas, methods, instructions or products referred to in the content.

Article

Thermally Assisted Optically Stimulated Luminescence (TA-OSL) from Commercial BeO Dosimeters

Georgios S. Polymeris

Laboratory of Archaeometry, Institute of Nanoscience and Nanotechnology, National Centre for Scientific Research "Demokritos", 15310 Agia Paraskevi, Greece; g.polymeris@inn.demokritos.gr

Abstract: BeO is another luminescent phosphor with very deep traps (VDTs) in its matrix that could not be stimulated using either thermal or conventional optical stimulations. The present study attempts to stimulate these traps using thermally assisted optically stimulated luminescence (TA-OSL), a combination of simultaneous thermal and optical stimulation that is applied to the material following a thermoluminescence measurement up to 500 °C. An intense, peak-shaped TA-OSL signal is measured throughout the entire temperature range between room temperature and 270 °C. This signal can be explained as the transfer of charges from VDTs to both dosimetric TL traps. Experimental features such as the peaked shape of the signal along with the presence of residual TL after the TA-OSL suggest that recombination of TA-OSL takes place via the conduction band. Isothermal TA-OSL is not effective for extending the maximum detection dose thresholds of BeO, unlike minerals such as quartz and aluminum oxide. Nevertheless, TA-OSL could be effectively used in order to either (a) control the occupancy of VDTs, circumventing the intense sensitivity changes induced by long-term uses and high accumulated dose to the VDTs, or (b) measure the total dose accumulated over a series of repetitive dose calculations.

Keywords: BeO; TL; TA-OSL; supralinearity; thermally assisted processes

Citation: Polymeris, G.S. Thermally Assisted Optically Stimulated Luminescence (TA-OSL) from Commercial BeO Dosimeters. *Materials* **2023**, *16*, 1494. https://doi.org/10.3390/ma16041494

Academic Editor: Dirk Poelman

Received: 8 January 2023
Revised: 5 February 2023
Accepted: 6 February 2023
Published: 10 February 2023

Copyright: © 2023 by the author. Licensee MDPI, Basel, Switzerland. This article is an open access article distributed under the terms and conditions of the Creative Commons Attribution (CC BY) license (https://creativecommons.org/licenses/by/4.0/).

1. Introduction

Beryllium oxide (BeO) could be considered as the luminescence dosimeter of the current decade. This material was very early suggested, during the 1960s, as a thermoluminescence (TL) dosimeter, due to a plethora of advanced properties. The most important include an easy TL glow curve with well-isolated, non-overlapping TL peaks, increased sensitivity to ionizing radiation, linearity of dose–response, and a near-tissue equivalent value of Z_{eff} = 7.14, i.e., very close to the value of Z_{eff} = 7.35–7.65 for biological tissues and the value of Z_{eff} = 7.51 for water [1]. This means that this material has a minimal (<20%) over- and under-response to low-energy photons, which is optimal for a wide range of applications in different fields of medicine, such as radiotherapy or radiation diagnostics [2–4]. A typical TL glow curve of a BeO dosimeter yields three TL glow peaks, with temperatures T_m corresponding to maximum intensities I_m at ~75, 190, and 317 °C for heating rate 1 °C/s [5].

Based on the number of available TL peaks within a TL glow curve and their corresponding peak positions, an experimental TL glow curve is usually divided into three main temperature regions [6]. Starting from room temperature (RT) up to around 150 °C, this region is considered the low-temperature region, including shallow traps. These traps are thermally unstable at ambient temperatures and thus inappropriate for dosimetric applications. The TL peak at 75 °C stands as a typical example of a TL peak at this specific region. Ranging between 150 and 300 °C, the second temperature region is considered the main dosimetric region. These traps are thermally stable at ambient temperatures and thus appropriate for dosimetry applications. All commercially available TL dosimeters, such as beryllium oxide, yield TL peaks at this specific temperature region [7]. The TL peak at

190 °C is the main dosimetric TL peak of BeO, as it is the TL peak that is widely used for dosimetric applications. Finally, the temperature region ranging between 300 and 500 °C is called the high-temperature region, corresponding to relatively deeper traps. These traps yield very prolonged lifetimes and are consequently more appropriate for dosimetry using natural phosphors, especially for dating applications; the TL peak with delocalization temperature T_m at 317 °C is a typical example.

However, the increased sensitivity of the material to light was firstly recognized as a major drawback that recently has resulted in a new advantage; BeO was suggested as a sensitive dosimeter for Optically Stimulated Luminescence (OSL). Bulur and Göksu [8] were the first who reported new (OSL) properties from an old dosimetric (TL) material before it was finally adopted in a commercial OSL dosimetry system [2,9–12]. The majority of the aforementioned TL advantages are relevant also for the OSL signal, in conjunction to a convenient OSL shape that yields a fast-decaying OSL component [13–15]. Nevertheless, the mechanism responsible for the OSL signal is not completely understood. Thus, conflicting results exist regarding the origin of the trapped charges responsible for the OSL signal. Recently, the low (5%), albeit substantial, fading in the OSL signal in the first 24 h of storage in the dark at room temperature, followed by the stability of the signal over many months, was recognized as another disadvantage. Among several other TL and OSL features, this latter quality was thoroughly discussed in the framework of the latest Solid-State Dosimetry conference that took place in Hiroshima, Japan in 2019. For an overview on both favorable and limiting properties of BeO, readers are referred to recent articles on the topic by Aşlar et al. [14], Polymeris et al. [15], and Yukihara [3,16,17].

Among these properties, intense transfer effects were often reported in the luminescence signal from BeO. Bulur and Göksu [8] were the first authors to report intense the OSL signal following TL without any intermediate irradiation. According to their terminology, they have measured "OSL after TL" at room temperature. This is exactly the thermally stimulated recuperation of the OSL signal that Yukihara [3,16] later termed thermally transferred OSL (TT-OSL hereafter). This signal was detected following previous TL within a narrow temperature range (225 to 310 °C), with a maximum intensity of barely 50% of the corresponding OSL intensity at 275 °C. Moreover, the TT-OSL curve indicates the typical decaying shape to that of a trivial OSL signal, which suggests that the same trapping centers are responsible for the OSL and the TT-OSL signal [3,16]. In an effort to exploit this specific signal for dosimetric applications, Yukihara has reported dose–response features for both low- [3] and high-dose [16] regions. Phototransfer is another proof of transfer effect that is being observed from stable trapping centers in BeO; it is stable enough to be occupied after heating to temperatures higher than 400 °C, leading to either TL (PTTL [18]) or OSL signals (PTOSL [19]).

Intense transfer effects were also reported by Aşlar et al. [14] in their effort to study possible correlations among the TL, OSL, and Electron Paramagnetic Resonance (EPR) signals in commercially available BeO dosimeters. Transfer was monitored either (a) between the two dosimetric traps or (b) from deeper than both traps responsible for the dosimetric TL peaks. Due to intense phototransfer effects from deeper to shallower traps, quantitative correlation using the signal intensities as a unique probe was not recommended. Moreover, following deconvolution of OSL decay curves and reconstruction, namely correction for thermal quenching, these authors have reported OSL components indicating an integrated signal that increases with increasing measurement temperature. This is a characteristic property of a thermally assisted process. Thus, Aşlar et al. [14] have suggested that indirect thermally assisted optically stimulated luminescence (TA-OSL) signals were monitored, indicating that transfer is more intense towards the trap corresponding to the TL peak at 317 °C. For this reason, even after depleting both TL peaks, intense OSL curves were still measured.

TA-OSL comprises simultaneous thermal and optical stimulation of traps that indicate delocalization temperatures T_m higher than 500 °C; these were termed very deep traps (VDTs hereafter). According to the early publication by Polymeris et al. [20], initially all conventional traps are emptied by a TL measurement up to 500 °C. Then, TA-OSL is measured in a continuous wave mode at a high temperature, promptly after the TL measurement and without any additional dose; it is the isothermal TA-OSL mode [6]. The signal registers the charge that was captured at traps with activation energies beyond 2 eV [6] that is transferred to shallower, albeit dosimetric, traps due to being empty because of the previous TL measurement. In other words, it is the most appropriate technique for stimulating VDTs, without heating the dosimeter to temperatures higher than those imposed by luminescence instrumental limitations [21,22]. The presence of such very deep traps is almost a ubiquitous feature of all wide-band semiconductors, either naturally occurring such as quartz, feldspars, and NaCl, or artificial such as Al_2O_3 with various dopants, $BaSO_4$:Eu, etc.; for an extended list of these materials, readers are referred to [22]. In this latter publication, just a selection of TA-OSL curves is presented for the case of BeO; nevertheless, a detailed study on the dosimetric properties of this signal is still missing from the literature.

TA-OSL was initially suggested as an alternative experimental method in order to not only measure the signal of the much deeper traps in Al_2O_3:C without heating the sample to temperatures greater than 500 °C, but also use this signal for high-dose-level dosimetry purposes as well [20]. Later on, TA-OSL was identified as a promising technique for extending both luminescence dosimetric [6] and dating limits [23], due to the following assets: (a) the lifetimes of such traps, of the order of 10^9 years or even more [6], combined with (b) insignificant athermal (anomalous) fading losses, even for the cases of reference materials such as feldspars or apatites [24,25], and (c) the higher charge capacity of these traps, resulting in dose–response curves of favorable features and larger dose saturation levels. Minerals such as quartz [26,27] and Al_2O_3:C [20–22] are being extensively studied, providing promising TA-OSL features and properties. Nevertheless, the favorable properties of TA-OSL for either dating or dosimetric applications are not reported for all materials. Typical examples include NaCl [28] and MgO (under preparation).

The present work attempts to fill in the hiatus of the scientific literature regarding the lack of an extended TA-OSL study in commercially available BeO samples. The aim of the present work is multifold, including (a) a detailed study on the TA-OSL features and properties of commercial BeO, (b) optimization of the measuring parameters, (c) study on the dose–response properties of the corresponding signal, and (d) comparison of the dose–response features of both conventional photo-transferred TL.

2. Materials and Methods

2.1. Materials and Apparatus

Commercially available BeO dosimeters in the form of pellets were used in the framework of the present study. The pellets were in square disc form with dimensions of 4 mm and thickness of 1 mm and were purchased from Thermalox 995, Brush Wellman Inc., USA. These detectors were annealed in an oven, inside an alumina crucible, at 900 °C for 60 min before the experiments, cooled immediately by putting the crucible in contact with a heat sink. The temperature was selected in order to empty all traps.

All stimulated luminescence measurements were performed using a Risø TL/OSL reader (model TL-OSL-DA-20, Reader ID: 267) equipped with a (0.12 ± 0.03) Gy/s ^{90}Sr/^{90}Y β-ray source. Unless otherwise stated, all TL and RTL measurements were performed up to a temperature of 500 °C. In all cases, a low heating rate of 2 °C/s was applied to avoid significant temperature lag [29]. The OSL measurements were performed in the continuous wave (CW-OSL) mode with blue LEDs (~470 nm) at constant stimulation intensity (90% of the maximum 36 mW/cm^2). All stimulated luminescence signals were recorded using a 7.5 mm thick Hoya U-340 filter (270–380 nm, FWHM 80 nm) in front of a bi-alkali EMI 9235QB photo multiplier tube (PMT). The test dose for each protocol is different.

2.2. Experimental Protocols

The experimental protocols that were applied constitute modified versions of the protocols that have been previously used for the study of TA-OSL in the cases of Al_2O_3:C [21] as well as of NaCl [28]. Two major protocols were applied, one for the selection of the appropriate stimulation temperature of the TA-OSL measurement, along with one that studies the dosimetric properties of the corresponding signal. In the former case, TA-OSL is measured for various stimulation temperatures according to protocol A (Table 1). In the latter case, the TA-OSL signal is measured for various doses to check the dose–response behavior of the signal. The two different protocols include steps for checking the sensitivity of the dosimetric traps of the material, following emptying the VDTs. Sensitivity is measured as the integrated TL peak signal following a steady test dose; a different test dose was used for each protocol. In addition to the action involved for each step, comments on the corresponding necessity of each step within the protocols are detailed in Tables 1 and 2. Each cycle of measurements was executed for both protocols using a different, freshly annealed BeO disk.

Table 1. Protocol A: TA-OSL for various stimulation temperatures for identifying the optimum stimulation temperature. The sample was cooled down to room temperature at the end of step 5 of each measurement cycle.

Step No.	Action	Comments and Technical Specifications
Step 1:	TL measurement	Remove any prior existing signal
Step 2:	Irradiation using 5 Gy dose	Populate all traps, including the VDTs
Step 3:	TL measurement	Empty shallow, dosimetric, and deep traps and monitor initial sensitivity S_0
Step 4:	Isothermal TL (ITL) (at room temperature for 60 s)	Check for possible overflow of the PMT
Step 5:	OSL measurement at temperatures T_{st} ranging from $\underline{30\ °C}$ up to $\underline{270\ °C}$ (in steps of 20 °C) over a period of 500 s	Measure the TA-OSL curves for various stimulation temperatures
Step 6:	TL measurement	Obtain the Residual TL (RTL) curves
Step 7:	Irradiation using 5 Gy dose	Populate all traps
Step 8:	TL measurement	Monitor final sensitivity S_f

Table 2. Protocol B: TA-OSL for various doses to check the dose–response behavior of the signal. The sample was cooled down to room temperature at the end of step 7 of each measurement cycle.

Step No.	Action	Comments and Technical Specifications
Step 1:	TL measurement	Remove any prior existing signal
Step 2:	Irradiation using 0.1 Gy dose	Populate shallow, dosimetric, and deep traps using the minimum dose
Step 3:	TL measurement	Empty shallow, dosimetric, and deep traps and monitor initial sensitivity S_k
Step 4:	Irradiation with a dose D_i *	Populate all traps, including the VDTs
Step 5:	TL measurement	Empty shallow, dosimetric, and deep traps and monitor high-dose sensitivity S_d
Step 6:	Isothermal TL (ITL) (at room temperature for 60 s)	Check for possible overflow of the PMT following the TL measurement

Table 2. Cont.

Step No.	Action	Comments and Technical Specifications
Step 7:	OSL measurement at the optimum stimulation temperature T_{st} over a period of 500 s	Measure the TA-OSL curves for various doses
Step 8:	Isothermal TL (ITL) (at room temperature for 60 s)	Check for possible overflow of the PMT following the TA-OSL measurement
Step 9:	TL measurement	Obtain the Residual TL (RTL) curves
Step 10:	Irradiation using 0.1 Gy dose	Populate shallow, dosimetric, and deep traps using the minimum dose
Step 11:	TL measurement	Monitor final sensitivity S_W

* The doses applied were 0.1, 0.25, 0.5, 1, 2, 4, 6.5, 8.5, 15, 21, 32, 43, 65, 85, and 110 Gy; each cycle of steps 1–11 was applied to different BeO disk.

2.3. RTL Deconvolution and PSM

All RTL glow curves were analyzed into individual TL peaks using the analytical expressions in the framework of the one trap one recombination center (OTOR) model [30]. The analytical expression given by Kitis and Vlachos [31] takes advantage of the Lambert W function $W(z)$ [32]:

$$I(T) = I_m \exp\left(-\frac{E}{kT} \cdot \frac{T_m - T}{T_m}\right) \cdot \frac{W(z_m) + W(z_m)^2}{W(z) + W(z)^2} \quad (1)$$

where z is now approximated by the expression:

$$z = \exp\left(\frac{R}{1-R} - \ln\left(\frac{1-R}{R}\right) + \frac{E \cdot \exp\left(\frac{E}{kT_m}\right)}{kT_m^2(1 - 1.05 \cdot R^{1.26})} \cdot F(T,E)\right) \quad (2)$$

and z_m is the value of z from Equation (2) for $T = T_m$.

Here R denotes the ratio A_n/A_m, with A_n and A_m being the re-trapping and recombination coefficients, respectively (in cm^3 s^{-1}). The $p(t,T)$ corresponds to the stimulation probability which is described by Equation (3):

$$p = \tau^{-1} = s \cdot \exp\left(-\frac{E}{kT}\right) \quad (3)$$

Equations (1)–(3) are used to fit the experimental TL glow curves, with the activation energy E, the temperature of maximum intensity T_m, and the ratio R (with R < 1) being the fitting parameters. The function $F(T,E)$ in this expression is the exponential integral appearing in TL models and is expressed in terms of the exponential integral function $E_i[-E/kT]$ as [30,32,33]:

$$F(T,E) = \int_{T_0}^{T} e^{-\frac{E}{kT}} dT = T \cdot \exp\left(-\frac{E}{kT}\right) + \frac{E}{k} \cdot E_i\left[-\frac{E}{kT}\right] \quad (4)$$

There are two numerical methods to evaluate the exponential integral function $E_i[z]$: (a) as an elementary function in commercially available software packages and (b) through approximate analytical expressions. For the majority of the cases, it has been implemented as an elementary built-in function in various software packages [30].

All curve fittings were performed using the software package Microsoft Excel, with the Solver utility [34], while the goodness of fit was checked using the Figure Of Merit (F.O.M.) [35]:

$$\text{F.O.M.} = \sum_i \frac{|Y_{\exp} - Y_{\text{Fit}}|}{A} \quad (5)$$

where $Y_{\exp}$ is the data point on the experimental curve, Y_{Fit} is the data point on the fitted curve, and A is the area of the fitted curve. A successful deconvolution process is obtained by optimizing fitting parameters so that F.O.M. values are as low as possible. For the present study, F.O.M. values were lower than 3.1%.

The peak shape methods are used in order to calculate the activation energy of the two RTL peaks, as these are not overlapping but isolated. The peak shape methods take advantage of the geometrical characteristics of the RTL glow curve, mainly the delocalization peak temperature T_m and the high- and low-temperature (T_1 and T_2, respectively) sides of the glow curve in the half maximum intensity. The approach of Kitis and Pagonis [36] was adopted, based on the general order kinetics model. Thus, the presented calculations in this work are based on the following equations:

$$C_\omega = \frac{\omega \cdot I_m}{\beta \cdot n_0} \quad (6)$$

$$C_\delta = \frac{\delta \cdot I_m}{\beta \cdot n_m} \quad (7)$$

$$C_\tau = \frac{\tau \cdot I_m}{\beta \cdot (n_0 - n_m)} \quad (8)$$

In these equations I_m expresses the maximum TL intensity, β is the heating rate, $n_m = \int_{T_m}^{\infty} I\, dt$ is the concentration of trapped electrons at maximum (half integral at high-temperature maximum, in units of cm^{-3}), $n_0 = \int_0^{T_{max}} I\, dt$ is the initial concentration of trapped electrons (total integral, also in units of cm^{-3}), and C_ω, C_δ, C_τ are the quantities that characterize the degree by which the area of a single glow curve approaches the area of a triangle. The following geometrical quantities can be defined:

$$\delta = T_2 - T_m, \quad \tau = T_m - T_1, \quad \omega = T_2 - T_1 \quad \mu_g = \delta/\omega \quad (9)$$

Here μ_g defines the symmetry (or geometrical) factor of the TL glow peak. The estimation of the activation energy is based on the ω and the integral symmetry factor, which can be defined as the ratio of the concentration of trapped electrons at maximum (n_m) over the initial concentration of trapped electrons (n_0), thus $\mu'_g = \frac{n_m}{n_0}$; therefore, according to [31,36,37], Equations (6)–(8) could be expressed as:

$$E_a = C_a \cdot \left(b^{\frac{b}{b-1}}\right) \cdot \frac{k \cdot T_m^2}{a} - 2kT_m \quad (10)$$

where α can be ω, δ, or τ.

In Equation (2) of the OTOR model, the value of the fitting parameter R denotes whether the re-trapping probability is significant; for values close to zero, the re-trapping probability is insignificant, while for values close to one, recombination and re-trapping probabilities are comparable. Similarly, in Equation (10), parameter b signifies the order of kinetics in the corresponding GOK models; for b values close to unity, the first order of kinetics implies negligible re-trapping, while for values close to two, the second order of kinetics suggests significant re-trapping probability.

2.4. Thermal Quenching and Reconstruction

As TA-OSL measurements include the combined action of thermal and optical stimulation, the presence of the thermal quenching effect is expected to substantially suppress the TA-OSL intensity. Thermal quenching describes the decrease in luminescence efficiency with increasing temperature [38] due to the increased probability of non-radiative transitions. The dependence of the luminescence efficiency, $\eta(T)$, on the temperature is given according to the following formula [38]:

$$\eta(T) = \frac{1}{1 + Ce^{-\frac{W}{kT}}} \quad (11)$$

where the parameters C and W are the so-called "quenching parameters", T is the temperature in units of K, and k is the Boltzmann constant. For more details regarding the physical meaning of the W parameter (in units of eV) as well as the dimensionless C parameter, readers are referred to [39]. For the case of isothermal TA-OSL, as the entire measurement takes place at a steady temperature T_{st}, the corrected unquenched integrated TA-OSL intensity $I_{uq}(T_{st})$ can be calculated using the following formula:

$$I_{uq}(Tst) = \frac{I_q(Tst)}{\eta(Tst)} \quad (12)$$

where $I_q(T_{st})$ and $I_{uq}(T_{st})$ are the integrated TA-OSL intensities which are quenched and unquenched, respectively, while $\eta(T_{st})$ corresponds to a single value of the thermal quenching efficiency at the specific measurement temperature. In this study, reconstruction was applied according to Equation (11), using the values of thermal quenching parameters W and C obtained by Aşlar et al. [40] for the Hoya U340 filter; the specific values used were $W = 0.62$ eV and $C = 1.3 \times 10^6$, namely, the values that were calculated for TL peak 2. These values were used because the Hoya U-340 filter was used for all measurements.

3. Results

3.1. Shapes of TL, TA-OSL, and RTL Glow Curves

TL glow curves of BeO indicate the same features as those reported by Aşlar et al. [5]; these are ubiquitously present in all TL glow curves of both protocols (steps 3 and 8 in protocol A, steps 3, 5, and 11 in protocol B) in the present study, as the inset of Figure 1 reveals. Nevertheless, there is a shift of the T_m towards higher values (203 and 329 °C for TL peaks 2 and 3, respectively). In the present study, the nomenclature will be the same as in the previous studies of our research group; the main dosimetric TL peak with delocalization temperature at 203 °C is termed TL peak 2 hereafter, while the TL peak with $T_m = 329$ °C is termed TL peak 3. Between these, TL peak 2 stands as the most intense, for a wide range of doses. For a detailed analysis on the kinetic parameters of the TL glow curves of this specific material, readers are referred to [5] and references therein.

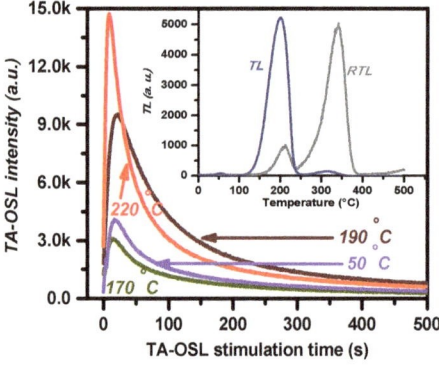

Figure 1. Examples of isothermal TA-OSL curves from BeO discs for four different measurement temperatures (protocol A); the inset presents a TL before and an RTL glow curve after TA-OSL at 190 °C. Intense TA-OSL signal is measured for all stimulation temperatures.

Examples of TA-OSL curves from BeO for a selection of stimulation temperatures (Figure 1) include one at low temperature along with three at higher counterparts, namely T_{st} = 50, 170, 190, and 220 °C. All curves correspond to TA-OSL measurements received promptly after the end of step 4 in protocol A. The signal in all cases emerges from VDTs, since each TA-OSL curve is monitored promptly after a TL measurement up to 500 °C. The shape of the TA-OSL does not resemble a classic background OSL signal. Instead, the intense TA-OSL signal is measured ubiquitously for all temperatures of stimulation according to step 5 of protocol A. All TA-OSL curves are bell-shaped, independent of the dose, similar to the TA-OSL curves reported for the case of aluminum oxide [20,21] and CaF_2:N [41]. The maximum intensity of the TA-OSL is monitored at stimulation time t_{max} (s); in general, this value is (a) lower than 50 s in all cases (thus, due to the 500 s of total stimulation, the peak-shaped TA-OSL signal yields a long tail at high stimulation times) and (b) shifted towards lower values as the stimulation temperature increases. Thus, as the t_{max} is temperature-dependent, the bell-shaped TA-OSL curve changes with stimulation temperature. This feature is strongly revealed from Figure 2a; t_{max} indicates an initial, wide peak with the maximum value attained at 110 °C, followed by an almost linear decrease for temperatures beyond. This behavior, being similar to the corresponding feature of the TA-OSL in aluminum oxide [20,21], provides an experimental argument for the presence of competition effects during TA-OSL recombination [42,43].

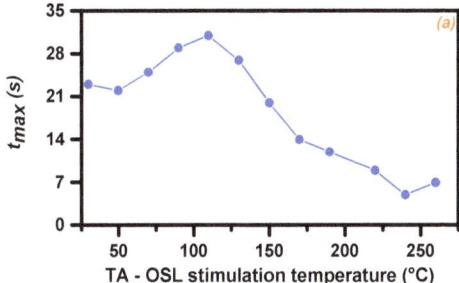

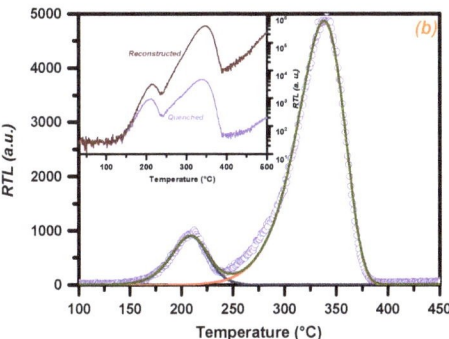

Figure 2. Plot (**a**) presents the time stimulation parameter t_{max} corresponding to the maximum intensity of the isothermal TA-OSL curve from BeO discs versus different stimulation temperatures (protocol A). Plot (**b**) presents the RTL glow curve after TA-OSL at 190 °C (experimental points as open dots) deconvolved into its two individual TL peaks; solid lines correspond to each individual TL peak along with the fitted TL glow curve. The fitting parameters for each RTL peak stand in good agreement with the corresponding values of TL peaks 2 and 3. Inset of plot (**b**) presents the same RTL glow curve before and after reconstruction; TL signal for temperatures higher than 375 °C suggests the presence of at least one more TL peak with delocalization temperature beyond 500 °C.

Residual TL (RTL) glow curves, namely TL signals promptly after the TA-OSL measurement, are measured during steps 6 and 9 in protocols A and B, respectively. RTL glow curves do not yield the unstable peak at ≈75 °C. Moreover, between TL peaks 2 and 3, RTL indicates the latter being more intense than the former for the majority of the stimulation temperatures. This feature is indicated by the inset of Figure 1, where both TL (from step 3) and RTL (from step 6) glow curves are presented for the case of the TA-OSL stimulation temperature of 190 °C. Figure 3 presents the intensities for both RTL peaks 2 and 3, plotted versus the TA-OSL measurement temperatures, in terms of integrated TL signal following deconvolution analysis. According to this figure, it becomes apparent that RTL peak 2 is more intense than RTL peak 3 only for a short range of stimulation temperatures (90 and 110 °C). Nevertheless, Figure 3 strongly supports the fact that intense transfer effects take place during the TA-OSL measurements.

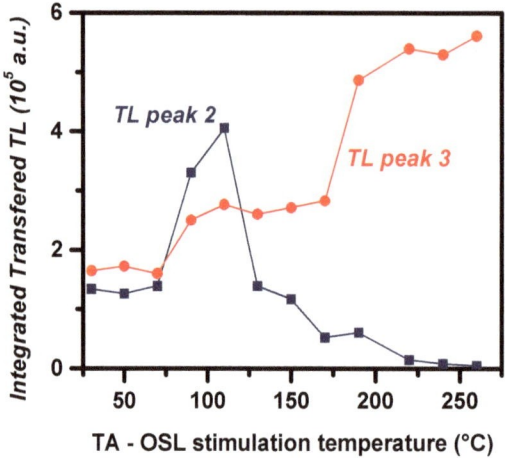

Figure 3. RTL signal (step 6 of protocol A) for both peaks 2 and 3 versus stimulation temperature of the TA-OSL signal (step 5), in terms of integrated TL signal following deconvolution analysis.

Within the temperature range 70–150 °C, intense transfer takes place to both RTL peaks; nevertheless, within this aforementioned temperature range, the transferred signal is much more intense for the case of RTL peak 2, as the stimulation temperature is lower than the delocalization temperature of this peak. For higher stimulation temperatures, which overlap with the temperature region of TL peak 2, intense transfer solely takes place to RTL peak 3.

It is worth emphasizing the asymmetry of the shape of the latter RTL glow peak, implying the fact that it could be described by the first order of kinetics. The enhanced sensitivity of RTL peak 3 provides a unique opportunity to study the corresponding structure. All RTL glow curves from protocol A were deconvolved. The aim of this deconvolution is twofold, namely to (a) calculate the integrated TL signal for both peaks and use it to construct Figure 3, as well as (b) to calculate the kinetic parameters of both RTL peaks. As RTL peaks 2 and 3 are not overlapping, kinetic analysis for both was studied using both deconvolution as well as PSM. Figure 2b presents an example of a deconvolved RTL glow curve, while Table 3 presents an outline of the results of both the deconvolution as well as the PSM analysis concerning the activation energies and the order of kinetics for either RTL peak.

Table 3. Kinetic parameters of the RTL glow peaks from both protocols according to the peak shape method (PSM) and deconvolution analysis. All symbols were explained inside the text and correspond to averages over the number of existing RTL glow curves for each protocol; E_{Dec} signifies the activation energy that was calculated according to the deconvolution analysis, while E_{PSM} is the respective value according to peak shape method analysis.

Protocol	RTL Peak T_m (°C)	E_{Dec} (eV)	R	ω (°C)	E_{PSM} (eV)	b
A	203.0 ± 1.0	1.21 ± 0.14	0.11 ± 0.03	43.0 ± 1.0	1.09 ± 0.09	1.15 ± 0.06
A	329.5 ± 0.5	1.32 ± 0.12	0.09 ± 0.02	54.0 ± 0.5	1.16 ± 0.11	1.09 ± 0.04
B	202.0 ± 1.5	1.18 ± 0.15	0.15 ± 0.02	41.5 ± 1.5	1.14 ± 0.12	1.14 ± 0.08
B	328.5 ± 1.5	1.27 ± 0.15	0.11 ± 0.01	54.5 ± 0.5	1.20 ± 0.12	1.07 ± 0.05

It is important to note that no correction for thermal quenching was applied before the RTL curves were analyzed using either the peak shape method or deconvolution analysis. The kinetic analysis on the RTL glow curves was performed to check whether the same trapping centers are responsible for the RTL and the ordinary TL signal. For such a comparison, reconstruction is not required. Moreover, Figure 2b presents an inset with an example of a reconstructed RTL glow curve. In both cases of reconstructed and quenched TL glow curves, the signal for temperatures beyond 375 °C justifies the presence of another TL glow peak with a delocalization temperature slightly larger than 500 °C. However, as this entire TL peak is not monitored, the deconvolution analysis of the entire TL glow curve will result in erroneous results due to resolution failure in the third peak. For a detailed analysis on the kinetic parameters of the reconstructed TL glow curves of this specific material, readers are referred to [5] and references therein.

3.2. TA-OSL versus Stimulation Temperature

Measuring TA-OSL for various incrementally increasing stimulation temperatures has become a standard routine in isothermal TA-OSL studies of various materials, with the aim being threefold [6]: (a) to verify the thermally assisted nature of the signal, (b) to determine the activation energy of the thermal assistance of the procedure, and (c) to identify the optimum stimulation temperature T_{st} of isothermal TA-OSL.

The integrated TA-OSL signal is plotted versus stimulation temperature in Figure 4. In all thermally assisted processes, unless thermal quenching is present, the signal intensity is expected to enhance with increasing stimulation temperature. Similar is the case for the thermally assisted OSL; the integrated OSL signal is expected to increase with increasing stimulation temperature, contrary to the signal of the conventional OSL, which is attributed to electrons from dosimetric traps. In the present case, the integrated TA-OSL signal (filled data points) indicates a smooth decrease for temperatures up to 170 °C, along with a mild increase for temperatures beyond. Nevertheless, these data were not corrected for the effect of thermal quenching. Figure 4 also presents the reconstructed data, namely following correction for compensation; these are presented as open data points. Such a correction was performed according to Equation (12). The behavior of the reconstructed data is similar to the behavior reported by Aşlar et al. [14] for the case of the component C_1 of their ordinary OSL signal when measured at elevated temperatures (please refer to their Figure 12B).

For the calculation of the thermal assistance activation energy, the natural logarithm of the corrected (reconstructed) integrated TA-OSL signal is plotted versus the 1/kT (where T is the absolute stimulation temperature in units of K). The Arrhenius plot that is presented in the inset of Figure 4 is derived. A clear linear region is yielded in this latter Arrhenius plot. The slope of linear fit corresponds to the thermally assisted activation energy that was calculated as (0.81 ± 0.06) eV. This value stands among the highest activation energies calculated for the thermal assistance on the TA-OSL signal. Higher values were calculated and reported only for the cases of Al_2O_3:C (1.08 eV [20,21]) and quartz (0.95 eV [26]).

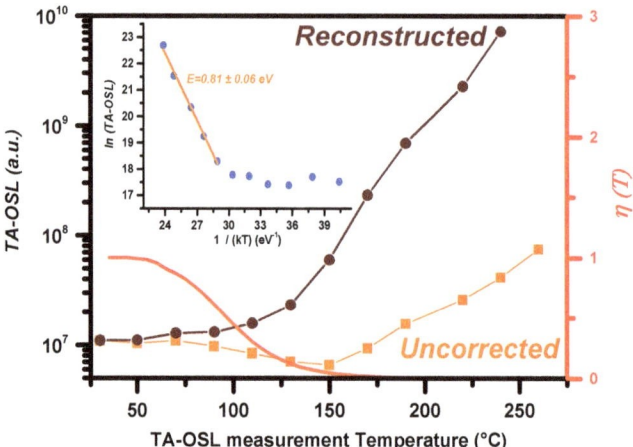

Figure 4. Integrated TA-OSL signals plotted versus the TA-OSL stimulation temperature (step 5 of protocol A). Both uncorrected data for thermal quenching as well as reconstructed data are presented. The same plot presents the thermal dependence of the thermal quenching efficiency $\eta(T)$ (red line, right-hand axis). Inset presents the corresponding Arrhenius plot of the reconstructed data.

Finally, the main criterion for the selection of the optimum TA-OSL measurement temperature deals with an enhanced signal-to-noise ratio. Especially in the presence of thermal quenching, a TA-OSL signal with high intensity becomes crucial. According to Figure 4, TA-OSL measurement at temperatures higher than 190 °C is proposed; this is the temperature region at which the integrated TA-OSL signal increases monotonically with increasing temperature. For reasons that include technical limitation on the measurement instrumentation, the optimum temperature is 220 °C.

3.3. TL Sensitivity Changes Following TA-OSL at Various Stimulation Temperatures

Protocol A consists of eight steps for each stimulation temperature. Among these, four involve TL measurements. TL of step 1 is used just for zeroing purposes, while the variation of either the shape or the intensity of the RTL glow curve of step 6 has already been discussed in Section 3.1. The other two TL measurements are quite important, as the comparison between the S_f and the S_0 glow curves (before and after the TA-OSL measurement, respectively) provides strong indications regarding possible changes in either TL glow curve shape or sensitivity. Despite the fact that TA-OSL was reported to induce shape changes in many luminescence phosphors such as Al_2O_3:C [21] and halites [28], the shape of the TL glow curve of BeO does not change after TA-OSL measurement. Moreover, the corresponding ratio of $\frac{S_f}{S_0}$ provides for each TL peak a measure of sensitization; this plot is presented in Figure 5 for both TL peaks. It is important to note that this ratio takes a value around unity (within statistical errors) for TL peak 2. For TL peak 3, this latter ratio takes values varying between 1.2 and 1.3, implying mild sensitization by 20–30%. There is not any specific behavior of this ratio with the TA-OSL stimulation temperature.

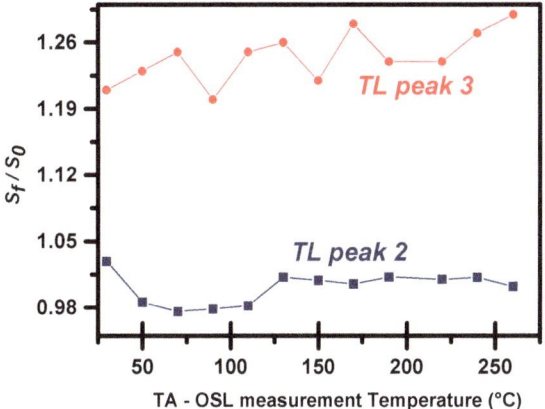

Figure 5. The ratio of the final (after measuring TA-OSL) over the initial (before measuring TA-OSL) TL sensitivity as a function of the stimulation temperature in the framework of protocol A.

3.4. TL and TA-OSL Dose–Responses

While the application of protocol A is required from a methodological point of view, for the optimization of the TA-OSL measurement sequence, protocol B will provide the most important information from a dosimetric point of view. Figure 6 presents the dose–response curves for both TA-OSL and TL signals from BeO. This material indicates favorable TL dose–response features, as TL peak 2 yields supralinear dose response throughout three decades of doses, namely from 0.1 to 100 Gy. The dose response of TL peak 3 is also supralinear, indicating quick saturation around 10 Gy and beyond. These features stand in good agreement with the TL dose–response features that were presented by Polymeris et al. [15]; in this recent paper, the dose–response curves were fitted using the formula:

$$TL, OSL = a \cdot D^m \qquad (13)$$

with the linearity index m being 1.71 for the case of TL peak 2 and 1.45 for the case of TL peak 3. According to Section 3.2, the most appropriate temperature for the TA-OSL measurement is 220 °C. The corresponding TA-OSL dose–response curve is presented in Figure 6 in terms of integrated signal, indicating supralinearity up to almost 30 Gy and a saturation sublinearity for larger doses. Nevertheless, due to the relative low intensity of the integrated TA-OSL signal compared to the corresponding integrated signal of TL peak 2, the TA-OSL dose response was measured at two more temperatures, namely 190 and 240 °C; these temperatures were selected as they surround 220 °C. All dose–response curves are presented in Figure 6. By comparing these three TA-OSL doses–response curves, not only among them, but also to the corresponding counterparts of TL peaks 2 and 3, one can easily conclude that independent of the stimulation temperature, all integrated (over 500 s) TA-OSL signals are more intense than the integrated TL signal of peak 3. Moreover, the intensity of the TA-OSL at T_{st} = 190 °C is almost of comparable magnitude to the integrated TL signal of peak 2. As the stimulation temperature increases, the integrated TA-OSL signal decreases, due to the effect of thermal quenching. Intense supralinearity features were yielded for all dose responses. Despite the fact that the dose responses are supralinear, no low-dose supralinearity is monitored for the TA-OSL at any stimulation temperature. Saturation sublinearity occurs for all three different stimulation temperatures. Moreover, as the stimulation temperature increases, so does the dose at which saturation takes place. Table 4 presents the saturation doses for each dose–response curve.

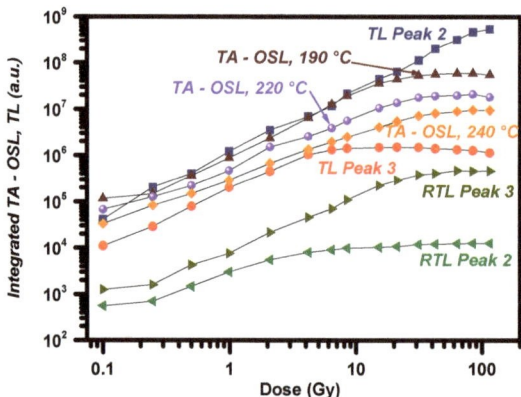

Figure 6. Dose–response curves in terms of integrated luminescence signals for TL and RTL glow peaks as well as TA-OSL signals presented in the same plot (protocol B).

Table 4. Saturation doses according to the corresponding dose–response curves.

Lum. Signal	TL Peak 2	TL Peak 3	TA-OSL, 190 °C	TA-OSL, 220 °C	TA-OSL, 240 °C	RTL Peak 2	RTL Peak 3
Saturation Dose (Gy)	-	13 ± 1	35 ± 3	52 ± 3	67 ± 4	7 ± 1	42 ± 4

Due to the large charge capacity of the VDTs, even after the end of the TA-OSL measurement of protocol B, a weak RTL signal is monitored. Two features are worth mentioning: (a) the shape of the RTL glow curve indicates similar features to the corresponding RTL of the previous protocol, namely the lack of the TL peak at 75 °C as well as the RTL peak 3 more intense than RTL peak 2, and (b) for both RTL peaks, the signal is almost stable within a wide dose region. Moreover, all RTL glow curves were subjected to both deconvolution and PSM analysis; the corresponding results are presented in Table 3. RTL integrated intensity versus initially attributed dose is also presented for both RTL peaks in Figure 6; it increases sublinearly with dose, while for larger doses it becomes stable and independent of dose. However, the integrated intensity is quite faint, almost three to four orders of magnitude lower than the TL and two to three less than the integrated TA-OSL signal.

3.5. TL Sensitivity Changes Following TA-OSL at Incremental Doses

Protocol B was designed in such way to include one TL measurement following a fixed test dose of 0.1 Gy before (step 3, S_k) and another one after the large dose and the TA-OSL (step 11, S_w). By comparing these two aforementioned TL measurements, possible changes on either TL glow curve shape or sensitivity could be easily monitored. The first obvious result deals with the lack of any modification of TL glow curve shape over the entire dose region. On the contrary, mild sensitivity changes (or alternatively, sensitization) were registered (Figure 7, filled data points for both TL peaks 2 and 3). Nevertheless, this change of sensitivity could be attributed to the combined action of large dose accumulation and TA-OSL measurement. In order to check the impact of solely the large dose on these sensitivity changes, protocol B was applied once again without steps 7–9, namely without the application of TA-OSL measurement.

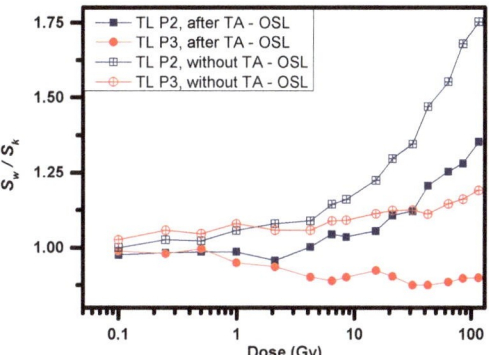

Figure 7. Sensitivity changes induced by the large dose accumulated (open data points) as well as by a combined action of large dose and a TA-OSL measurement (filled data points) for both TL peaks of BeO.

These results are also presented in Figure 7 as open data points for both TL peaks. Insignificant sensitivity changes are monitored for doses as large as 1–5 Gy. For larger doses, mild sensitization is prominent for both TL peaks. This sensitization is dose-dependent, indicating an increasing trend with dose. This change in the TL sensitivity is larger in the absence of the TA-OSL measurement for both TL peaks. Moreover, between these two, sensitization is always larger for TL peak 2.

4. Discussion

BeO, as a wide-band semiconductor, is another luminescent phosphor with very deep traps in its matrix. These traps could be effectively stimulated using TA-OSL, yielding an intense, bell-shaped signal. TA-OSL was used as an indirect verification for the presence of either very deep traps or transferred signal from these, solely by monitoring the increase in the signal versus increasing stimulation temperature. This feature is the main experimental evidence for a thermally assisted process and at the same time the main experimental difference between the TA-OSL and the conventional OSL signals. Of course, this increase was monitored even for the quenched signal, for stimulation temperatures above 170 °C. Nevertheless, following reconstruction, namely correction and compensation due to thermal quenching, this indication is prominent even from lower stimulation temperatures. According to [3,7,16,17,44], overlapping TL peaks were found to be located in the region 400–650 °C of the corresponding TL glow curve. For the case of the present study, the TA-OSL signal originates from TL traps with delocalization temperatures within the range from 500 to 650 °C. Moreover, both quenched but mostly the reconstructed RTL glow curves in the inset of Figure 2b provide strong experimental evidence supporting the existence of such traps. These delocalization temperatures are well below the corresponding temperatures for the VDTs of the Al_2O_3:C, with the latter being around 750–900 °C; thus, these traps could be easily stimulated via a direct TL measurement using the new luminescence instrumentation. Nevertheless, the interference from a high infrared background signal at such high temperatures, in conjunction with the thermal quenching effect, makes TL measurements up to temperatures as high as 600–650 °C quite problematic.

The presence of RTL signals in both TA-OSL protocols indicates that the main measurement in TA-OSL protocols is able to create PTTL by liberating electrons from VDTs. When PTTL is observed, then the TA-OSL possibly originates either directly from the VDTs or indirectly from the shallower trap, to which electrons are photo-transferred [6]. It is important to note that the RTL signal measured after the TA-OSL includes the afterglow emission due to radiative relaxation of unstable centers, in addition to the high-temperature blue stimulated OSL component. For the majority of the RTL glow curves, TL peak 3 not only exists, but also dominates in the RTL spectra after the step of TA-OSL. According to protocol A, the TA-OSL measurement temperature is always lower than the delocalization temperature of TL peak 3; nevertheless, it gets larger than the delocalization temperature of TL peak 2. Thus, during the TA-OSL measurement, the stimulation temperature is always lower than the T_m of TL peak 3. Therefore, the corresponding trap accumulates charge that recombines totally only during the following step of the RTL measurement.

The kinetic analysis that was performed in the RTL glow curves of both protocols provides solid proof regarding the dominance of the first order of kinetics in the mechanism of the transferred luminescence. This conclusion could imply a lack of re-trapping during the transfer effects. However, following a TA-OSL measurement for 500 s, the RTL signal registers the remaining electrons that were transferred to the main traps. These electrons are quite few while all VDTs are empty, inducing strong competition; the re-trapping probability is quite high and the number of re-trappings for every electron is large. According to Kitis and Pagonis [45], in such cases, the large number of re-trappings indicates TL glow curves that are also dominated by the first order of kinetics. The activation energies are slightly larger than 1 eV, in good agreement with the results of the activation energies reported by our group for both TL peaks 2 and 3 [4,5,14,15,40]. Moreover, following TA-OSL, there is no change in the TL glow curve shape of BeO; such a change was quite obvious in both cases of Al_2O_3:C and NaCl. All aforementioned features stand as arguments for the fact that TL, PTTL, and TA-OSL use the same traps, being more than one, in addition to the main dosimetric trap that is responsible for TL peak 2.

The bell peak shape of the TA-OSL signal is quite important, as it can provide strong hints for the identification of the recombination pathways of TA-OSL. The presence of the peak-shaped isothermal TA-OSL curve, in conjunction with a strong PTTL signal, could be considered as being typical of transfer effects taking place via the conduction band. This conclusion is further supported by the model adopted for the fitting of these curves in the case of CaF_2:N [41]. According to a recent paper by Polymeris et al. [46], the bell shape is typical for the cases where recombination takes place via the conduction band. This was verified using Time-Resolved TA-OSL measurements for the case of Al_2O_3:C [47]; TR–TA-OSL measurements will provide stronger arguments for the case of BeO as well. The shape of the TA-OSL curves could be used as a probe in order to distinguish the recombination pathways of these signals.

As far as the dosimetric properties of BeO's TA-OSL are concerned, the present study revealed that the TA-OSL signal could be effectively used for dosimetry purposes. In fact, independent of the stimulation temperature, the dose response of the TA-OSL signal indicates supralinearity features, similar to the counterparts of either TL or the component C_2 of the ordinary OSL signal [15]. Unfortunately, the capacity of these traps in BeO is not high enough to enable extension of the detectable dose limit of the material beyond 35–67 Gy, being the current saturation doses of the TA-OSL, depending on the stimulation temperature. Furthermore, the dose response of the (main dosimetric) TL peak 2, despite being supralinear, extends the detectable dose up to almost 100 Gy. Thus, the dosimetric properties of the TA-OSL signal are not as favorable for being exploited for high-dose-level dosimetry purposes as was suggested for Al_2O_3:C [20,21].

Nevertheless, the results have obvious implications for practical dosimetry using BeO. As was also the case for TT-OSL [3,16], TA-OSL also enables re-estimation of the doses in order to perform a cross-check in cases where the detectors are accidentally exposed to light. Moreover, the standard procedure for measuring doses using BeO includes a short optical stimulation that is followed by an intense optical zeroing of the optically sensitive traps. As the photo-ionization cross-section of the VDTs is quite small and these traps are slightly affected by light at ambient temperatures, in a series of consecutive and repetitive dose estimations using the same BeO disk, TA-OSL could be effectively used for the calculation of the cumulative dose over the entire number of consecutive uses. Experimental studies on BeO crystals have revealed several features which do not follow from the known models describing the luminescence kinetics in solids. Such features include supralinear dose responses along with sensitivity changes of both TL dosimetric peaks; these features are enhanced following long-term use of the same BeO disk due to the large accumulated dose to the VDTs. Even though these features are still not thoroughly understood, our experience gained from similar studies on Al_2O_3:C indicated that a considerable advancement in the understanding of the mechanism of these observed phenomena was the establishment of the participation of very deep traps in these processes [48–50]. The occupancy of the VDTs, a measure of which could be provided by the cumulative dose accumulated in them, could be the most important parameter affecting these features. The occupancy of VDTs is controlled by selectively emptying them using TA-OSL measurements. The less occupied these traps are, the more intense competition effects take place. The present study has revealed that TA-OSL is able to limit the sensitivity changes that might be induced due to long-term uses and large accumulated doses to the very deep traps.

5. Conclusions

TA-OSL is an alternative experimental method to not only measure the signal from very deep traps in BeO without heating the sample to temperatures greater than 500 °C, but also exploit the properties of this signal for dosimetric purposes as well. The intense, peak-shaped TA-OSL signal is measured throughout the entire temperature range. This signal is attributed to intense transfer effects from VDTs to both TL peaks 2 (main dosimetric) and TL peak 3. The increase in the integrated signal for increasing stimulating temperatures indicates the thermally assisted nature of the signal. A temperature of 220 °C was selected as the optimum stimulation temperature, mostly for signal-to-noise ratio reasons. The dose response of the TA-OSL integrated signal is sublinear, indicating practically saturation levels at relatively low doses. Isothermal TA-OSL is not effective for extending the maximum detection dose thresholds of BeO. Nevertheless, TA-OSL could be effectively used in order to either (a) control the occupancy of VDTs and thus get rid of intense sensitivity changes induced by long-term uses and high accumulated dose to the VDTs or (b) measure the dose accumulated over a series of repetitive dose calculations.

Funding: This research received no external funding.

Institutional Review Board Statement: Not applicable.

Informed Consent Statement: Not applicable.

Data Availability Statement: Data are available upon request.

Conflicts of Interest: The author declares unanimously no conflict of interest.

References

1. Bos, A.J.J. High Sensitivity thermoluminescence dosimetry. *Nucl. Instrum. Methods Phys. Res. Sect. B* **2001**, *184*, 3–28. [CrossRef]
2. Sommer, M.; Henniger, J. Investigation of a BeO-based optically stimulated luminescence dosimeter. *Radiat. Protect. Dosim.* **2006**, *119*, 394–397. [CrossRef]
3. Yukihara, E.G. Observation of strong thermally transferred optically stimulated luminescence (TT-OSL) in BeO. *Radiat. Meas.* **2019**, *121*, 103–108. [CrossRef]
4. Aşlar, E.; Şahiner, E.; Polymeris, G.S.; Meriç, N. Feasibility of determining Entrance Surface Dose (ESD) and mean glandular dose (MGD) using OSL signal from BeO dosimeters in mammography. *Radiat. Phys. Chem.* **2020**, *177*, 109151. [CrossRef]
5. Aşlar, E.; Şahiner, E.; Polymeris, G.S.; Meriç, N. Determination of trapping parameters in BeO ceramics in both quenched as well as reconstructed thermoluminescence glow using various analysis methods. *Appl. Radiat. Isot.* **2012**, *129*, 142–151. [CrossRef]
6. Polymeris, G.S.; Kitis, G. Thermally Assisted Optically Stimulated Luminescence (TA-OSL) from Very Deep Traps. In *Advances in Physics and Applications of Optically and Thermally Stimulated Luminescence*; Chen, R., Pagonis, V., Eds.; World Scientific: Singapore, 2019; Chapter 4.
7. McKeever, S.W.S.; Moscovitch, M.; Townsend, P.D. *Thermoluminescence Dosimetry Materials: Properties and Uses*; Nuclear Technology Publishing: Ashford, UK, 1995.
8. Bulur, E.; Göksu, H.Y. OSL from BeO ceramics: New observations from an old material. *Radiat. Meas.* **1998**, *29*, 639–650. [CrossRef]
9. Sommer, M.; Freudenberg, R.; Henniger, J. New aspects of a BeO-based optically stimulated luminescence dosimeter. *Radiat. Meas.* **2007**, *42*, 617–620. [CrossRef]
10. Sommer, M.; Jahn, A.; Henniger, J. Beryllium oxide as optically stimulated luminescence dosimeter. *Radiat. Meas.* **2008**, *43*, 353–356. [CrossRef]

11. Jahn, A.; Sommer, M.; Ulrich, W.; Wickert, M.; Henniger, J. The BeOmax system—Dosimetry using OSL of BeO for several applications. *Radiat. Meas.* **2013**, *56*, 324–327. [CrossRef]
12. Jahn, A.; Sommer, M.; Henniger, J. Environmental dosimetry with the BeO OSL personal dosimeter—State of the art. *Radiat. Meas.* **2014**, *71*, 438–441. [CrossRef]
13. Bulur, E.; Saraç, B.E. Time-resolved OSL studies on BeO ceramics. *Radiat. Meas.* **2013**, *59*, 129–138. [CrossRef]
14. Aşlar, E.; Meriç, N.; Şahiner, E.; Erdem, O.; Kitis, G.; Polymeris, G.S. A correlation study on the TL, OSL and ESR signals in commercial BeO dosimeters yielding intense transfer effects. *J. Lumin.* **2019**, *24*, 116533. [CrossRef]
15. Polymeris, G.S.; Çoskun, S.; Tsoutsoumanos, E.; Konstantinidis, P.; Aşlar, E.; Şahiner, E.; Meriç, N.; Kitis, G. Dose response features of quenched and reconstructed, TL and deconvolved OSL signals in BeO. *Results Phys.* **2021**, *25*, 104222. [CrossRef]
16. Yukihara, E.G. Characterization of the thermally transferred optically stimulated luminescence (TT-OSL) of BeO. *Radiat. Meas.* **2019**, *126*, 106132. [CrossRef]
17. Yukihara, E.G. A review on the OSL of BeO in light of recent discoveries: The missing piece of the puzzle? *Radiat. Meas.* **2020**, *134*, 106291. [CrossRef]
18. Bulur, E. Photo-transferred luminescence from BeO ceramics. *Radiat. Meas.* **2007**, *42*, 334–340. [CrossRef]
19. Yukihara, E.G.; Andrad, A.B.; Eller, S. BeO optically stimulated luminescence dosimetry using automated research readers. *Radiat. Meas.* **2016**, *94*, 27–34. [CrossRef]
20. Polymeris, G.S.; Raptis, S.; Afouxenidis, D.; Tsirliganis, N.C.; Kitis, G. Thermally assisted OSL from deep traps in Al_2O_3: C. *Radiat. Meas.* **2010**, *45*, 519–522. [CrossRef]
21. Polymeris, G.S.; Kitis, G. Thermally assisted photo transfer OSL from deep traps in Al_2O_3:C grains exhibiting different TL peak shapes. *Appl. Radiat. Isot.* **2012**, *70*, 2478–2487. [CrossRef]
22. Polymeris, G.S. Thermally assisted OSL (TA-OSL) from various luminescence phosphors; an overview. *Radiat. Meas.* **2016**, *90*, 145–152. [CrossRef]
23. Wintle, A.G.; Adamiec, G. Optically stimulated luminescence signals from quartz: A review. *Radiat. Meas.* **2017**, *98*, 10–33. [CrossRef]
24. Polymeris, G.S.; Giannoulatou, V.; Sfampa, I.K.; Tsirliganis, N.C.; Kitis, G. Search for stable energy levels in materials exhibiting strong anomalous fading: The case of apatites. *J. Lumin.* **2014**, *153*, 245–251. [CrossRef]
25. Polymeris, G.S.; Sfampa, I.K.; Niora, M.; Stefanaki, E.C.; Malletzidou, L.; Giannoulatou, V.; Pagonis, V.; Kitis, G. Anomalous fading in TL, OSL and TA-OSL signals of Durango apatite for various grain size fractions; from micro to nano scale. *J. Lumin.* **2018**, *195*, 216–224. [CrossRef]
26. Polymeris, G.S.; Şahiner, E.; Meriç, N.; Kitis, G. Experimental features of natural thermally assisted OSL (NTA-OSL) signal in various quartz samples; preliminary results. *Nucl. Instrum. Methods Phys. Res. Sect. B* **2015**, *349*, 24–30. [CrossRef]
27. Şahiner, E.; Polymeris, G.S.; Meriç, N. Thermally assisted OSL application for equivalent dose estimation; comparison of multiple equivalent dose values as well as saturation levels determined by luminescence and ESR techniques for a sedimentary sample collected from a fault gauge. *Nucl. Instrum. Methods Phys. Res. Sect. B* **2017**, *392*, 21–30. [CrossRef]
28. Majgier, R.; Biernacka, M.; Palczewski, P.; Mandowski, A.; Polymeris, G.S. Investigation on thermally assisted optically stimulated luminescence (TA-OSL) signal in various sodium chloride samples. *Appl. Radiat. Isot.* **2019**, *143*, 98–106. [CrossRef]
29. Kitis, G.; Kiyak, N.G.; Polymeris, G.S. Temperature lags of luminescence measurements in a commercial luminescence reader. *Nucl. Instrum. Methods Phys. Res. Sect. B* **2015**, *359*, 60–63. [CrossRef]
30. Kitis, G.; Polymeris, G.S.; Pagonis, V. Stimulated luminescence emission: From phenomenological models to master analytical equations. *Appl. Radiat. Isot.* **2019**, *153*, 108797. [CrossRef]
31. Kitis, G.; Vlachos, N.D. General semi-analytical expressions for TL, OSL and other luminescence stimulation modes derived from the OTOR model using the Lambert W-function. *Radiat. Meas.* **2012**, *48*, 47–54. [CrossRef]
32. Corless, R.M.; Gonnet, G.H.; Hare, D.G.E.; Jeffrey, D.J.; Knuth, D.E. On the Lambert W function. *Adv. Comput. Math.* **1996**, *5*, 329–359. [CrossRef]
33. Sadek, A.M.; Eissa, H.M.; Basha, A.M.; Kitis, G. Development of the peak fitting and peak shape methods to analyze the thermoluminescence glow-curves generated with exponential heating function. *Nucl. Instrum. Methods Phys. Res. B* **2014**, *330*, 103–107. [CrossRef]
34. Konstantinidis, P.; Kioumourtzoglou, S.; Polymeris, G.S.; Kitis, G. Stimulated luminescence; Analysis of complex signals and fitting of dose response curves using analytical expressions based on the Lambert W function implemented in a commercial spreadsheet. *Radiat. Phys. Chem.* **2021**, *176*, 109870. [CrossRef]
35. Balian, H.G.; Eddy, N.W. Figure-of-merit (FOM), an improved criterion over the normalized chi-squared test for assessing goodness-of-fit of gamma-ray spectral peaks. *Nucl. Instrum. Methods* **1977**, *145*, 389–395. [CrossRef]
36. Kitis, G.; Pagonis, V. Peak shape methods for general order thermoluminescence glow-peaks: A reappraisal. *Nucl. Instrum. Methods Phys. Res. B* **2007**, *262*, 313–322. [CrossRef]
37. Kitis, G.; Chen, R.; Pagonis, V. Thermoluminescence glow-peak shape methods based on mixed order kinetics. *Phys. Stat. Solidi (a)* **2008**, *205*, 1181–1189. [CrossRef]
38. Curie, D. *Luminescence in Crystals*; Methuen: London, UK, 1963.
39. Pagonis, V.; Ankjærgaard, C.; Murray, A.S.; Jain, M.; Chen, R.; Lawless, J.; Greilich, S. Modelling the thermal quenching mechanism in quartz based on time-resolved optically stimulated luminescence. *J. Lumin.* **2010**, *130*, 902–909. [CrossRef]

40. Aşlar, E.; Şahiner, E.; Polymeris, G.S.; Meriç, N. Calculation of thermal quenching parameters in BeO ceramics using solely TL measurements. *Radiat. Meas.* **2017**, *103*, 13–25. [CrossRef]
41. Polymeris, G.S.; Tsirliganis, N.C.; Kitis, G. TL and OSL properties of CaF_2:N. *Nucl. Instrum. Methods Phys. Res. B* **2016**, *251*, 133–142. [CrossRef]
42. Nikiforov, S.V.; Pagonis, V.; Merehnikov, A.S. Sublinear dose dependence of thermoluminescence as a result of copetition between electron and hole trapping centers. *Radiat. Meas.* **2017**, *105*, 54–61. [CrossRef]
43. Chen, R.; McKeever, S.W.S. *Theory of Thermoluminescence and Related Phenomena*; World Scientific: Singapore, 1997.
44. Yukihara, E.; McKeever, S.W.S. *Optically Stimulated Luminescence: Fundamentals and Applications*; John Wiley and Sons: Hoboken, NJ, USA, 2011.
45. Kitis, G.; Pagonis, V. On the Need for Deconvolution Analysis of Experimental and Simulated Thermoluminescence Glow Curves. *Materials* **2023**, *16*, 871. [CrossRef]
46. Polymeris, G.S.; Şahiner, E.; Aşlar, E.; Kitis, G.; Meriç, N. Deconvolution of isothermal TA-OSL decay curves from sedimentary quartz using combinations of various contemporary models. *Radiat. Meas.* **2018**, *119*, 93–101. [CrossRef]
47. Nyirenda, A.N.; Chithambo, M.L.; Polymeris, G.S. On luminescence stimulated from deep traps in α-Al_2O_3:C. *Radiat. Meas.* **2016**, *90*, 109–112. [CrossRef]
48. Milman, I.I.; Kortov, V.S.; Nikiforov, S.V. An interactive process in the mechanism of the thermally stimulated luminescence of anion-defective α-Al_2O_3 crystals. *Radiat. Meas.* **1998**, *29*, 401–410. [CrossRef]
49. Kortov, V.S.; Nikiforov, S.V.; Milman, I.I.; Moyseikin, E.V. Specific features of luminescence of radiation-colored α-Al_2O_3 single crystals. *Radiat. Meas.* **2004**, *38*, 451–454. [CrossRef]
50. Meriç, N.; Şahiner, E.; Polymeris, G.S. Thermally assisted OSL (TA-OSL) reproducibility in Al_2O_3:C and its implication on the corresponding thermoluminescence (TL) reproducibility. *Radiat. Meas.* **2016**, *90*, 269–273. [CrossRef]

Disclaimer/Publisher's Note: The statements, opinions and data contained in all publications are solely those of the individual author(s) and contributor(s) and not of MDPI and/or the editor(s). MDPI and/or the editor(s) disclaim responsibility for any injury to people or property resulting from any ideas, methods, instructions or products referred to in the content.

Article

Trap Parameters for the Fast OSL Signal Component Obtained through Analytical Separation for Various Quartz Samples

Magdalena Biernacka [1,*], Alida Timar-Gabor [2,3], Zuzanna Kabacińska [3], Piotr Palczewski [1] and Alicja Chruścińska [1]

1. Institute of Physics, Faculty of Physics, Astronomy and Informatics, Nicolaus Copernicus University in Toruń, ul. Grudziądzka 5, 87-100 Toruń, Poland
2. Faculty of Environmental Science and Engineering, Babeș-Bolyai University, Fântânele 30, 400294 Cluj-Napoca, Romania
3. Interdisciplinary Research Institute on Bio-Nano-Sciences, Babeș-Bolyai University, Treboniu Laurian 42, 400271 Cluj-Napoca, Romania
* Correspondence: m.biernacka@umk.pl

Citation: Biernacka, M.; Timar-Gabor, A.; Kabacińska, Z.; Palczewski, P.; Chruścińska, A. Trap Parameters for the Fast OSL Signal Component Obtained through Analytical Separation for Various Quartz Samples. *Materials* 2022, 15, 8682. https://doi.org/10.3390/ma15238682

Academic Editor: Dirk Poelman

Received: 9 November 2022
Accepted: 2 December 2022
Published: 6 December 2022

Publisher's Note: MDPI stays neutral with regard to jurisdictional claims in published maps and institutional affiliations.

Copyright: © 2022 by the authors. Licensee MDPI, Basel, Switzerland. This article is an open access article distributed under the terms and conditions of the Creative Commons Attribution (CC BY) license (https://creativecommons.org/licenses/by/4.0/).

Abstract: Trap stability is essential in luminescence dating and thermochronometry. Trap depth and frequency factors determining the stability of the fast component of optically stimulated luminescence (OSL) in quartz, which is the most important in dating, have yet to be uniquely determined, especially for samples with an OSL signal not dominated by this component. One can determine them in OSL thermal depletion curve (OTDC) experiments. The separation of the fast OSL signal undisturbed by other OSL components is vital for obtaining accurate parameters for the traps of interest. This work presents a method of simultaneous thermal and optical stimulation using red light (620 nm) to separate the fast OSL component (the thermally modulated OSL method—TM-OSL). The OTDC experiment with the TM-OSL stimulation was used for the trap parameter determination on a variety of quartz samples, leading us to report for the first time, the trap parameters for the fast OSL component analytically separated in quartz from rock samples. The results obtained for these samples with the fast component of low intensity are consistent with those with an intensive fast OSL component. Results of OTDC measurements for all investigated quartz samples were tested for a wide range of irradiation doses.

Keywords: traps; thermal stability; optically stimulated luminescence; quartz

1. Introduction

In this study, we work towards an accurate determination of the values of trap parameters (trap depth, denoted by E, and frequency factor, denoted by s) obtained in optically stimulated luminescence (OSL) isothermal experiments for quartz of various origins. Knowledge about the stability of investigated traps in quartz is crucial for applying the OSL method in dating and thermochronometry.

OSL dating method using quartz has an effective dating range of 10^1–10^6 years (e.g., [1]) and low-temperature sensitivity of ~35–60 °C (e.g., [2]) which makes it especially well-suited for Quaternary geochronology. For quartz, it was identified that the OSL signal comprises at least three components called, depending on the rate of decay, "fast", "medium" and "slow" [3,4]. The primary condition for the OSL dating of the sediment is resetting its OSL signal by sunlight before the sediment layer formation. The fastest decaying OSL component is recognized as the one that guarantees accurate age estimation. That made the fast OSL component an object of extensive studies (e.g., Refs. [5,6]). The continuous-wave OSL (CW-OSL) method using blue (470 nm) stimulation light is commonly used to investigate the thermal stability of OSL traps in quartz (e.g., [7]). In the CW-OSL method, optical stimulation is performed with constant stimulation energy at a constant temperature. During such stimulation, the OSL signal from various optically active traps is depopulated simultaneously and the observed OSL signal represents the overall existing OSL components in the sample. The OSL signal decay rate for a particular electron trap depends on its optical cross-section (OCS), which stays constant at fixed stimulation energy and temperature. Thus, for two kinds of traps with different OCSs, the ratio of OSL decay rates remains unchanged; therefore, the ability of the CW-OSL method to separate the signal related to various traps is limited. However, in many studies, the trap parameters obtained in this way for sedimentary quartz samples of various geological origins show variability in E and s values, for example, $E = 1.65 \pm 0.02$ eV and $s = 5.01 \cdot 10^{13}$ s^{-1} [8], $E = 1.41 \pm 0.13$ eV and $s = 4.25 \cdot 10^{11}$ s^{-1} [9], $E = 1.70 \pm 0.01$ eV and $s = 7.76 \cdot 10^{13}$ s^{-1} [10]. Besides the variability in the obtained parameters, deviation from first-order kinetics was reported for the thermal depletion curves (e.g., Refs. [10–12]). These observations may be explained by the different contributions of the OSL components in the depleted OSL signal. The charge carrier transitions between traps, such as thermal transfer, as well as the significant variation of OSL components contributions from sample to sample can negatively affect the accuracy of the obtained values. Therefore, the isolation of the fast OSL component from other components is of the utmost importance. Previous works [13–15] proposed thermally modulated OSL to achieve such a separation and showed that the fast OSL component in quartz could be measured using stimulation at 620 nm. The TM-OSL method uses the dependency of the OCS on the temperature to separate the signal from various types of traps more effectively. Differences between the OCSs of various traps are more significant if the light of a longer wavelength is used for stimulation, and the OCS increases more dynamically with temperature. The stimulation light effectively empties the traps with the highest OCS if a suitable wavelength is applied. Moreover, the rising temperature causes a continuous OCS increase and the time necessary to depopulate the investigated traps is shortened. This ensures that the traps, which are not intended to be bleached, are emptied to a negligible extent. In order to observe a pure OSL signal, the TM-OSL stimulation should be accomplished in the temperature range where the thermal activation of investigated traps is minimal.

Recently OSL thermal depletion curves (OTDC) were constructed using TM-OSL$_{620\text{ nm}}$ instead of CW-OSL$_{470\text{ nm}}$ for analyzing the thermal stability of traps responsible for the fast OSL component in sedimentary quartz. The trap parameters found in this way were compared with those obtained using other methods such as initial rise, variable heating rate and TL peak fitting [16].

While determining these values for sedimentary quartz is important as they partially dictate the age range of dating, the situation is much more complex when rock quartz samples are used for thermochronometry. Thermal stability analysis of the OSL signal is crucial for the evaluation of the "closure temperature" [2,17]. Using erroneous values for these parameters to calculate the lifetime of a charge carrier in a trap leads to a false reconstruction of the time-temperature history of rocks in thermochronometry. Moreover, in the case of rock quartz, the OSL signal usually has a much lower intensity than in sedimentary quartz and the contribution of the fast OSL component is significantly lower (e.g., [10]). Despite this, the common trap parameters generally used in thermochronometry applications (see e.g., Refs. [18,19]) are those obtained for sedimentary samples dominated by the fast OSL component, namely $E = 1.59$ eV and $s = 10^{12.9}$ s^{-1} obtained on the WIDG8 sample by Murray and Wintle [20] or $E = 1.59$ eV and $s = 2.8 \cdot 10^{12}$ s^{-1} obtained on sample LW 94/1 by Spooner and Questiaux [21]. Keeping in mind the variability observed in sediments, one should be cautious when E and s values obtained for the fastest blue—OSL component in sedimentary quartz are applied to thermochronometry analysis for quartz from rocks. For practical application in this field, rock samples need to be characterized.

Here, the thermally modulated method (TM-OSL$_{620\text{ nm}}$) where the optical stimulation is conducted during linear heating of the sample was applied for investigating the thermal stability of the OSL signal of quartz extracted from granites. The results are compared to those obtained on sedimentary quartz. The influence of measurement parameters such as (i) preheat temperature, or in other words, the initial occupation of traps involved in the isothermal process, and (ii) the impact of irradiation dose on the isothermal analysis results are thoroughly investigated. Based on these investigations we report for the first time the analytical separation of the fast OSL signal component in quartz extracted from plutonic rocks and present recommendations for improving the accuracy of determination of trap parameters for the defect responsible for this signal in such problematic samples.

2. Materials and Methods

2.1. Samples

The following four samples of sedimentary quartz (coarse grain) were investigated:

- 2 MV 570—dia. 63–90 µm, a loess sediment from Mircea Voda, Romania. A detailed description of the preparation of sample 2 MV 570 can be found elsewhere [22], with references therein (in this work subsample 2 MV 570A was used);
- QC (calibration quartz)—dia. 180–200 µm, aeolian sediments, Jutland, the Radiation Research Division of the Technical University of Denmark in Risø (Risø DTU/RRD). A detailed description of the preparation of the QC sample can be found elsewhere [23,24].
- MR—dia. 150–250 µm, from ~8 to 9 ka old deposits 'Silver Sands of Morar', reworked intensely by fluvial and marine processes. The preparation of sample MR is described in detail by Schmidt et al. [25].
- FB—dia. 150–250 µm, Oligocene coastal sand from the Fontainebleau Sand Formation, from the time of the last marine intrusion into the Paris Basin (the Stampian Sea) ~35 Ma ago. The preparation of sample FB is described in detail in Kreutzer et al. [26] (in this work batch FB, subsample FB3A was used);

Two investigated samples of igneous (rock) quartz:

- GC—Catalina granite, dia. 250–500 µm, Cenozoic Granite with an age of about 26 Ma, Tucson, southeast Arizona, USA [27,28].
- GO—Oracle granite, dia. 500–1000 µm, Proterozoic granite with a crystallization age of about 1.4 Ga, Tucson, southeast Arizona, USA [27,28].

Quartz has been extracted by crushing, light mineral fraction separation by Wilfley table followed by centrifugation in diluted $Na_6[H_2W_{12}O_{40}] \times H_2O$ (2.62 g/cm^3 and 2.75 g/cm^3) and 40 % hydrofluoric acid etching for 60 min followed by rinsing with 10% HCl to remove any acid-soluble fluoride precipitates.

2.2. Instrumentation

The OTDC experiment using TM-OSL$_{620\,nm}$ protocol was carried out using equipment set up based on a Risø TL/OSL- DA-20 System equipped with an additional stimulation light source. In the system, the signal was detected by an EMI 9235QB photomultiplier under a 7.5 mm Hoya U-340 bandpass filter. Optical stimulation was conducted using a dedicated module made of a single high-power LED integrated with an optical adapter inserted into the Risø reader. The LED spectral band was 620 ± 12 nm. The stimulation power used was ~33 mW/cm^2, and heating was performed in an Ar atmosphere. Irradiation was performed using beta radiation delivered by ^{90}Sr/^{90}Y β source with a dose rate of 112 mGy s^{-1}. The 6 mm diameter mask was used for quartz grains deposition on stainless steel discs delivered by Risø. A silicon spray assured a single-grain layer. The masses of all the used aliquots were comparable (about 5 mg).

2.3. Trap Parameter Determination

The values of trap parameters determine the lifetime (τ) of charge carriers in a trap, i.e., the mean duration that an electron can be expected to remain trapped. The lifetime for a trap of the depth (or the thermal activation energy) E (eV) can be calculated according to the following equation:

$$\tau = s^{-1} \exp\left(\frac{E}{kT}\right), \tag{1}$$

where s is the frequency factor (s^{-1}) described as the number of escape attempts of an electron from a trap per second [29], T is the temperature (K) and k is the Boltzmann constant (eV K^{-1}).

The OTDC is constructed by the repeated measurement of the remaining OSL signal after keeping the sample at a specific temperature T (°C), hereinafter referred to as holding temperature for a different time t (from a few to many thousands of seconds), hereinafter referred to as holding time (see Table 1). During such an experiment the thermal release of the electrons from traps related to the OSL is observed. The series of measurements is terminated when the OSL signal is decreased to the PMT background level. One repeats the OTDC measurement for several holding temperatures. Assuming the most basic first-order kinetics and a single kind of depopulated traps, the shape of the OTDC can be described by a simple exponential function (e.g., Refs. [29–31]):

$$I(t) = I_0 \exp\left(\frac{-t}{\tau}\right), \quad (2)$$

where I_0 is the initial OSL signal for holding time $t = 0$ and τ (s) is the lifetime of the electron in the trap at temperature T (Equation (1)). In this study, the trap parameters (E and s) have been derived using the TM-OSL$_{620\,nm}$ protocol (see Table 1). The lifetime was estimated for each holding temperature by attempting to fit a single exponential function to the OTDC first. The values of E and s were derived by regression method from the Arrhenius plot (obtained from Equation (1)):

$$\ln \tau = \frac{E}{kT} - \ln s. \quad (3)$$

Table 1. OTDC protocol used in the experiments.

	Protocol TM-OSL$_{620\,nm}$
1	Irradiation (D) *
2	Preheat to **150 °C or 280 °C**, 2 Ks^{-1}, 0 s
3	Heating the sample to T ** with the heating rate of 1 Ks^{-1} and holding at T during t *** seconds
4	TM-OSL$_{620\,nm}$ from 40 °C to 200 °C, 2 Ks^{-1}
5	Heating to 500 °C, 5 Ks^{-1}
6	Go to step 1

Sample measured: 2 MV 570A, GO, GC; QC, FB, MR; *—10 Gy, 100 Gy, 1000 Gy; **—sample holding temperatures in °C: 240, 250, 260, 270, 280; ***—sample holding times in seconds: 0, 3, 10, 32, 100, 316, 1000, 3160, 10,000.

The plot of $\ln \tau$ versus $1/T$ yields a straight line of slope E/k, and intercept $\ln s$ on the ordinate axis (e.g., [30]).

Protocol

The OTDC protocol used in experiments are shown in Table 1. Samples measured are listed in the legend of Table 1. The OTDC protocol using the TM-OSL$_{620\,nm}$ method was described in detail previously [16]. The modification of this protocol applied here consisted in replacing the preheat up to 150 °C with the preheat to the highest holding temperature of those used in the protocol (in this case it was 280 °C). This was conducted in order to remove electrons from shallow traps and to preserve the most similar initial filling of traps for all isothermal decays used in the experiment. The optical stimulation with 620 nm was initiated at 40 °C and continued up to 200 °C whereas linear heating was performed with the rate of 2 Ks^{-1}. The heating up to 500 °C was conducted in step 5 to remove all electrons from the traps responsible for the TL signal observed below this temperature. The TM-OSL$_{620\,nm}$ signal was taken as the integral under the TM-OSL curve in the range of 40–150 °C. The sum of counts from 21 channels at the maximum of TL peak

110 °C (measurement in step 2 of the protocol in Table 1) was used for correction of the TM-OSL$_{620\,nm}$ signal for luminescence sensitivity changes [32]. The OTDCs were normalized to the signal measured for $tj = 0$ s (thus, the y-axis caption is as follows: TM-OSL/TM-OSL (0 s).

In the TM-OSL$_{620\,nm}$ protocol, at least two aliquots of each sample were used. One of the aliquots was tested in all the OTDC experiments, and the second aliquot was used to verify the reproducibility of results for selected points of the procedures used.

3. Results

3.1. Initial Characterization of Luminescence Properties: TL, CW-OSL, TM-OSL

The luminescence properties of the samples were investigated to demonstrate the basic differences between them. TL, CW-OSL$_{470\,nm}$ as well as TM-OSL$_{620\,nm}$ were detected in the UV detection range (Hoya U-340 filter) after the beta irradiation with a dose of 1000 Gy.

Figure 1 clearly shows the differences in the shapes of TL curves. Three temperature ranges can be distinguished above 110 °C. In the first one, below 160 °C, all samples except QC have similar shapes of the TL curves. For some of them (GC, FB and MR), the peak at approximately 130 °C is more pronounced, but the overall trend of a slight decrease in TL with temperature is visible. From 160 °C to 240 °C, there is a further general trend of TL decrease, resulting from thermal quenching [33–36] in which for samples GC, FB and 2 MV 570A, a peak around 200 °C is clearly marked. Between 240 °C and 280 °C, the TL in rock samples QC and GO is distinguished by its stronger intensity (maximum about 260 °C) in contrast to the clear lowering in the signal of other samples. Above 280 °C, the signal of samples MR and QC stands out, with TL curves having a broad TL maximum centered at 340 °C. In the rest of the samples, TL in this range decays slowly, and no particular peaks can be distinguished.

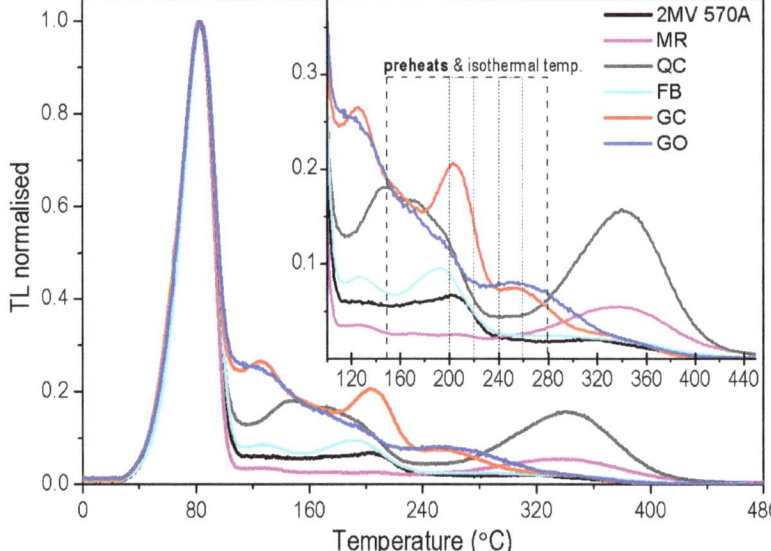

Figure 1. Normalized TL glow curves measured for one aliquot of each sample after irradiation with a dose of 1000 Gy. The inset shows the same data for the high-temperature range with the marked preheat (dashed lines) and storage temperatures (dotted line) used in experiments. TL glow curves were measured immediately after irradiation with heating rate of 2 Ks^{-1} and standard background subtraction was performed for each aliquot.

The complex shapes of TL glow curves indicate a multi-trap structure in the investigated quartz samples. The obtained diversity in the TL curve shapes shows that the concentrations of the defects responsible for the individual electron traps in the samples are different.

Based on the CW-OSL$_{470\,nm}$ decays (Figure 2) obtained for the same quartz samples, it is possible to distinguish two qualitatively different sample groups with two samples (GC and MR) displaying a slower decay rate than the others. It means that the fast OSL component contribution in the total OSL signal for GC and MR samples is lower than for FB, QC, GO, and 2 MV 570A samples. The observations that the OSL signal originating from FB is dominated by the fast OSL component whereas in the case of the MR sample the total OSL signal is evidently composed of the other OSL components agrees with LM-OSL results for these samples published recently [25]. The LM-OSL curves for both samples were decomposed into several first-order kinetics LM-OSL components. The fast component observed in sample FB was not present in the decomposition results obtained for sample MR.

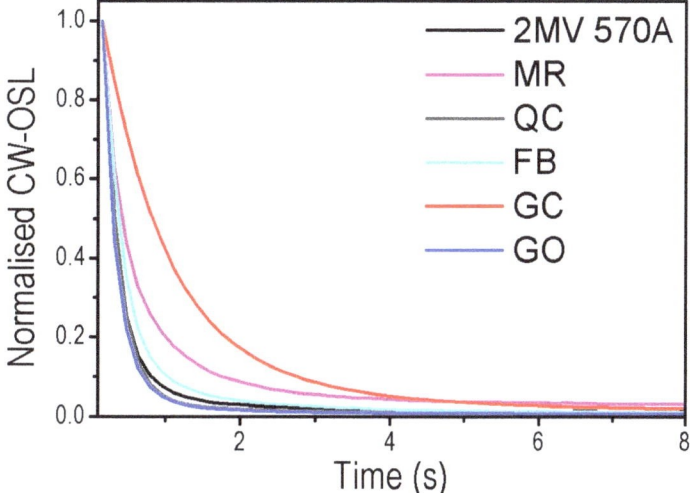

Figure 2. CW-OSL$_{470\,nm}$ curves measured for one aliquot of each sample after irradiation with a dose of 1000 Gy and preheat up to 240 °C. Data set after normalization to the initial intensity. The blue LEDs were used for the optical stimulation with 90% of their maximal power ~80 mW/cm^2.

The shape of TM-OSL$_{620\,nm}$ curves depends on the population of the trap responsible for the fast OSL component. When the fast OSL component is present in the sample, the TM-OSL$_{620\,nm}$ curves have a broad-peak shape as presented in Figure 3a, where a comparison of the TM-OSL$_{620\,nm}$ curves measured for each of the quartz samples is shown. The brightest in TM-OSL$_{620\,nm}$ among the investigated samples are sedimentary quartz FB and QC. Rock quartz samples are less sensitive; however, two sedimentary samples MR and 2 MV 570A are comparable with them in intensities. Differences are visible in the shape of TM-OSL curves for GC and MR samples, where the maximum of the TM-OSL peak is shifted in the higher temperature region; in the case of the MR sample, this shift is slight, whereas for GC is evident (Figure 3b). This behavior agrees with CW-OSL$_{470\,nm}$ decays indicating the small contribution of the fast OSL components in these samples. Nevertheless, in the case of these two samples, another OSL component can be observed in the TM-OSL$_{620\,nm}$ curve. It manifests itself in a TM-OSL peak at about 150–160 °C, for a heating rate of 2 Ks^{-1}. In the case of bright samples FB and QC, the predominance of the fast component over the rest of the OSL signal is vast. In this way, the slower component (medium) is not easy for direct observation in the TM-OSL$_{620\,nm}$.

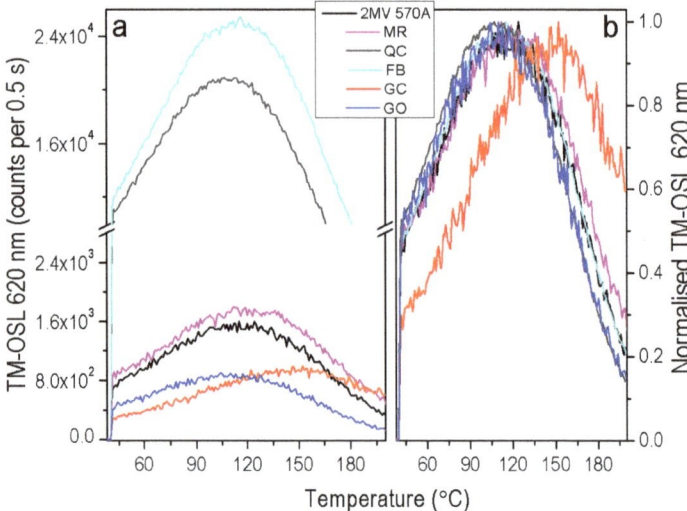

Figure 3. (**a**) TM-OSL$_{620\,nm}$ curves measured for one aliquot of each sample after irradiation with a dose of 1000 Gy and preheat up to 280 °C, (**b**) the same data set normalized to the curve maximum. TM-OSL signals were measured after irradiation and preheat up to 280 °C with a heating rate of 2 Ks^{-1}. During linear heating with a rate of 2 Ks^{-1}, the optical stimulation was carried out using 620 nm LEDs with power ~33 mW/cm^2.

3.2. TM-OSL Method Used for the Fast OSL Component Separation

As was recently shown [37], the preheat applied immediately after irradiation before starting the storage at a certain holding temperature (step 2 in Table 1) is important not only to remove carriers from shallow traps but also, to ensure a similar level of trap filling for each value of the holding temperature applied in the experiment. In the first stage of the OTDC experiment, the effect of changing the preheat temperature from 150 °C to 280 °C in the TM-OSL$_{620\,nm}$ procedure was tested using samples FB and 2 MV 570. The OTDCs obtained for the FB sample (the same aliquot) for the preheat temperatures of 280 °C and 150 °C are shown in Figures 4a and 4b, respectively. For this sample, additional holding temperatures (250 and 270 °C) next to the standard 240, 260, and 280 °C values were applied. Additional holding temperatures were chosen in order to follow thermal depletion processes more carefully. Finally, the best OTDC fits were obtained for holding temperatures: 240, 260, and 280 °C when the preheat temperature of 150 °C was used and for 250, 260, 270 and 280 °C in the case of preheat to 280 °C. When preheat temperature of 150 °C is applied, the experimental points deviate from the exponential curve. The improvement in isothermal decay fitting quality (compare the obtained FOM values in Figure 4a,b) is visible for the preheat to 280 °C. For both samples tested here, when one achieves the better first-order fitting result, i.e., for the higher preheat temperature, the obtained trap depths are higher than for the lower preheat temperature.

Figures 5 and 6 show, for four samples, the sets of OTDCs created for the OSL signal measured by the TM-OSL$_{620\,nm}$ method. The protocol applied, in this case, is shown in Table 1. Results were selected for samples that originate both from sediments and rock and whose contribution of the fast component is different.

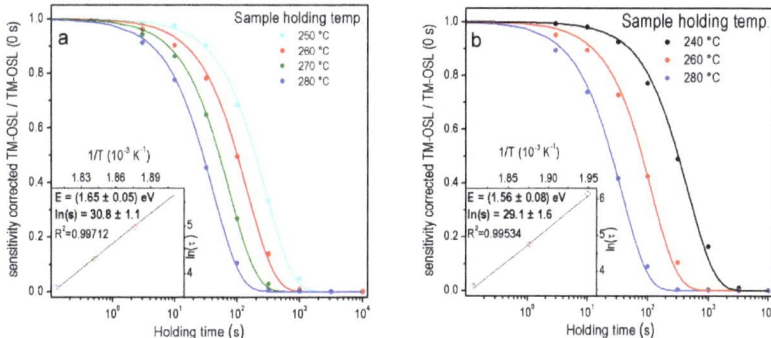

Figure 4. The OTDCs for one aliquot of the FB sample using TM-OSL$_{620\,nm}$ procedure after irradiation with dose 1000 Gy and preheating: (**a**) 280 °C, FOM$_{250}$ = 1.89%, FOM$_{260}$ = 3.17%, FOM$_{270}$ = 2.03%, FOM$_{280}$ = 2.33%, (**b**) 150 °C, FOM$_{240}$ = 2.75%, FOM$_{260}$ = 4.13%, FOM$_{280}$ = 3.91%.

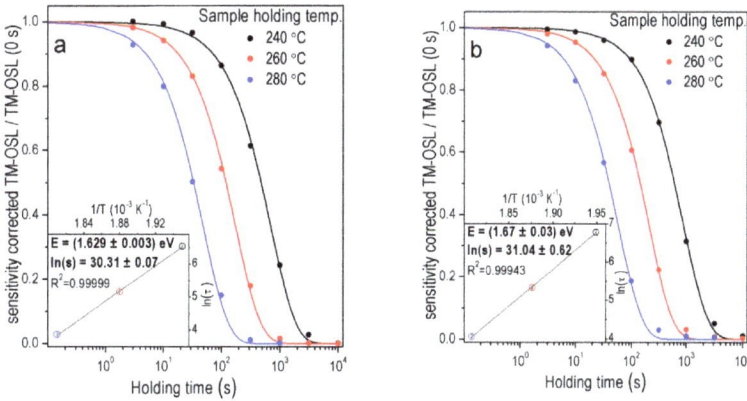

Figure 5. The OTDCs of the quartz samples (**a**) 2 MV 570A for dose 1000 Gy, FOM$_{240}$ = 1.57%, FOM$_{260}$ = 1.53%, FOM$_{280}$ = 2.99%, (**b**) MR for dose 100 Gy, FOM$_{240}$ = 1.17%, FOM$_{260}$ = 1.65%, FOM$_{280}$ = 2.98% for one aliquot and preheating 280 °C.

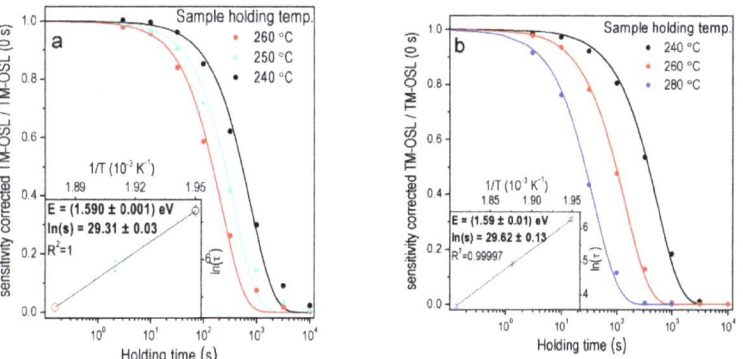

Figure 6. The OTDCs of the granite samples (**a**) Catalina FOM$_{240}$ = 4.46% FOM$_{250}$ = 4.70% FOM$_{260}$ = 5.23% (**b**) Oracle FOM$_{280}$ = 3.57% FOM$_{260}$ = 1.90% FOM$_{240}$ = 2.72% for one aliquot after dose 1000 Gy and preheating 280 °C.

Sample 2 MV 570 (Figure 5a) represents the case of sediment quartz with a substantial share of the fast component in the OSL signal. In this case, the first-order curves correctly reproduce the OTDCs. Estimated trap parameters $E = 1.629 \pm 0.003$ eV and $s = (1.46 \pm 0.11) \times 10^{13}$ s^{-1} have a low value of standard uncertainties. The investigation of OSL components in the sample MR was performed previously [25] and showed that this sample has a very low intensity of the fast OSL component. As shown in Figure 5b, the first-order decay curves can be fitted to OTDCs obtained in these experiments with excellent accuracy. The trap depth and frequency factor estimated for sample MR equal $E = 1.67 \pm 0.03$ eV and $s = (3.0 \pm 1.9) \times 10^{13}$ s^{-1} are consistent with the values for the rest of the samples.

The TM-OSL method, whose outcomes are shown in Figure 6a,b, enables the construction of OTDCs having regular first-order shapes for both the GO sample with a significant share of the fast component and the GC sample characterized by a low percentage of the fast component in the total OSL signal. It is worth mentioning that the first-order decay fitting for holding a temperature equal to 280 °C for sample granite Catalina is worse than for other temperatures. Therefore, the OTDC for 280 °C was replaced by OTDC measured at 250 °C, for which the first-order fitting gives accurate results. The E and s ($E = 1.59 \pm 0.01$ eV; $s = (7.3 \pm 0.9) \times 10^{12}$ s^{-1} for GO and $E = 1.590 \pm 0.001$ eV; $s = (5.3 \pm 0.17) \times 10^{12}$ s^{-1} for GC) were determined with good precision and agree with those obtained for sediment samples. Summing up, in all the above-presented cases, the first-order decay can be fitted with good quality when the TM-OSL$_{620\,nm}$ procedure is applied.

3.3. Influence of the Irradiation Dose on the OTDC Measurement Results

The final results of experiments for three different irradiation doses—10 Gy, 100 Gy and 1000 Gy are presented in Table 2, while representative TM-OSL signals for different doses are presented in Figure 7. The first-order kinetics TL peaks are shown in Figure 8 for the E and s values obtained in these experiments. They are calculated for a heating rate of 2 Ks^{-1}. Assuming a linear rate of heating the sample, b (Ks^{-1}), the TL intensity for the first-order kinetic traps can be expressed as follows [38]:

$$I_{TL}(T) = sn_0\, exp\left(\frac{-E}{kT}\right) exp\left[-\frac{s}{b}\int_{T_0}^{T} exp\left(-\frac{E}{kT'}\right) dT'\right] \tag{4}$$

n_0 (m^{-3}) is the number of trapped electrons at the beginning of heating.

When applying the TM-OSL$_{620\,nm}$ protocol with a preheat temperature of 280 °C, one can observe some slight variation in parameters with the change of dose. However, no general trend of these changes can be seen. The thermal activation energy varies from 1.54 ± 0.02 eV and 1.54 ± 0.08 eV (the lowest values obtained, respectively, for GC and FB samples, both for dose 10 Gy) up to 1.67 ± 0.31 eV and 1.67 ± 0.03 eV (the highest values obtained for 2 MV 570 and MR samples, both for dose 100 Gy). The frequency factor s varies from $(5.52 \pm 2.33) \times 10^{11}$ s^{-1} (the lowest value obtained for GC when dose 10 Gy was applied) up to $(3.14 \pm 21.3) \times 10^{13}$ s^{-1} (the highest value obtained for 2 MV 570 when dose 100 Gy was applied). Although the values of frequency factors differ significantly, they are consistent within their uncertainties. The precision of trap parameters also varies from experiment to experiment carried out for one aliquot. The good precision of the majority of E results and very high one estimated in a few cases (MR for dose 10 Gy, GC and 2 MV 570 for dose 1000 Gy) contrast with very low precision obtained for samples 2 MV 570 and MR when 100 Gy and 1000 Gy were applied, respectively (Table 2). The low precision of the determined parameters results from the quality of fitting of the exponential curve. Although small, the observed differences in energy values and significant differences in the precision of the parameters determined in OTDC measurements confirm the importance of the dose selection also in the TM-OSL$_{620\,nm}$ protocol. Results in Table 2 seem to indicate that the trap parameters obtained for rock samples are slightly lower than those for sediments.

Table 2. Trap parameters values (*E* and *s*) obtained for various irradiation doses in OTDC experiment using TM-OSL$_{620\,nm}$ protocol for preheat 280 °C. T$_{max}$ values are the maxima of TL peak positions simulated using first-order kinetic for adequate *E*, *s* and heating rate of 2 K/s.

		Rocks			Sediments		
		GO	GC	2 MV 570	FB	QC	MR
10 [Gy]	E [eV]	1.60 ± 0.11	1.54 ± 0.02	1.56 ± 0.04	1.54 ± 0.08	1.64 ± 0.05	**1.653 ± 0.004**
	s [s^{-1}]	(8.3 ± 19.0) × 10^{12}	(5.52 ± 2.33) × 10^{11}	(2.96 ± 2.51) × 10^{12}	(2.21 ± 3.88) × 10^{12}	(1.98 ± 2.02) × 10^{13}	**(2.28 ± 0.19) × 10^{13}**
	T$_{max}$ [°C]	307.7	335.3	312.1	309.3	307.3	309.3
100 [Gy]	E [eV]	1.56 ± 0.03	1.63 ± 0.03	1.67 ± 0.31	1.64 ± 0.03	1.65 ± 0.01	1.67 ± 0.03
	s [s^{-1}]	(3.43 ± 2.15) × 10^{12}	(1.4 ± 0.9) × 10^{13}	(3.14 ± 21.3) × 10^{13}	(2.15 ± 1.38) × 10^{13}	(2.50 ± 0.32) × 10^{13}	(3.0 ± 1.9) × 10^{13}
	T$_{max}$ [°C]	308.9	309.4	310.0	305.8	306.4	310.6
1000 [Gy]	E [eV]	1.59 ± 0.01	1.590 ± 0.001	1.629 ± 0.003	1.65 ± 0.05	**1.65 ± 0.04**	1.63 ± 0.15
	s [s^{-1}]	(7.3 ± 0.9) × 10^{12}	(5.3 ± 0.17) × 10^{12}	(1.46 ± 0.11) × 10^{13}	(2.47 ± 2.72) × 10^{13}	**(2.60 ± 2.14) × 10^{13}**	(4.51 ± 14.4) × 10^{12}
	T$_{max}$ [°C]	306.5	312.4	308.5	306.9	306.1	329.0

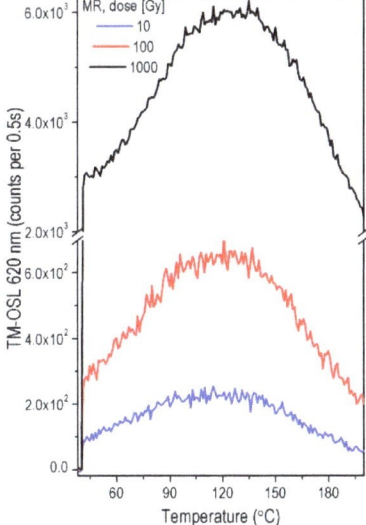

Figure 7. TM-OSL$_{620\,nm}$ curves measured for one aliquot of MR sample after irradiation with a dose: 10 Gy (blue), 100 Gy (red) and 1000 Gy (black) and preheat up to 280 °C.

It is worth noting that the positions of the first-order TL peaks obtained for all the results presented in Table 2 (Figure 8) do not differ significantly, except in two cases: sample GC measured with dose 10 Gy (red line) and MR with 1000 Gy (magenta dot line). We attribute this behavior to the dependence of OTDC results on the initial filling of traps active in the processes taking place in the OTDC measurement, as indicated earlier in the simulation study [37] (see Section 4).

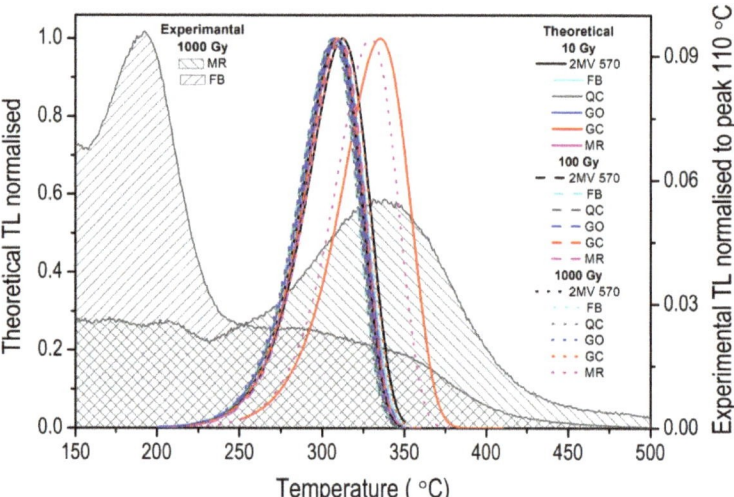

Figure 8. The theoretical single first-order kinetic TL peaks calculated for trap parameters E and s obtained in OTDC experiments set in Table 2 according to Equation (4) (left vertical axis), heating rate of 2 Ks^{-1}. Experimental TL curves (area filled with grey lines) measured using heating rate of 2 Ks^{-1} for MR and FB samples after irradiation with a dose of 1000 Gy and normalisation to peak 110 °C (right vertical axis)—the same data as shown in Figure 1.

4. Discussion

Recently, the processes occurring during OTDC measurements in a simple OSL model were studied by simulations using the luminescence kinetics model [37]. The shape of OTDC was presented for a few cases when the measurement for a single OSL trap kind is disturbed by the contribution of other traps in the thermal decay or OSL process. The signal of the investigated OSL traps was assumed to decay according to the first-order kinetics when the process is independent of other traps. These basic investigations allowed us to distinguish two main reasons for OTDCs deformation (1) the optical release of carriers from disturbing traps and their participation in the emission during the OSL measurement, and (2) the thermal transfer of carriers to OSL traps from disturbing traps during the isothermal holding. Both disturbing processes strongly depend on the trap filling at the start of the isothermal holding, so they depend on the irradiation dose and preheat temperature applied in the OTDC protocol.

In light of the research mentioned above, the fundamental problem in OTDC experiments is separating the fast OSL component from other components for investigations. As previously shown, the TM-OSL method with 620 nm wavelength stimulation enables the separation of the fast OSL component in quartz [13–16]. It is important to note that the method allows for detecting the fast component even in samples in which its intensity is very low. Such a case is shown in Figure 7 for the MR sample for which it was previously shown that the fast component is about 100 times less intense than in well-behaved samples such as sample FB. This component was not detected in sample MR by the LM-OSL measurement with blue light (see [25], (Figure 9) therein).

The OTDC measurement simulations proved that the constant preheat temperature should be used in the whole OTDC protocol and it should be not lower than the highest used holding temperature [37]. This ensures that the filling of traps at the beginning of the isothermal holding is constant in the procedure. However, this is not the only condition that must be met when selecting the correct temperature of the preheat. The preheat should be high enough to empty the shallower traps participating in the OSL and others participating in the thermal transfer during the isothermal holding. This problem, however, also applies to the TM-OSL$_{620\text{ nm}}$ protocol, even though the signal from the fast component is well separated. When initially a preheat to 150 °C was used, then the shape of the OTDCs was usually distorted. It is demonstrated in Section 3.2 for samples with the fast OSL component of relatively high intensity. As mentioned, the E and s values determined with too low preheat temperature are lower than those established with properly selected preheat for protocol TM-OSL$_{620\text{ nm}}$ (compare Figure 4a,b) for sample FB. It is worth noting that the reduction in values E and s means a significant reduction in a trap lifetime. For example, the trap lifetimes in the case of sample 2 MV 570 for the preheat to 280 °C is $(2.14 \pm 0.04) \times 10^5$ ka, for preheat 150 °C it is $(1.58 \pm 1.70) \times 10^4$ ka (calculated for 10 °C).

On the other hand, the selected preheat temperature cannot lead to excessive depopulation of the tested OSL traps. When empty, the shallow traps effectively retrap charge carriers thermally released during the isothermal holding from a deeper trap. Such a pure effect of thermal transfer causes OTDC deformation, especially at longer holding times (see [37], (Figure 13d) therein). In other cases, significant depletion of the tested OSL traps may emphasize in the measured OSL signal the previously negligible contribution from other slower decaying OSL traps in the sample. Then a lowering of trap parameters from OTDC measurement was observed. In the presented measurement results, such a situation seems to occur for sample GC with a low intensity of the fast OSL component. The OTDC experiment carried out using the TM-OSL protocol with the irradiation dose of 10 Gy led to the E and s values of 1.54 ± 0.02 eV and $(5.52 \pm 2.33) \times 10^{11}$ s^{-1}, respectively. These values are lower than those obtained for the same aliquot of this sample but for higher doses (Table 2). The first-order peak for these parameters is shifted into high temperatures from the peak obtained for other doses in the case of sample GC and the rest of the samples by more than 20 °C.

The above example indicates the significance of the irradiation dose selection in the OTDC protocol. This has been demonstrated by the simulation study mentioned above. When a preheat temperature that ensures that the shallower traps are emptied is applied before isothermal holding, it may turn out that the initial occupation of the investigated OSL traps is very low. This can lead to a low signal-to-background ratio. Then, using a higher irradiation dose in the protocol helps obtain better precision in the OSL measurements. However, the higher dose used in the measurements does not always guarantee the correct determination of the parameters of the tested OSL traps. As shown, the crucial condition is the significant dominance of the signal from these traps in the total OSL signal measured in the OTDC protocol. Applying a higher dose does not necessarily make the condition true. It is the case, for example, when other slower decaying, thermally deeper OSL traps have a significantly higher concentration and, due to a lower retrapping factor, they fill up more slowly. For the doses for which the tested OSL traps are full, the share of the total OSL signal of the slower traps begins to play an increasingly important role. Hence, when a high dose that is too high is applied, the stipulation of the tested trap dominance in the total OSL signal is weakened and the resulting OTDCs have a slower decay.

In Figure 8, one can notice another example of the TL peak shifting from the others to higher temperatures. It is the curve calculated for the E and s obtained for sample MR, hence again for a sample with a relatively low intensity of the fast OSL component. The case corresponds to the above-mentioned situation when a too-high dose is used in the OTDC protocol. One can fit the OTDCs by an exponential decay with reasonable accuracy, but the frequency factor assuring this result is much lower than for smaller doses ($s = (4.51 \pm 14.4) \times 10^{12}$ s^{-1} versus $s = (2.28 \pm 0.19) \times 10^{13}$ s^{-1} for doses 1000 Gy and 10 Gy, respectively). Its low value determines the position of the calculated TL peak on the temperature axis. In the case of other samples, such a high dose did not cause a similarly severe parameter deviation, indicating that the effects are sample-dependent. Intuitively, they should be such when one explains deformations of OTDCs by processes resulting from the competition of various traps. The latter is controlled by trap concentrations, which may vary significantly from sample to sample and, in the case of sediments, from grain to grain. It means that one should not arbitrarily apply a once-selected OTDC procedure to all tested samples but check how the results change with the preheat temperature and dose. The shape of OTDCs and the quality of curve fitting should be carefully viewed and selected for trap parameter estimation.

5. Conclusions

Trap parameters estimated by OTDC measurements may depend on various measurement parameters applied. The significant discrepancy in the values of the trap depth and the frequency factor of traps responsible for the fast OSL component in quartz previously presented in the literature is most probably a consequence of the complexity of the processes occurring in quartz during OTDC measurements. It was demonstrated that a careful selection of the measurement parameters and the protocols used is needed. To sum up:

- Both in samples with a low share and with a dominant share of the fast component in the total OSL signal, the proper separation of the fast OSL component in the OSL measurement is crucial for obtaining reliable values of trap parameters.
- A procedure using the TM-OSL method for measurement of the isothermal depletion curve with preheat 280 °C seems to be the best solution, at least in the case of the investigated samples. This is supported by the fact that a single exponential function is sufficient to obtain a good reconstruction of OTDC even for quartz samples of low OSL fast component intensity.
- For samples not dominated by the fast OSL component, the OTDCs can be more stretched and the fit quality can decrease. However, if the TM-OSL signal is measurable, probably after finding the proper dose, preheat and thermal holding temperatures, the trap stability may be successfully designated still using the first-order kinetic decay.
- The trap responsible for the fast OSL component in quartz is independent of the quartz sample type (sediment or rock). The mean trap depth and s factor (calculated based on bold data from Table 2) are: $E = 1.63 \pm 0.01$ eV, $s = (1.83 \pm 0.37) \times 10^{13}$ s^{-1} and are close to these established earlier for sedimentary quartz that had the fast component dominating in OSL signal by Murray and Wintle [20]: $E = 1.66 \pm 0.03$ eV, $s = (1.00 \pm 1.99) \times 10^{13}$ s^{-1}.

Overall, we showed that the TM-OSL$_{620\,nm}$ method used in the isothermal procedure can be used independently of quartz origin even for low sensitivity samples if the TM-OSL signal is measurable and experiments are carried out for proper values of dose, preheat, and isothermal holding temperatures.

Author Contributions: Conceptualization, A.C., A.T.-G. and M.B.; methodology, A.C., A.T.-G. and M.B.; validation, M.B., Z.K. and P.P.; formal analysis, M.B., Z.K. and P.P.; investigation, M.B., Z.K. and P.P.; resources, A.C. and A.T.-G.; data curation, M.B. and Z.K.; writing—original draft preparation, M.B., A.C. and A.T.-G.; writing—review and editing, A.T.-G., M.B. and Z.K.; visualization, M.B.; supervision, A.C. and A.T.-G.; project administration, A.C. and A.T.-G.; funding acquisition, A.C. and A.T.-G. All authors have read and agreed to the published version of the manuscript.

Funding: The research was financed by the grant of the National Science Centre, Poland, No. 2018/31/B/ST10/03917. A.T.-G. and Z.K. are funded by the European Research Council (ERC) under the European Union's Horizon Europe research and innovation program ERC-2021-COG (grant agreement No. 101043356).

Institutional Review Board Statement: Not applicable.

Informed Consent Statement: Not applicable.

Data Availability Statement: All data was made open access on Zenodo and is digitally identified by DOI https://doi.org/10.5281/zenodo.7383041.

Acknowledgments: The experiments were realised using the equipment belonging to the Centre for Modern Interdisciplinary Technologies, Nicolaus Copernicus University in Torun, ul. Wilenska 4, 87-100 Torun, Poland (e-mail: icnt@umk.pl).

Conflicts of Interest: The authors declare unanimously no conflict of interest.

References

1. Rhodes, E.J. Optically stimulated luminescence dating of sediments over the past 200,000 years. *Annu. Rev. Earth Planet Sci.* **2011**, *39*, 461–488. [CrossRef]
2. Guralnik, B.; Jain, M.; Herman, F.; Paris, R.B.; Harrison, T.M.; Murray, A.S.; Valla, P.G.; Rhodes, E.J. Effective closure temperature in leaky and/or saturating thermochronometers. *Earth Planet. Sci. Lett.* **2013**, *384*, 209–218. [CrossRef]
3. Smith, B.W.; Rhodes, E.J. Charge movements in quartz and their relevance to optical dating. *Radiat. Meas.* **1994**, *23*, 329–333. [CrossRef]
4. Bailey, R.M.; Smith, B.W.; Rhodes, E.J. Partial bleaching and the decay form characteristics of quartz OSL. *Radiat. Meas.* **1997**, *27*, 123–136. [CrossRef]
5. Murray, A.S.; Wintle, A.G. The single aliquot regenerative dose protocol: Potential for improvements in reliability. *Radiat. Meas.* **2003**, *37*, 377–381. [CrossRef]
6. Li, B.; Li, S.H. Comparison of De estimates using the fast component and the medium component of quartz OSL. *Radiat. Meas.* **2006**, *41*, 125–136. [CrossRef]
7. Wang, X.L.; Wintle, A.G.; Lu, Y.C. Thermally transferred luminescence in fine-grained quartz from Chinese loess: Basic observations. *Radiat. Meas.* **2006**, *41*, 649–658. [CrossRef]
8. Wu, T.S.; Jain, M.; Guralnik, B.; Murray, A.S.; Chen, Y.G. Luminescence characteristics of quartz from Hsuehshan Range (Central Taiwan) and implications for thermochronometry. *Radiat. Meas.* **2015**, *81*, 104–109. [CrossRef]
9. Durcan, J.A. Assessing the reproducibility of quartz OSL lifetime determinations derived using isothermal decay. *Radiat. Meas.* **2018**, *120*, 234–240. [CrossRef]
10. Mineli, T.D.; Sawakuchi, A.O.; Guralnik, B.; Lambert, R.; Jain, M.; Pupim, F.N.; del Rio, I.; Guedes, C.C.F.; Nogueira, L. Variation of luminescence sensitivity, characteristic dose and trap parameters of quartz from rocks and sediments. *Radiat. Meas.* **2021**, *144*, 106583. [CrossRef]
11. King, G.E.; Herman, F.; Lambert, R.; Valla, P.G.; Guralnik, B. Multi-OSL-thermochronometry of feldspar. *Quat. Geochronol.* **2016**, *33*, 76–87. [CrossRef]
12. Faershtein, G.; Guralnik, B.; Lambert, R.; Matmon, A.; Porat, N. Investigating the thermal stability of TT-OSL main source trap. *Radiat. Meas.* **2018**, *119*, 102–111. [CrossRef]
13. Chruścińska, A.; Kijek, N. Thermally modulated optically stimulated luminescence (TM–OSL) as a tool of trap parameter analysis. *J. Lumin.* **2016**, *174*, 42–48. [CrossRef]
14. Chruścińska, A.; Kijek, N.; Topolewski, S. Recent development in the optical stimulation of luminescence. *Radiat. Meas.* **2017**, *106*, 13–19. [CrossRef]
15. Chruścińska, A.; Szramowski, A. Thermally modulated optically stimulated luminescence (TM-OSL) of quartz. *J. Lumin.* **2018**, *195*, 435–440. [CrossRef]
16. Biernacka, M.; Chruścińska, A.; Palczewski, P.; Derkowski, P. Determining the kinetic parameters of traps in quartz using the thermally modulated OSL method. *J. Lumin.* **2022**, *252*, 119289. [CrossRef]
17. Dodson, M.H. Closure temperature in cooling geochronological and petrological systems. *Contrib. Mineral. Petrol.* **1973**, *40*, 259–274. [CrossRef]
18. Herman, F.; Rhodes, E.J.; Braun, J.; Heiniger, L. Uniform erosion rates and relief amplitude during glacial cycles in the Southern Alps of New Zealand, as revealed from OSL-thermochronology. *Earth Planet. Sci. Lett.* **2010**, *297*, 183–189. [CrossRef]
19. Li, B.; Li, S.H. Determining the cooling age using luminescence-thermochronology. *Tectonophysics* **2012**, *580*, 242–248. [CrossRef]
20. Murray, A.S.; Wintle, A.G. Isothermal decay of optically stimulated luminescence in quartz. *Radiat. Meas.* **1999**, *30*, 119–125. [CrossRef]
21. Spooner, N.A.; Questiaux, D.G. Kinetics of red, blue and UV thermoluminescence and optically-stimulated luminescence from quartz. *Radiat. Meas.* **2000**, *32*, 659–666. [CrossRef]
22. Groza-Săcaciu, Ș.M.; Panaiotu, C.; Timar-Gabor, A. Single Aliquot Regeneration (SAR) Optically Stimulated Luminescence Dating Protocols Using Different Grain-Sizes of Quartz: Revisiting the Chronology of Mircea Vodă Loess-Paleosol Master Section (Romania). *Methods Protoc.* **2020**, *3*, 19. [CrossRef] [PubMed]
23. Madsen, A.T.; Murray, A.S.; Andersen, T.J. Optical dating of dune ridges on Rømø, a barrier island in the Wadden Sea, Denmark. *J. Coast. Res.* **2007**, *23*, 1259–1269. [CrossRef]
24. Hansen, V.; Murray, A.; Buylaert, J.P.; Yeo, E.Y.; Thomsen, K. A new irradiated quartz for beta source calibration. *Radiat. Meas.* **2015**, *81*, 123–127. [CrossRef]
25. Schmidt, C.; Chruścińska, A.; Fasoli, M.; Biernacka, M.; Kreutzer, S.; Polymeris, G.S.; Sanderson, D.C.W.; Cresswell, A.; Adamiec, A.; Martini, M. A systematic multi-technique comparison of luminescence characteristics of two reference quartz samples. *J. Lumin.* **2022**, *250*, 119070. [CrossRef]
26. Kreutzer, S.; Friedrich, J.; Sanderson, D.; Adamiec, G.; Chruścińska, A.; Fasoli, M.; Martini, M.; Polymeris, G.S.; Burbidge, C.I.; Schmidt, C. Les sables de fontainebleau: A natural quartz reference sample and its characterization. *Anc. TL* **2017**, *35*, 21–31.
27. Fornash, K.F.; Patchett, P.J.; Gehrels, G.E.; Spencer, J.E. Evolution of granitoids in the Catalina metamorphic core complex, southeastern Arizona: U Pb, Nd, and Hf isotopic constraints. *Contrib. Mineral. Petrol.* **2013**, *165*, 1295–1310. [CrossRef]

28. Ducea, M.N.; Triantafyllou, A.; Krcmaric, J. New timing and depth constraints for the Catalina metamorphic core complex, southeast Arizona. *Tectonics* **2020**, *39*, e2020TC006383. [CrossRef]
29. Aitken, M.J. *Thermoluminescence Dating*; Academic Press: London, UK, 1985.
30. McKeever, S.W.S. *Thermoluminescence of Solids*; Cambridge University Press: Cambridge, UK, 1985.
31. Chen, R.; McKeever, S.W.S. *Theory of Thermoluminescence and Related Phenomena*; World Scientific: Singapore, 1997.
32. Wintle, A.G.; Murray, A.S. A review of quartz optically stimulated luminescence characteristics and their relevance in single-aliquot regeneration dating protocols. *Radiat. Meas.* **2006**, *41*, 369–391. [CrossRef]
33. Wintle, A.G. Thermal quenching of thermoluminescence in quartz. *Geophys. J. R. Astron. Soc.* **1975**, *41*, 107–113. [CrossRef]
34. Subedi, B.; Oniya, E.; Polymeris, G.S.; Afouxenidis, D.; Tsirliganis, N.C.; Kitis, G. Thermal quenching of thermoluminescence in quartz samples of various origin. *Nucl. Instrum. Methods Phys. Res. B* **2011**, *269*, 572–581. [CrossRef]
35. Subedi, B.; Polymeris, G.S.; Tsirliganis, N.C.; Pagonis, V.; Kitis, G. Reconstruction of thermally quenched glow curves in quartz. *Radiat. Meas.* **2012**, *47*, 250–257. [CrossRef]
36. Friedrich, J.; Kreutzer, S.; Schmidt, C. Radiofluorescence as a detection tool for quartz luminescence quenching processes. *Radiat. Meas.* **2018**, *120*, 33–40. [CrossRef]
37. Pawlak, N.K.; Chruścińska, A.; Biernacka, M.; Palczewski, P. Thermal stability assessment of OSL signal by measuring the OSL thermal depletion curves. *Measurement* **2022**, *199*, 111505. [CrossRef]
38. Randall, J.T.; Wilkins, M.H.F. Phosphorescence and electron traps. I. The study of trap distributions. *Proc. R. Soc. Lond. Ser. A* **1945**, *184*, 365–389.

Article

Study of Spectrally Resolved Thermoluminescence in Tsarev and Chelyabinsk Chondrites with a Versatile High-Sensitive Setup

Alexander Vokhmintsev [1], Ahmed Henaish [1,2], Taher Sharshar [3], Osama Hemeda [2] and Ilya Weinstein [1,4,*]

1. NANOTECH Center, Ural Federal University, 620002 Ekaterinburg, Russia; a.s.vokhmintsev@urfu.ru (A.V.); a.Henaish@urfu.ru (A.H.)
2. Physics Department, Faculty of Science, Tanta University, Tanta 31527, Egypt; omhemeda@science.tanta.edu.eg
3. Physics Department, Faculty of Science, Kafrelsheikh University, Kafr El-Shaikh 33735, Egypt; taher.sharshar@hotmail.com
4. Institute of Metallurgy of the Ural Branch of the Russian Academy of Sciences, 620016 Ekaterinburg, Russia
* Correspondence: i.a.weinstein@urfu.ru; Tel.: +7-343-375-9374

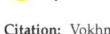

Citation: Vokhmintsev, A.; Henaish, A.; Sharshar, T.; Hemeda, O.; Weinstein, I. Study of Spectrally Resolved Thermoluminescence in Tsarev and Chelyabinsk Chondrites with a Versatile High-Sensitive Setup. *Materials* **2021**, *14*, 6518. https://doi.org/10.3390/ma14216518

Academic Editors: Dirk Poelman and Valery V. Tuchin

Received: 19 September 2021
Accepted: 28 October 2021
Published: 29 October 2021

Publisher's Note: MDPI stays neutral with regard to jurisdictional claims in published maps and institutional affiliations.

Copyright: © 2021 by the authors. Licensee MDPI, Basel, Switzerland. This article is an open access article distributed under the terms and conditions of the Creative Commons Attribution (CC BY) license (https://creativecommons.org/licenses/by/4.0/).

Abstract: Thermoluminescence (TL) research provides a powerful tool for characterizing radiation-induced processes in extraterrestrial matter. One of the challenges in studying the spectral features of the natural TL of stony meteorites is its weak intensity. The present work showcases the capabilities of a high-sensitive original module for measuring the spectrally resolved TL characteristics of the Chelyabinsk and Tsarev chondrites. We have analyzed the emission spectra and glow curves of natural and induced TL over the 300–650 nm and RT–873 K ranges. A quasi-continuous distribution of traps active within the 350–650 K range was found in the silicate substructure of both meteorites under study. Based on the general order kinetic formalism and using the natural TL data, we also estimated the activation energies of E_A = 0.86 and 1.08 eV for the Chelyabinsk and Tsarev chondrites, respectively.

Keywords: TL spectroscopy; ordinary chondrite; Chelyabinsk LL5; Tsarev L5; activation energy

1. Introduction

Thermoluminescence (TL) spectroscopy is a well-proven experimental method for studying the spectral characteristics and kinetic mechanisms of radiation-stimulated processes in irradiated materials [1,2]. In practice, there are different TL-based techniques to apply in archaeological and geological dating for dose exposure and radiation contamination monitoring by commercial systems, and different techniques correspond to specific targets within solid-state dosimetry using luminescent technologies [3]. Moreover, TL research provides a large-yield stomping ground for characterizing thermal history and describes various impact events, metamorphic processes, and features of the inorganic composition in extraterrestrial materials [4]. Catching the low intensity of natural TL is an arduous enough issue in exploring the spectral properties of meteorites. Usually, an integral luminescent response in a wide wavelength range is detected only. However, this method of studying meteorites that exhibit simultaneous emission in different bands is not always effective [5,6]. Therefore, one does not succeed in analyzing the spectral peculiarities of the natural and induced TL at a required spectral resolution. Previously, we designed and put in practice a high-temperature module as a supplementary unit for commercial fluorescence spectrometers. Additionally, its performance was tested within the range of up to 773 K for temperature, using wide-gap nitrides as examples [7,8]. In this paper, we demonstrate the capabilities of an original high-temperature module by measuring the spectrally resolved thermoluminescence of the Chelyabinsk and Tsarev meteorites. The goal is to evaluate spectral and energy parameters of thermally stimulated processes in the chondrites exposed by the high-dose irradiation.

2. Materials and Methods

Several fragments of the Chelyabinsk LL5 (fall date is 15 February 2013) and Tsarev L5 (fall date is 6 December 1922) chondrites have been studied. The meteorite cores were refined from the fusion crust, followed by grinding into a micro-sized powder. Later on, we applied hydrochloric acid treatment to the latter to remove metal particles (see Figure 1).

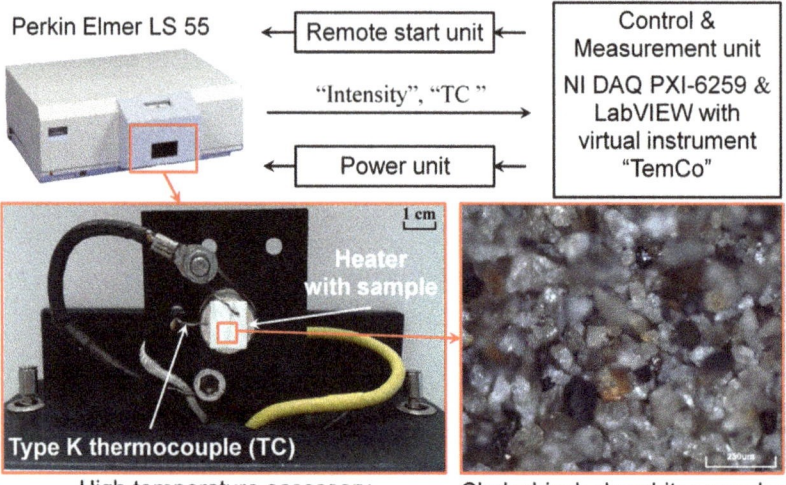

Figure 1. Thermoluminescent spectrometer.

The thermoluminescence measurements of the samples were carried out in the phosphorescence regime (12.5 ms of gate time and 20 ms of cycle time) using a LS 55 Perkin Elmer spectrometer with an original heating accessory module [7]. Figure 1 presents a block diagram of the developed TL spectrometer. The latter includes four main parts: a high-temperature accessory with a heating stage and a thermocouple, a power unit, a remote start unit, and a control and measurement unit. A detailed configuration and the operating regimes of the experimental setup, as well as its abilities to study thermally activated luminescence processes in wide-gap materials, were described in References [7,8].

The natural and induced TL glow curves for both meteorites were recorded for the 440 ± 20 nm band within the RT–873 K range with a linear heating rate of r = 2 K/s. An UELR-10-15S linear accelerator with 10 MeV electrons was utilized for the irradiation of the samples and excited an induced TL response. The radiation doses amounted to 9.1–36.4 kGy. For subsequent numerical processing, 4 measurements of the TL glow curves were performed for each value of the dose.

The TL spectra ranged from 300 to 650 nm and RT–873 K were analyzed with a scanning speed of 700 nm/min and r = 0.5 K/s. About 15 spectra were recorded during the single heating process, while the temperature of the sample changed by 15 K within one measurement of the spectral dependence. In this case, a starting temperature of the recording was assigned to each spectrum. Figures with TL spectra and TL glow curves show the selection of the measured dependencies, accounting for the clarity and completeness of the experimental data. The spectral parameters of natural TL for the Tsarev meteorite could not be analyzed due to its very low emission intensity.

3. Results and Discussion

Figure 2 shows experimental spectra of the natural and induced TL for the Chelyabinsk and Tsarev chondrites, respectively. A wide structureless band in the visible spectral range was observed in the TL emission for both meteorites under investigation at the indicated temperatures. All the TL spectra could be approximated with high accuracy (coefficient of determination is $R^2 > 0.993$) by a superposition of two G_1 and G_2 Gaussians; see Figure 3. For the appropriate temperatures, the G1 dominated in the TL emission, as its intensity was four to six times higher than that of the G_2 peak. The values of the E_{max} maximum energies and ω_E half-widths for the Gaussian bands are presented in Table 1 in comparison with independent data on spectral parameters of photo (PL) and cathodoluminescence (CL) for the same chondrites.

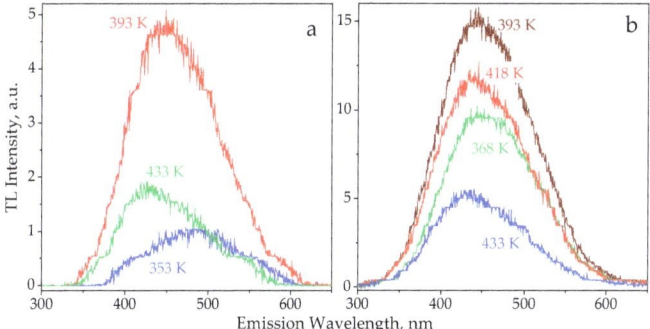

Figure 2. Emission spectra of natural TL in the Chelyabinsk chondrite (**a**) and induced TL in the Tsarev chondrite (**b**).

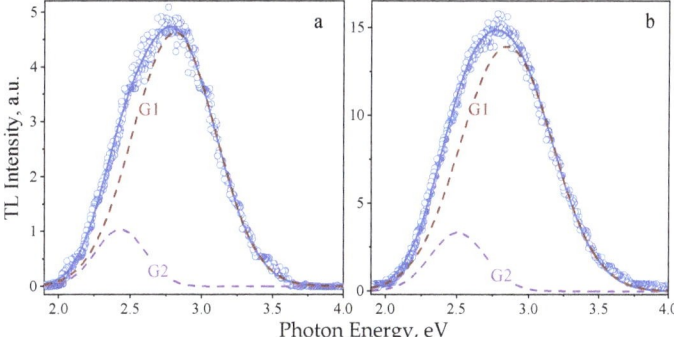

Figure 3. Deconvolution of TL spectra measured at 393 K for the Chelyabinsk (**a**) and Tsarev (**b**) chondrites. For both plots: circles—experimental data; colored dashed lines—the corresponding Gaussian components; and solid blue lines—resulting approximation curves.

Previous studies of the Chelyabinsk [6] and Tsarev [9] meteorites have shown that the observed PL emission spectra are also characterized by two Gaussians. The shape of the PL bands did not change, while their intensity decreased with varying excitation photon energy within the 6.2–4.5 eV range. The values of the spectral parameters obtained for the G_1 and G_2 components suggest that the PL and TL processes are due to the same recombination centers in both meteorites. In turn, the CL spectrum for the Chelyabinsk meteorite contained only a single Gaussian-shaped band and its parameters were consistent with the G_1 component; see Table 1. It can be concluded that the TL, CL, and PL spectra have the same emission composition, which indicates the similarity of the recombination centers involved in the mechanisms of the luminescence of the investigated meteorites exposed by UV, electrons, and space irradiation.

Analysis of the data obtained and independent studies allows one to conclude that the observed features of the luminescence can be associated with defective recombination centers in the structure of forsterite [10–12] or enstatite [13]. In the works mentioned above, the shown spectra exhibited a broad band in the 350–550 nm range. It should be noted that the 2.75 eV (450 nm) emission is well known for α-quartz and thought to be an intrinsic property of SiO_4 tetrahedrons, which are the main structural motifs in the olivines and pyroxenes [14].

Table 1. Spectral parameters of luminescence for chondrites.

Chondrite	Method	T, K	E_{max}, ±0.05 eV	ω_E, ±0.05 eV	Reference
Chelyabinsk	TL	393	2.81 2.43	0.68 0.40	This work
	PL	RT	2.80 2.45	0.70 0.37	[6]
	CL	RT	2.68	0.75	[15]
Tsarev	TL	368	2.81 2.47	0.70 0.42	This work
		393	2.85 2.50	0.72 0.44	
		418	2.87 2.50	0.73 0.42	
		443	2.87 2.48	0.74 0.33	
	PL	RT	2.90 2.48	0.85 0.42	[9]

Figure 4 shows natural and induced TL curves for the studied samples. Regarding the Chelyabinsk LL5 chondrite, a maximum level of the natural TL was visible at 400–520 K. Apart from this, a high-temperature shoulder at 520–750 K was revealed. It is worth noting that the Dhajala meteorite showcases similar parameters for the main TL peak [5]. The emission demonstrates two ten-fold intensity changes, as the maximum temperature of the TL peak shifts from T_{max} = 427 to 487 K for different samples. This can be due to inhomogeneous mineral phases or mineral compositions, or due to differences in irradiation space doses [4]. The estimated values of the shape parameters for the TL curves measured for the chondrites under study are presented in Table 2. It can be noted that the high temperature shift of the induced TL peaks with increased doses was observed within the interval of T_{max} = (390–430) ± 10 K, while the halfwidths change was observed at ω_T = 61–83 K. The maximum temperatures of the natural TL peaks are noticeably higher for both meteorites.

The observed dependencies of the TL curve parameters on the irradiation dose cannot be described under the assumption made for independent charge-capturing centers, particularly in the frame of the "one trap—one recombination center" model [16]. The high values calculated for a geometric factor of μ_g = 0.58–0.64 indicate the processes with the kinetics order of b > 2 and are consistent with the estimates performed earlier in [17]. These facts evince the possible presence of a quasi-continuous system of capturing levels, which are active and interact in the investigated temperature range. Such a situation is quite typical for silicates (pyroxene, olivine, and others), in which various structural defects form luminescent complexes to be responsible for the processes analyzed.

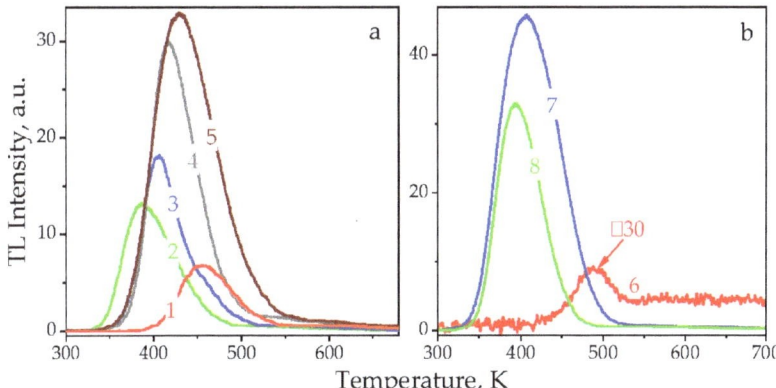

Figure 4. TL glow curves for meteorites irradiated with various doses. (**a**) The Chelyabinsk chondrite: 1—natural, 2—9.1 kGy, 3—18.2 kGy, 4—27.3 kGy, and 5—36.4 kGy. (**b**) The Tsarev chondrite: 6—natural, 7—9.1 kGy, and 8—9.1 kGy after 3 days of storage in the dark.

Table 2. TL parameters of chondrites after irradiation.

Chondrite	Dose, kGy	T_{max}, K	ω_T, K	μ_g
Chelyabinsk	9.1	387	70	0.64
	18.2	402	66	0.62
	27.3	418	61	0.59
	36.4	429	83	0.58
	natural	456	72	0.56
Tsarev	9.1	408	86	0.53
	9.1 [1]	394	62	0.58
	natural	490	60	0.50

[1] after 3 days of storage in the dark.

The natural TL glow curve for the Tsarev L5 chondrite was found to contain a low intensity peak with a maximum temperature of T_{max} = 490 K and a halfwidth of w_T = 60 K (see Figure 4b and Table 2). In addition, storage time affects the induced TL maximum; it shifts to lower temperatures from 408 ± 5 to 394 ± 5 K. In this case, the halfwidth narrows from 86 ± 5 to 62 ± 5 K. The observed TL faded away by 45% after a 3-day holding period. For drawing a more reliable conclusion concerning the number of different capturing centers emptied within the RT–500 K temperature range, it is necessary to use additional TL techniques, such as dose or heating rate variation, step pre-heating, etc. [2,18].

In the case of measuring the natural TL, the equilibrium signal with information about the accumulated dose is assumed to be read from a trap, specifically the last emptying and deepest one. Accordingly, the natural glow curves obtained were analyzed in terms of the peak shape formalism for the general order kinetics [18]:

$$E_A = \left[0.976 + 7.3\left(\mu_g - 0.42\right)\right] \frac{kT_{max}^2}{\delta}, \quad (1)$$

Here, μ_g is the geometrical factor; δ is the high temperature halfwidth of the TL peak; k is the Boltzmann constant; and J·K^{-1} and E_A are the activation energies in eV. The values of E_A = 0.86 ± 0.10 and 1.08 ± 0.10 eV were calculated using the natural TL curves for the Chelyabinsk and Tsarev chondrites, respectively. The data obtained are in satisfactory agreement with E_A = 0.9–1.6 eV, which was taken from an analysis of the Dhajala meteorite thermoluminescence [5]. For the induced TL approximation, a superposition of several glow peaks, which characterized the presence of a quasi-continuous system of traps, should be used. We have no sufficient information to choose the number of the kinetic components.

4. Conclusions

The emission spectra, specifically the natural and induced TL glow curves in the 300–650 nm range, were measured for the Chelyabinsk and Tsarev stony meteorites using a luminescent spectrometer with a developed high-temperature appliance. All the TL spectra were approximated by a superposition of two Gaussians with maximum energies near 2.8 and 2.5 eV. The 2.8 eV band dominated in the TL emission and had an intensity four to six times higher than that of the 2.5 eV band. The conducted analysis of the obtained and independent data on the spectral parameters of PL and CL allows one to conclude that the observed features of the luminescence can be caused by defective recombination centers in the structure of forsterite in the meteorite composition.

The study of induced TL has shown that a high-temperature shift of $\approx$40 K is observed for the TL peak maximum as the dose increases within the 9.1–36.4 kGy range. We have revealed that a quasi-continuous traps distribution is active at 350–650 K. Based on the general order kinetics and using natural TL data for the Chelyabinsk and Tsarev meteorites, we also estimated the values of activation energies $E_A = 0.86 \pm 0.10$ and 1.08 ± 0.10 eV. This work has demonstrated that thermoluminescence processes in the Chelyabinsk LL5 and Tsarev L5 chondrites are characterized by similar spectral and kinetic peculiarities.

Author Contributions: Conceptualization, I.W. and A.V.; formal analysis, A.H.; investigation, A.V. and A.H.; resources, I.W. and O.H.; writing—original draft preparation, A.V. and A.H.; writing—review and editing, I.W., T.S. and O.H.; visualization, A.V. and I.W.; supervision, I.W. and T.S.; funding acquisition, I.W. and T.S. All authors have read and agreed to the published version of the manuscript.

Funding: This work was supported by the Minobrnauki research project, number FEUZ-2020-0059.

Institutional Review Board Statement: Not applicable.

Informed Consent Statement: Not applicable.

Data Availability Statement: The data presented in this study are available on request from the corresponding author.

Acknowledgments: Authors thank O. Ryabukhin and A. Ishchenko for their help in the irradiation of the samples. The fragment of the Tsarev chondrite was provided for study by V. Grokhovsky.

Conflicts of Interest: The authors declare no conflict of interest. The funder had no role in the design of the study.

References

1. Bos, A.J.J. Thermoluminescence as a research tool to investigate luminescence mechanisms. *Materials* **2017**, *10*, 1357. [CrossRef] [PubMed]
2. Sunta, C.M. *Unraveling Thermoluminescence*; Springer: New Delhi, India, 2015; 188p.
3. Murthy, K.V.R. Thermoluminescence and its Applications: A Review. *Defect Diffus. Forum* **2013**, *347*, 35–73. [CrossRef]
4. Sears, D.W.G.; Ninagawa, K.; Singhvi, A.K. Luminescence studies of extraterrestrial materials: Insights into their recent radiation and thermal histories and into their metamorphic history. *Chem. Erde-Geochem.* **2013**, *73*, 1–37. [CrossRef]
5. Biswas, R.H.; Morthekai, P.; Gartia, R.K.; Chawla, S.; Singhvi, A.K. Thermoluminescence of the meteorite interior: A possible tool for the estimation of cosmic ray exposure ages. *Earth Plan. Sci. Lett.* **2011**, *304*, 36–44. [CrossRef]
6. Popova, O.P.; Jenniskens, P.; Emel'yanenko, V.; Kartashova, A.; Biryukov, E.; Khaibrakhmanov, S.; Shuvalov, V.; Rybnov, Y.; Dudorov, A.; Grokhovsky, V.I.; et al. Chelyabinsk airburst, damage assessment, meteorite recovery, and characterization. *Science* **2013**, *342*, 1069–1073. [CrossRef] [PubMed]
7. Vokhmintsev, A.S.; Minin, M.G.; Chaykin, D.V.; Weinstein, I.A. A high-temperature accessory for measurements of the spectral characteristics of thermoluminescence. *Instr. Exp. Tech.* **2014**, *57*, 369–373. [CrossRef]
8. Vokhmintsev, A.S.; Minin, M.G.; Henaish, A.M.A.; Weinstein, I.A. Spectrally resolved thermoluminescence measurements in fluorescence spectrometer. *Measurement* **2015**, *66*, 90–94. [CrossRef]
9. Vokhmintsev, A.S.; Weinstein, I.A.; Grokhovsky, V.I. Luminescence characterization of Tsarev L5 chondrite. *Meteorit. Planet. Sci.* **2015**, *50* (Suppl. 1), 5200.

10. Benstock, E.J.; Buseck, P.R.; Steele, I.M. Cathodoluminescence of meteoritic and synthetic forsterite at 296 and 77 K using TEM. *Am. Mineral.* **1997**, *82*, 310–315. [CrossRef]
11. Jones, R.H.; Carey, E.R. Identification of relict forsterite grains in forsterite-rich chondrules from the Mokoia CV3 carbonaceous chondrites. *Am. Mineral.* **2006**, *91*, 1664–1674. [CrossRef]
12. Nishido, H.; Endo, T.; Ninagawa, K.; Kayama, M.; Gucsik, A. Thermal effects on cathodoluminescence in forsterite. *Geochronometria* **2013**, *40*, 239–243. [CrossRef]
13. Zhang, Y.; Huang, S.; Schneider, D.; Benoit, P.H.; DeHart, J.M.; Lofgren, G.E.; Sears, D.W. Pyroxene structures, cathodoluminescence and the thermal history of the enstatite chondrites. *Meteorit. Planet. Sci.* **1996**, *31*, 87–96. [CrossRef]
14. Waychunas, G.A. Luminescence, X-ray emission and new spectroscopies. *Rev. Mineral.* **1988**, *18*, 639–698.
15. Weinstein, I.A.; Vokhmintsev, A.S.; Ishchenko, A.V.; Grokhovsky, V.I. Luminescence characterization of different lithologies in Chelyabinsk LL5 chondrite. *Meteorit. Planet. Sci.* **2014**, *49* (Suppl. 1), A428.
16. Pagonis, V.; Kitis, G.; Furetta, C. *Numerical and Practical Exercises in Thermoluminescence*; Springer: New York, NY, USA, 2006; 225p.
17. Weinstein, I.A.; Vokhmintsev, A.S.; Ishchenko, A.V.; Grokhovsky, V.I. Spectral and kinetic features of thermoluminescence in Chelyabinsk LL5 chondrite. *Meteorit. Planet. Sci.* **2013**, *48* (Suppl. 1), A368.
18. McKeever, S.W.S. *Thermoluminescence of Solids*; Cambridge University Press: London, UK, 1988; 386p.

Charge Transfer and Charge Trapping Processes in Ca- or Al-Co-doped Lu$_2$SiO$_5$ and Lu$_2$Si$_2$O$_7$ Scintillators Activated by Pr^{3+} or Ce^{3+} Ions

Valentyn Laguta [1,*], Lubomir Havlak [1], Vladimir Babin [1], Jan Barta [1,2], Jan Pejchal [1] and Martin Nikl [1]

[1] Institute of Physics of the Czech Academy of Sciences, Cukrovarnicka 10/112, 16200 Prague, Czech Republic; jan.barta@fjfi.cvut.cz (J.B.)
[2] Faculty of Nuclear Sciences and Physical Engineering, Czech Technical University in Prague, Břehová 7, 11519 Prague, Czech Republic
* Correspondence: laguta@fzu.cz

Abstract: Lutetium oxyorthosilicate Lu$_2$SiO$_5$ (LSO) and pyrosilicate Lu$_2$Si$_2$O$_7$ (LPS) activated by Ce^{3+} or Pr^{3+} are known to be effective and fast scintillation materials for the detection of X-rays and γ-rays. Their performances can be further improved by co-doping with aliovalent ions. Herein, we investigate the Ce^{3+}(Pr^{3+}) → Ce^{4+}(Pr^{4+}) conversion and the formation of lattice defects stimulated by co-doping with Ca^{2+} and Al^{3+} in LSO and LPS powders prepared by the solid-state reaction process. The materials were studied by electron paramagnetic resonance (EPR), radioluminescence spectroscopy, and thermally stimulated luminescence (TSL), and scintillation decays were measured. EPR measurements of both LSO:Ce and LPS:Ce showed effective Ce^{3+} → Ce^{4+} conversions stimulated by Ca^{2+} co-doping, while the effect of Al^{3+} co-doping was less effective. In Pr-doped LSO and LPS, a similar Pr^{3+} → Pr^{4+} conversion was not detected by EPR, suggesting that the charge compensation of Al^{3+} and Ca^{2+} ions is realized via other impurities and/or lattice defects. X-ray irradiation of LPS creates hole centers attributed to a hole trapped in an oxygen ion in the neighborhood of Al^{3+} and Ca^{2+}. These hole centers contribute to an intense TSL glow peak at 450–470 K. In contrast to LPS, only weak TSL peaks are detected in LSO and no hole centers are visible via EPR. The scintillation decay curves of both LSO and LPS show a bi-exponential decay with fast and slow component decay times of 10–13 ns and 30–36 ns, respectively. The decay time of the fast component shows a small (6–8%) decrease due to co-doping.

Keywords: scintillation material; luminescence; EPR; lattice defect; radioluminescence

1. Introduction

Scintillation materials are currently widely used for radiation detection in many fields, such as medical imaging, high energy physics calorimetry, bolometry for rare events searches, industrial control, safety and homeland security, and others [1]. Among them, wide bandgap oxide dielectrics with a high degree of structural perfection are the most suitable for such purposes [2]. In general, for scintillation applications, the material must accomplish fast and efficient transformations of incoming high energy photons/particles (or energy arising in a nuclear reaction with neutrons) into a number of electron–hole pairs collected in the conduction and valence bands, respectively, and their radiative recombination at suitable luminescence centers in the material. Therefore, most of the applications using scintillation materials are based on the density, scintillation, and time response performances. Based on these parameters, an impressive number of heavy cation-based hosts (particularly lutetium/yttrium/gadolinium) doped with Ce^{3+} or Pr^{3+} have been developed. Among them, a promising family of scintillation crystals is based on Lu$_2$SiO$_5$ (LSO), (Lu,Y)$_2$SiO$_5$ (LYSO) oxyorthosilicates, and Lu$_2$Si$_2$O$_7$ (LPS) and recently (Gd,La)$_2$Si$_2$O$_7$ pyrosilicates doped by Ce^{3+} or Pr^{3+} ions (see, e.g., review paper [2] and refs. [3–6]). All the abovementioned materials are excellent candidates for the detection of gamma rays in both positron emission tomography (PET), a very powerful medical imaging method to monitor metabolism, blood flow, or neurotransmission [7], and high energy calorimetry [2,8]. LYSO:Ce crystals are currently used in scintillation detectors in PET scanners, and various co-doping schemes have been reported in the last decade to further improve their performance [2,9,10]. In particular, co-doping LSO:Ce with divalent Ca^{2+} or Mg^{2+} ions has been shown to eliminate shallow electron traps and decrease the scintillation decay time from ~43 ns to ~30 ns while maintaining a high light output [8,11,12]. This improvement is at least partially related to the effective $Ce^{3+} \rightarrow Ce^{4+}$ conversion, where the stable Ce^{4+} ion creates an additional fast radiative recombination pathway, which efficiently competes in electron trapping from the conduction band with any other electron traps [11]. Alternatively, co-doping with a trivalent metal ion (Al, Ga, or In) that substitutes a tetravalent Si ion has been proposed [13,14]. It was assumed that such co-doping creates a positive charge deficit that limits the trapping of electrons responsible for afterglow.

The same positive effect of $Ca^{2+}(Mg^{2+})$ or Al^{3+} co-doping on decay time improvements is expected in Ce- or Pr-doped $Lu_2Si_2O_7$ pyrosilicates [15]. They show even better scintillation characteristics than oxyorthosilicates. In particular, with nearly the same light yield, energy resolution, and scintillation decay time as reported for LSO:Ce, Ce-doped pyrosilicates are free from the intense afterglow as reported in [16,17]. However, in general, the effect of Ca^{2+}, Mg^{2+}, or Al^{3+} co-doping on charge trapping processes has not been investigated practically, especially in Pr-doped oxyorthosilicates and pyrosilicates.

In oxyorthosilicates as well as in scintillating garnets, in addition to the $Ce^{3+} \rightarrow Ce^{4+}$ recharge that improves the timing characteristics, the creation of additional charge trapping sites due to co-doping with aliovalent ions is important as well. Such traps will decrease the light yield and can also contribute to delayed luminescence or afterglow. This phenomenon can seriously limit the time response of a scintillator, which is crucial for PET scanners and other time-of-flight applications where a sub-100 ps time resolution is desirable. Although charge traps have been extensively studied in both LSO and LPS activated by Ce^{3+} and Pr^{3+} ions using the thermally stimulated luminescence (TSL) and the local probe electron paramagnetic resonance (EPR) methods [18–21], not much is known about charge traps in $Ca^{2+}(Mg^{2+})$ or Al^{3+} co-doped materials.

In this paper, we present the results of a detailed EPR study of polycrystalline Ce- and Pr-doped Lu_2SiO_5 and $Lu_2Si_2O_7$ co-doped with Ca^{2+} and Al^{3+} with the aim of clarifying the $Ce^{3+}(Pr^{3+}) \rightarrow Ce^{4+}(Pr^{4+})$ conversion and to study the formation of lattice defects stimulated by co-doping with aliovalent ions. Our EPR study is also accompanied by the TSL and scintillation decay measurements of the synthesized materials.

2. Materials and Methods

The powder samples of LSO and LPS doped with 2000 ppm Ce (or Pr)/Lu and co-doped with 5000 ppm Al/Si or 5000 ppm Ca/Lu (see Table 1) were prepared by a conventional solid-state reaction process [22] consisting of several periods of annealing in air up to 1500–1600 °C/72 h and remixing in an agate mortar. The starting materials were 5N Lu_2O_3, 4N8 SiO_2, 5N Al_2O_3, CaO grade I, 4N CeO_2, or 5N Pr_6O_{11}. The weights of the starting oxides were corrected for the moisture content in the base materials determined after annealing at 1200 °C/12 h. Annealing was carried out in corundum boats with lids, which were washed and then annealed up to 1600 °C/12 h before their first use. It should be noted that due to segregation during the growth of crystallites, the final contents of the dopants in the lattice could be smaller by up to a factor of 0.5 [8] than those indicated in Table 1.

Table 1. Nominal composition of samples.

Theoretical Sample Formula	Designation in the Text
$Lu_{1.996}Ce_{0.004}SiO_5$	LSO:Ce
$Lu_{1.986}Ce_{0.004}Ca_{0.01}SiO_5$	LSO:Ce,Ca
$Lu_{1.996}Ce_{0.004}Si_{0.995}Al_{0.005}O_5$	LSO:Ce,Al
$Lu_{1.996}Ce_{0.004}Si_2O_7$	LPS:Ce
$Lu_{1.986}Ce_{0.004}Ca_{0.01}Si_2O_7$	LPS:Ce,Ca
$Lu_{1.996}Ce_{0.004}Si_{1.99}Al_{0.01}O_7$	LPS:Ce,Al
$Lu_{1.996}Pr_{0.004}SiO_5$	LSO:Pr
$Lu_{1.986}Pr_{0.004}Ca_{0.01}SiO_5$	LSO:Pr,Ca
$Lu_{1.996}Pr_{0.004}Si_{0.995}Al_{0.005}O_5$	LSO:Pr,Al
$Lu_{1.996}Pr_{0.004}Si_2O_7$	LPS:Pr
$Lu_{1.986}Pr_{0.004}Ca_{0.01}Si_2O_7$	LPS:Pr,Ca
$Lu_{1.996}Pr_{0.004}Si_{1.99}Al_{0.01}O_7$	LPS:Pr,Al

The phase purity of all synthesized LSO and LPS powders was evaluated by X-ray powder diffraction (XRPD) using a Rigaku MiniFlex 600 diffractometer equipped with a Cu X-ray tube and a NaI:Tl scintillation detector. The diffraction patterns exactly matched the following LSO and LPS database records in the ICDD PDF-2 database: 01-070-9485 and 01-071-3309, respectively (see Supplementary Materials). This proved that all samples after the final annealing step were phase-pure materials with a phase composition corresponding to the desired one; LSO samples contained just Lu_2SiO_5 (space group $C2/c$ [23]) and LPS samples contained just $Lu_2Si_2O_7$ (space group $C2/m$ [24]).

EPR spectra were measured using a commercial Bruker EMX plus spectrometer at X-band (microwave frequency 9.25–9.5 GHz) within the temperature range of 10–290 K. An X-ray tube operating at a voltage and current of 55 kV and 30 mA, respectively, with a Co anode (ISO-DEBYEFLEX 3003 Seifert Gmbh., Ahrensburg, Germany) was used as the source of X-ray irradiation for LSO and LPS powders.

The radioluminescence (RL) and thermally stimulated luminescence measurements were performed in the spectral range 200–800 nm using the Horiba Jobin-Yvon 5000M spectrometer with an Oxford liquid nitrogen cryostat and a TBX-04 (IBH) photomultiplier (Glasgow, Scotland). The spectral bandwidth of the monochromator was 8 nm. The RL spectra were recorded at 295 K. The spectrally unresolved TSL glow curves were recorded in the temperature range of 77–500 K with a heating rate of 0.1 K/s. Irradiation of the samples was performed at 77 K via a Seifert X-ray tube operating at 40 kV with a tungsten target; the dose was estimated to be about 450 Gy. All the spectra were corrected for spectral distortions caused by the experimental setup.

Scintillation decays with ultra-high time resolution under the pulsed X-ray excitation were measured using a picosecond (ps) X-ray tube N5084 (Hamamatsu, 40 kV, Shizuoka, Japan). The X-ray tube was driven by a ps pulsed laser at a repetition rate of up to 1 MHz. The signal was detected by a hybrid picosecond photon detector and a Fluorohub unit (Horiba Scientific). The instrumental response function FWHM of the setup is about 75 ps. Spectrally unresolved luminescence decay curves were detected from the surface excited by X-rays. The convolution procedure of the instrumental response and fit function was applied to fit the decay curves and determine the true decay times (SpectraSolve™ software package for Windows, Ames Photonics, Hurst, TX, USA).

3. Results

3.1. EPR Spectra in Ca and Al Co-doped LSO:Ce and LSO:Pr

A detailed investigation of the $Ce^{3+} \to Ce^{4+}$ charge conversion that plays an important role in the acceleration of the Ce^{3+} scintillation decay was performed on LSO:Ce, LSO:Ce,Al, and LSO:Ce,Ca samples. The corresponding EPR spectra are presented in Figure 1. The spectra contain two spectral lines from Ce^{3+} ions ($S = 1/2$, $4f^1$) corresponding to two principal g factors: $g_1 = 2.262$ and $g_2 = 1.686$. The third Ce^{3+} spectral line at $g_3 = 0.563$ ($B_r = 16.676$ kG) is outside of the magnetic field range. These g factors coincide with those measured previously in crystals [25]. Note that the second Ce^{3+} center detected in crystals was not resolved in the powder spectrum due to the low intensity of its spectral lines (the population of the second Ce^{3+} center is only about 5% [25]).

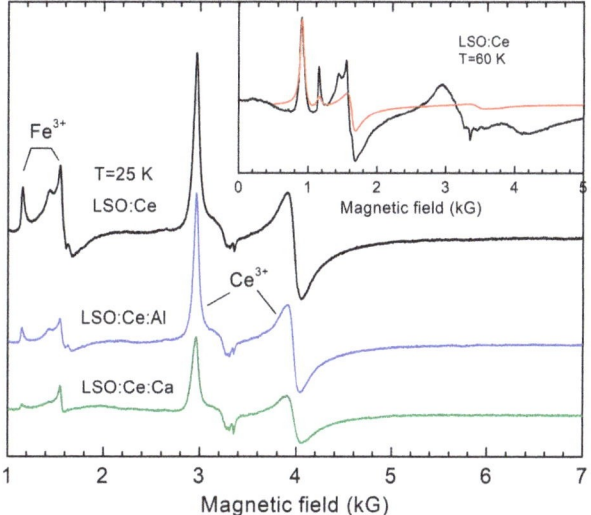

Figure 1. EPR spectra measured in LSO:Ce showing the change in the Ce^{3+} and Fe^{3+} concentration (EPR intensity is directly proportional to concentration) under co-doping with Ca^{2+} and Al^{3+} ions. The signal at low magnetic fields is assigned to Fe^{3+} ions. The inset shows a comparison of the Fe^{3+} simulated spectrum (red line) with the measured spectrum (black line). All spectra are normalized to the same sample volume.

There are other spectral lines at low magnetic fields, which we assign to Fe^{3+} accidental impurities. The spectrum is typical for this ion ($S = 5/2$, $3d^5$) in a low-symmetry crystal field with large zero-field splitting of energy levels [26]. The corresponding simulated spectrum is shown in the inset of Figure 1 (red line), and the spin Hamiltonian parameters are listed in Table 2. For the fit, the conventional spin Hamiltonian was used:

$$\mathbf{H} = \beta \mathbf{S} \cdot \mathbf{g} \cdot \mathbf{B} + \frac{1}{3}(b_2^0 O_2^0 + b_2^2 O_2^2),$$

where β is the Bohr magneton, $\mathbf{g}$ is the g-tensor, and b_l^m and O_l^m are crystal field parameters and Stevens spin operators [27], respectively. The fourth-rank crystal field terms were neglected as they are much smaller than the other terms and could not be determined from powder spectra. The roughly estimated concentration of Fe^{3+} ions from the EPR intensity is only 5–10 ppm. These ions, most probably, come from the raw materials.

Table 2. Spectral parameters of Ce^{3+}, Fe^{3+}, and O^- centers in LSO and LPS powders.

Material	Spin Hamiltonian Parameters	HF Constant	Reference
LSO:Ce LSO:Pr	Ce^{3+}: $g_1 = 2.262$ $g_2 = 1.686$ $g_3 = 0.563$		This paper, [25]
	Fe^{3+}: $g = 1.99$ $b_2^0 = 0.165$ cm^{-1} $b_2^2 = 0.165$ cm^{-1}		This paper
LPS:Ce LPS:Pr	Ce^{3+}: $g_1 = 3.000$ $g_2 = 0.705$ $g_3 \approx 0.10$		This paper, [25,28]
	Fe^{3+}: $g = 2.00$ $b_2^0 = 0.700$ cm^{-1} $b_2^2 = 0.380$ cm^{-1}		This paper
LPS:Ce,Al	O^-: $g = 2.012$	Not determined	This paper
LPS:Ce,Ca	O^-: $g = 2.011$	^{175}Lu: $A_{00} = 4.0 \times 10^{-4}$ cm^{-1}	This paper

A substantial (three-fold) decrease in Ce^{3+} concentration in Ca-co-doped samples was observed, confirming the effective $Ce^{3+} \rightarrow Ce^{4+}$ conversion, while the effect of Al co-doping (when Al^{3+} is substituted for Si^{4+}) was much smaller, at only about 15–20%. The Fe^{3+} EPR intensity also decreases under Ca and Al co-doping, indicating the recharge of the Fe^{3+} ions to a valence state invisible in EPR.

We also measured the $Pr^{3+} \rightarrow Pr^{4+}$ conversion in LSO:Pr samples stimulated by Al and Ca co-doping. The EPR spectrum of Pr^{3+} ($S = 1$, $4f^2$) is not visible at 9–35 GHz which is available in our spectrometer due to a large splitting of the $4f^2$ energy levels already in zero field (non-paramagnetic ground state is not excluded as well). Moreover, even Pr^{4+}, which has the same electron shell as Ce^{3+} and should be easily detectable in EPR, was not detected in the EPR spectra even at measurement temperatures down to 3.5 K. This suggests a much lower concentration of Pr^{4+} ions compared to Pr^{3+} ions. Therefore, the $Pr^{3+} \rightarrow Pr^{4+}$ conversion stimulated by Al and Ca co-doping of LSO is much less effective than the $Ce^{3+} \rightarrow Ce^{4+}$ conversion. Instead, as in the case of LSO:Ce, we observed a pronounced change in the Fe^{3+} concentration (Figure 2) induced by Al and Ca co-doping of LSO:Pr. However, here, the Fe^{3+} concentration increases, suggesting that the $Fe^{2+} \rightarrow Fe^{3+}$ conversion takes place to compensate, at least partly, the excess negative charge caused by replacement of Lu^{3+} by Ca^{2+}.

3.2. EPR Spectra in Ca- and Al-Co-doped LPS:Ce and LPS:Pr

The EPR spectra measured in LPS:Pr, LPS:Pr,Al, and LPS:Pr,Ca powders are presented in Figure 3. As in the case of Pr-doped LSO samples, only Fe^{3+} ions were detected in EPR spectra. The spectral parameters of these ions in the LPS lattice (obtained from the simulation of the Fe^{3+} powder spectrum) are listed in Table 2. In contrast to the Pr-doped LSO, no change in the Fe^{3+} concentration was observed in LPS. However, in Ce-doped LPS (Figure 4), the Fe^{3+} concentration markedly increased with Ca co-doping. The Ce^{3+} spectral lines were identified according to the data published in [25,28]. The Ce^{3+} concentration markedly decreases with Al and Ca co-doping. The effect of the co-doping is especially large for the Ca co-dopant, similar LSO:Ce, indicating an effective $Ce^{3+} \rightarrow Ce^{4+}$ conversion.

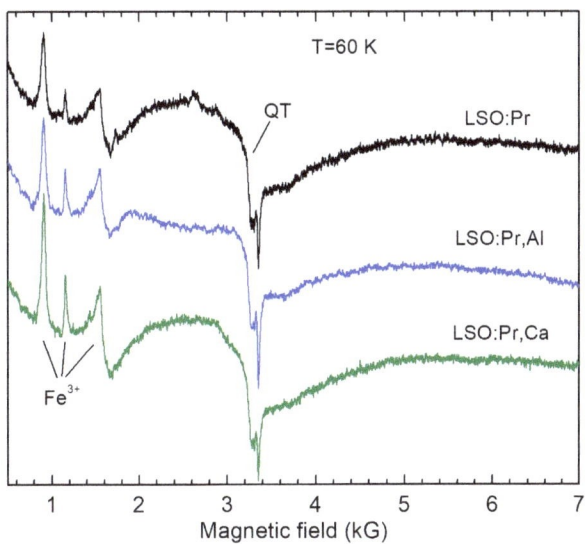

Figure 2. EPR spectra measured in LSO:Pr, LSO:Pr,Al, and LSO:Pr,Ca powders showing the change in the Fe^{3+} concentration under Al and Ca co-doping. The broad signal denoted as QT belongs to quartz tubes of LHe cryostat. All spectra are normalized to the same sample volume.

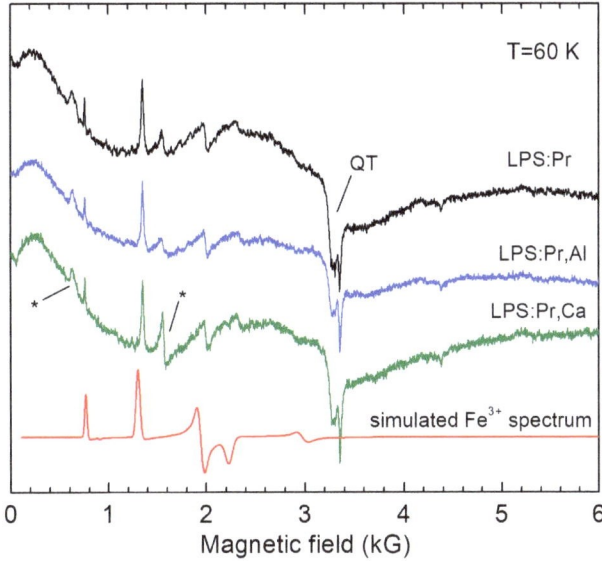

Figure 3. EPR spectra measured in LPS:Pr, LPS:Pr,Al, and LPS:Pr,Ca powders. The broad signal denoted as QT belongs to quartz tubes of the LHe cryostat and asterisks denote unidentified spectral lines. The simulated Fe^{3+} spectrum (red line) is shown as well. All spectra are normalized to the same sample volume.

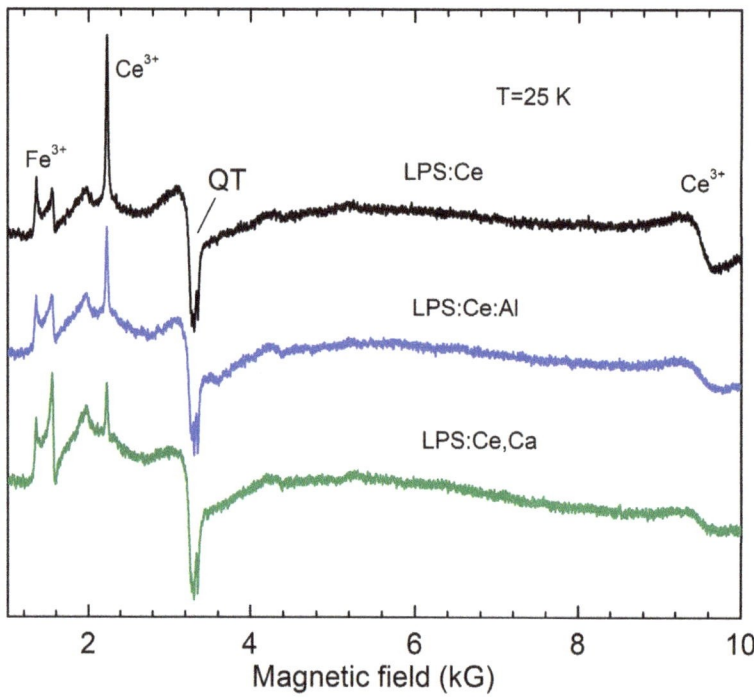

Figure 4. EPR spectra measured in LPS:Ce showing the change in the Ce^{3+} concentration under co-doping with Ca^{2+} and Al^{3+} ions. All spectra are normalized to the same sample volume.

3.3. EPR Spectra Created by X-ray Irradiation in LPS:Ce, LPS:Pr, LSO:Ce, and LSO:Pr Co-doped with Ca^{2+} or Al^{3+} and Comparison with TSL

The effect of Al and Ca co-doping on the hole and electron trapping processes was investigated in X-ray-irradiated LPS:Ce(Pr) and LSO:Ce(Pr) samples via EPR measurements of X-ray-irradiation-induced paramagnetic centers and with TSL measurements. These measurements showed formation of O^- hole centers (a hole trapped at an oxygen ion) stabilized by Al^{3+} or Ca^{2+} ions at Si^{4+} or Lu^{3+} sites in LPS samples, respectively. Figure 5 illustrates the O^- EPR spectra created by X-ray irradiation at 295 K for selected samples.

The interpretation of these O^- spectra is based on our detailed study of the O^- centers in LPS:Ce and LPS:Pr crystals [21]. In particular, the O^- EPR spectra are mainly constructed from unresolved hyperfine components from ^{175}Lu isotopes (nuclear spin $I = 7/2$, 97.4% natural abundance). In single crystal samples, the hyperfine structure of ^{175}Lu isotopes is clearly observed [21]. As can be seen from Figure 5a,b, X-ray irradiation creates the same spectrum in both LPS:Ce and LPS:Ce,Al samples. However, the spectral intensity is much higher in the LPS:Ce,Al sample, suggesting that O^- centers are created near Al^{3+} impurities (Al^{3+} at Si^{4+} sites stabilizes the trapped hole in the neighboring oxygen lattice ion). Such hole centers are usually called bound small polarons [29]. The nominally Al undoped LPS:Ce probably also contains small amounts of the same O^-–Al centers, as powders were synthesized in an Al_2O_3 boat at 1500–1600 °C. It is assumed that the trapped hole interacts with the nuclear spins of the two nearest Lu ions and the nuclear spin of Al impurities. Together with the g factor anisotropy, this creates quite a complex spectral pattern which cannot realistically be simulated due to many unknown parameters. Therefore, this center was characterized in this work only by its average g factor value measured at the center of gravity of the spectral line (Table 2).

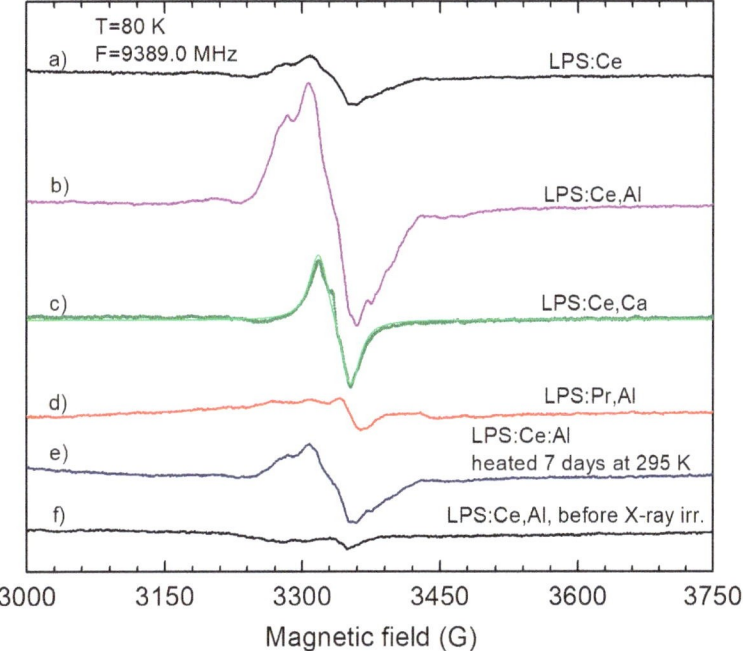

Figure 5. O⁻ EPR spectra created by X-ray irradiation at room temperature in (**a**) LPS:Ce; (**b**) LPS:Ce,Al; (**c**) LPS:Ce,Ca; and (**d**) LPS:Pr,Al. (**e**) EPR spectrum after heating of the X-ray irradiated LPS:Ce,Al powder at 295 K for 7 days. (**f**) EPR spectrum measured in LPS:Ce,Al before X-ray irradiation. Simulated O⁻ EPR spectrum is shown by the green solid line for LPS:Ce,Ca in spectrum (**c**). The simulated spectrum coincides with the measured one (black solid line).

In the case of Ca co-doping, the Ca^{2+} ion at the Lu^{3+} site also serves as a stabilizing defect for the hole trapped at an oxygen ion. In the O⁻–Ca centers, trapped holes interact only with the nuclear spin of one Lu ion (Ca has no isotopes with non-zero nuclear spins and the ^{29}Si isotope only has a small natural abundance of 4.67%) and the O⁻ spectral line in the LPS:Ce,Ca sample is narrow (Figure 5c). Its shape can be easily simulated including the ^{175}Lu hyperfine interaction (green solid line in Figure 5c and parameters in Table 2).

X-ray irradiation of Pr-doped LPS also created O⁻ hole centers. However, their EPR spectrum is broad (Figure 5d) and could not be qualitatively analyzed. O⁻ centers in LPS created by X-ray irradiation at room temperature are thermally stable to about 350 K, but their concentration slowly decreases even at room temperature (Figure 5e). They completely disappear at annealing temperatures of ~450–500 K (see also ref. [21]). This correlates well with the main TSL peak at 440–480 K created by X-ray irradiation measured in the same samples (Figure 6a,b). This TSL peak is broad and contains contributions from several traps [19,30], not all of them being paramagnetic. A complete fitting of the TSL peaks is presented in refs. [19,20,30]. Therefore, in general, the intensity of TSL does not correlate exactly with the intensity of the EPR spectra. On the other hand, one can notice that the TSL intensity is much higher in the Al-co-doped LPS:Pr sample as compared to the Al-free sample (Figure 6b).

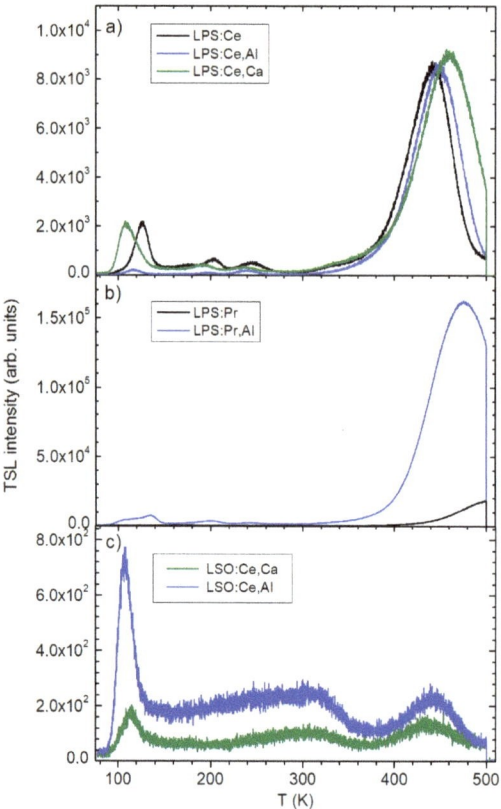

Figure 6. TSL glow curves of: (**a**) LPS:Ce, LPS:Ce,Al, and LPS:Ce,Ca; (**b**) LPS:Pr and LPS:Pr,Al; and (**c**) LSO:Ce,Al and LSO:Ce,Ca. TSL is created by X-ray irradiation at 77 K.

Surprisingly, no X-ray-created paramagnetic active centers were detected in LSO samples even after irradiation at 77 K. TSL measurements revealed a weak glow peak at ≈110 K (Figure 6c), indicating an effective (practically full) recombination of electron–hole pairs at activator ions. This contradicts the data obtained in measurements of LSO:Ce single crystals, where a sequence of intense glow peaks was observed at 350–600 K [11,20]. To explain the nature of the corresponding charge traps, the authors of ref. [20] proposed a model where Ce^{3+} serves as the charge donor, the recombination center, and also the trap creating center. The traps are related to specific configurations of oxygen ions around the central Ce^{3+} ion. Each configuration is able to trap the 5d electron in a metastable electronic state, with mixed Ce^{3+} 5d and O^{2-} orbitals at the same time creating a hole state near or at the Ce ion. According to our EPR data, the hole state is the O^- ion rather than the Ce^{4+} ion, but both may exist depending on the presence of other defects in the surrounding lattice.

3.4. Scintillation Decay Time Measurements

For selected samples, radioluminescence and scintillation decay curves within two time windows of 50 ns and 2 μs were measured at 296 K. The emission spectra of LPS:Ce samples (Figure 7a) are characterized by intense Ce-related bands with maxima located at about 380 nm in all samples. The emission intensity is largest in the non-co-doped sample and decreases by about 1.2-fold in the Al-co-doped sample and by 1.4-fold in the Ca-co-doped sample. In LSO:Ce samples, the Ce-related emission bands have maxima at about 405 nm; the intensity is the highest in the Al-co-doped sample while that of the Ca-co-doped it is about 2 times lower.

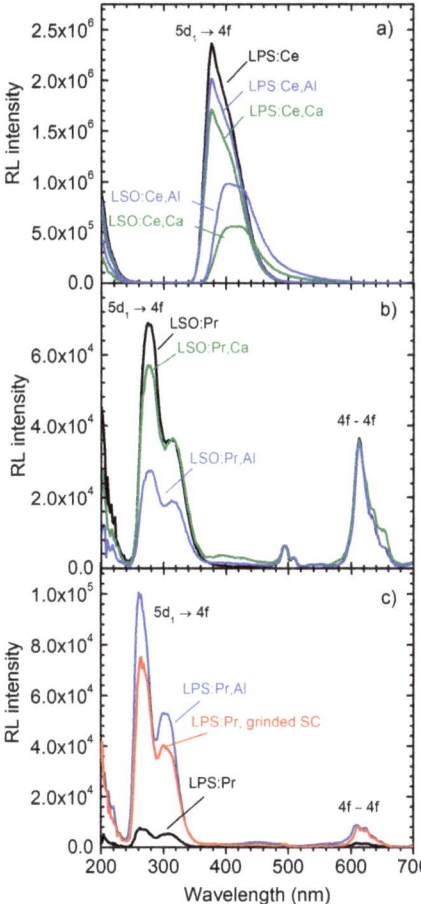

Figure 7. X-ray excited luminescence spectra measured at 296 K in (**a**) LPS:Ce, LPS:Ce,Al, LPS:Ce,Ca, LSO:Ce,Al, and LSO:Ce:Ca powders; (**b**) LSO:Pr, LSO:Pr,Ca, and LSO:Pr,Al powders; and (**c**) LPS:Pr and LPS:Pr,Al powders and LPS:Pr grinded SC.

The spectra of LSO:Pr (Figure 7b) are dominated by the 5d–4f doublet emission peak with maxima at about 280 nm. Weaker 4f–4f transitions can be observed at longer wavelengths at around 610 nm. The emission spectra of LPS:Pr samples (Figure 7c) are similar to those of LSO:Pr, but the main 5d–4f emission peak is shifted to 260 nm and the 4f–4f transitions are much weaker. Although co-doping does not influence the spectral position of the emission bands, it markedly changes the RL emission intensity.

In order to study the influence of co-doping on scintillation kinetics characteristics, the ps-pulsed X-ray excited decay curves were measured with an ultra-high time resolution, and the results are shown in Figure 8.

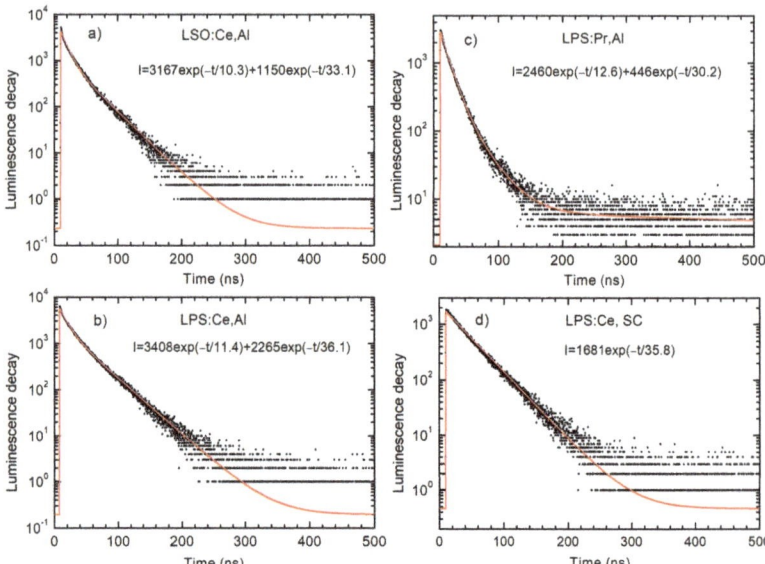

Figure 8. Luminescence decay of Ce^{3+} and Pr^{3+} emissions excited by X-ray pulses at 296 K in (**a**) LSO:Ce,Al, (**b**) LPS:Ce,Al, and (**c**) LPS:Pr,Al powder samples. For comparison, the luminescence decay in the LPS:Ce single crystal grinded in powder is presented in graph (**d**). The fit of the experimental data (black dots) is shown by the red solid lines.

All experimental decay curves were fitted using a bi-exponential function and the calculated parameters are presented in Table 3. The measurements showed, along with the conventional 33–38 ns decay component in the Ce-doped samples, a second fast component with the decay time of only about 11 ns. On the contrary, in the Pr-doped samples, along with the conventional fast decay component 13–14 ns, a second slower component (30–31 ns) is presented. The origin of the second components is unclear. To the best of our knowledge, the fast 11 ns scintillation decay component in Ce-doped LSO and LPS has never been observed before and this result needs further detailed investigation. For comparison, the decay curve of the LPS:Ce single crystal grinded into a powder can be fitted using a single exponential function (Figure 8d).

Table 3. Luminescence decay times and relative intensities of components in LPS and LSO powders doped with Ce or Pr and co-doped with Ca or Al, approximated by the bi-exponential function $I(t) = \sum A_i \exp(-t/\tau_i) + b$, where $i = 1, 2$, to fit the scintillation decay curves. The relative intensity of each component is calculated as $I_i = (A_i \tau_i / \sum A_i \tau_i) \times 100\%$. For comparison, the luminescence decay time of the LPS:Ce single crystal and published data for decay times in LSO and LPS under γ-ray excitation are listed as well.

Composition	τ_1 (ns)	I_1 (%)	τ_2 (ns)	I_2 (%)
LPS:Ce	11.5	31.3	36.3	68.7
LPS:Ce,Al	11.4	32.2	36.1	67.8
LPS:Ce,Ca	10.8	32.5	35.5	67.5
LPS:Ce, SC			35.8, 38 *	100
LSO:Ce,Al	10.3	46.1	33.1	53.9
LSO:Ce,Ca	9.8	50.8	32.9	49.2
LSO:Ce, SC			40 *	100

Table 3. Cont.

Composition	τ_1 (ns)	I_1 (%)	τ_2 (ns)	I_2 (%)
LPS:Pr	13.6	69.0	30.8	31.0
LPS:Pr,Al	12.6	69.7	30.2	30.3
LPS:Pr, SC	20 **, 15 ***	100		

* Decay time in SC under γ-ray excitation [16]; ** photoluminescence decay time in SC under excitation at 256 nm [28]; *** decay time in SC under γ-ray excitation [31].

The changes in the decay time values due to co-doping are almost negligible. However, some common tendencies could be observed. The values of both fast and slow decay components demonstrate a mild decrease with the following sequence: non-co-doped → Al-co-doped → Ca-co-doped samples (e.g., compare τ_1 component in LPS:Ce: 11.5 → 11.4 → 10.8 ns).

4. Conclusions

Single-phase powder samples of Lu_2SiO_5 (space group $C2/c$) and $Lu_2Si_2O_7$ (space group $C2/m$) doped with 2000 ppm Ce (or Pr) and co-doped with 5000 ppm Al/Si or 5000 ppm Ca/Lu were prepared by the conventional solid-state reaction process. The phase purity was confirmed by X-ray diffraction measurements.

Detailed EPR measurements of LSO:Ce revealed a substantial (three-fold) decrease in the Ce^{3+} concentration in Ca-co-doped samples, confirming the effective $Ce^{3+} \rightarrow Ce^{4+}$ conversion, while the effect of Al co-doping (Al^{3+} substituted for Si^{4+}) was much smaller, at only a 15–20% decrease.

On the contrary, EPR measurements in LPS:Ce, LPS:Ce,Al, and LPS:Ce,Ca samples revealed approximately the same decrease in Ce^{3+} concentration (2–3 times) in both the Al- and Ca-co-doped samples (again with a stronger effect for Ca-co-doped sample), suggesting that the crystal structure and different number of chemical bonds with oxygen ions in LSO and LPS influences the charge balance and charge transfer between doping ions.

In both LSO and LPS samples, accidental Fe^{3+} impurities were detected by EPR. The Fe^{3+} EPR intensity changed under Ca and Al co-doping, indicating participation of the Fe^{3+} ions in the charge transfer processes induced by Ca and Al co-doping.

EPR measurements in the Pr-doped LSO and LPS did not reveal possible $Pr^{3+} \rightarrow Pr^{4+}$ conversions stimulated by Al and Ca co-doping, despite the fact that the Pr^{4+} ion has the same electron shell as Ce^{3+} and, consequently, it should be clearly observed by EPR. This suggests that the charge compensation of Al^{3+} and Ca^{2+} ions is mainly realized via participation of other impurities and/or lattice defects.

X-ray irradiation of Ce- and Pr-doped LPS creates O^- centers attributed to a hole trapped at an oxygen ion in the neighborhood of Al^{3+} and Ca^{2+} ions. These hole centers are reasonably thermally stable up to about 400 K. They contribute to intense TSL glow peaks at 450–470 K. On the other hand, only a weak TSL peak was visible in all LSO samples at T ≈ 110 K and no paramagnetic active centers were detected via EPR after X-ray irradiation despite the intense radioluminescence. This suggests effective (practically full) recombination of electron–hole pairs at activator ions.

The X-ray-excited scintillation decay curves were measured for both LSO and LPS samples and could be approximated by a bi-exponential function. The measurements showed, along with the conventional dominating 33–38 ns decay component in Ce-doped LSO and LPS, the presence of a second fast component with a decay time of only about 11 ns. The origin of the latter component is unclear and needs further detailed investigations. On the contrary, in the Pr-doped samples, along with the usual fast decay component of 13–15 ns, a second slower component of 30–31 ns is present. Only a weak acceleration of the scintillation decay (6–8%) is found for the fast component in the Ca- and Al-co-doped samples. For the slower component, the acceleration is negligibly small.

Supplementary Materials: The following supporting information can be downloaded at: https://www.mdpi.com/article/10.3390/ma16124488/s1.

Author Contributions: Conceptualization, J.P. and M.N.; Methodology, J.B.; Validation, L.H. and J.P.; Formal analysis, V.B. and J.B.; Investigation, V.L., L.H. and V.B.; Writing—original draft, V.L.; Writing—review & editing, M.N. All authors have read and agreed to the published version of the manuscript.

Funding: This research was funded by the Czech Science Foundation project GA20-12885S.

Institutional Review Board Statement: Not applicable.

Informed Consent Statement: Not applicable.

Data Availability Statement: Data are contained within the article.

Acknowledgments: This work was supported by the Czech Science Foundation project GA20-12885S.

Conflicts of Interest: The authors declare no conflict of interest.

References

1. Rodnyi, P.A. *Physical Processes in Inorganic Scintillators*; CRC Press: Boca Raton, FL, USA, 1997.
2. Dujardin, C.; Auffray, E.; Bourret-Courchesne, E.; Dorenbos, P.; Lecoq, P.; Nikl, M.; Vasil'ev, A.N.; Yoshikawa, A.; Zhu, R.-Y. Needs, trends, and advances in inorganic scintillators. *IEEE Trans. Nukl. Sci.* **2018**, *65*, 1977–1997. [CrossRef]
3. Wei, Q.; Shi, H.; Zhou, Z.; Liu, G.; Chen, Z.; Qin, L.; Shu, K.; Liu, Q. A study on the structure, luminescence and thermos-stability of polycrystalline $Gd_2Si_2O_7$:Ce and $(Gd,La)_2Si_2O_7$:Ce. *J. Mater. Chem. C* **2017**, *5*, 1443–1451. [CrossRef]
4. Kantuptim, P.; Akatsuka, M.; Nakauchi, D.; Kato, T.; Kawaguchi, N. Scintillation properties of Pr-doped $Lu_2Si_2O_7$ single crystal. *Radiat. Meas.* **2020**, *134*, 106320. [CrossRef]
5. Lopez, L.; Pichon, P.; Loiseau, P.; Viana, B.; Mahiou, R.; Druon, F.; Georges, P.; Balembois, F. Ce:LYSO, from scintillator to solid-state lighting as a blue luminescent concentrator. *Sci. Rep.* **2023**, *13*, 7199. [CrossRef]
6. Kantuptim, P.; Fukushima, H.; Kimura, H.; Nakauchi, D.; Kato, T.; Koshimizu, M.; Kawaguchi, N.; Yanagida, T. VUV- and X-ray-induced properties of $Lu_2Si_2O_7$, $Y_2Si_2O_7$, and $Gd_2Si_2O_7$ single crystals. *Sens. Mater.* **2021**, *33*, 2195–2201. [CrossRef]
7. Van Eijk, C.W.E. Inorganic scintillators in medical imaging. *Phys. Med. Biol.* **2002**, *47*, R85–R106. [CrossRef] [PubMed]
8. Blahuta, S.; Bessière, A.; Viana, B.; Dorenbos, P.; Ouspenski, V. Evidence and consequences of Ce^{4+} in LYSO:Ce,Ca and LYSO:Ce,Mg single crystals for medical imaging applications. *IEEE Trans. Nukl. Sci.* **2013**, *60*, 3134–3141. [CrossRef]
9. Wu, Y.; Peng, J.; Rutstrom, D.; Koschan, M.; Foster, C.; Melcher, C.L. Unraveling the critical role of site occupancy of lithium codopants in Lu_2SiO_5:Ce^{3+} single-crystalline scintillators. *ACS Appl. Mater. Interfaces* **2019**, *11*, 8194–8201. [CrossRef]
10. Tian, J.; Xie, J.; Zhuang, W. Recent advances in multi-site luminescence materials: Design, identification and regulation. *Materials* **2023**, *15*, 2179. [CrossRef]
11. Yang, K.; Melcher, C.L.; Rack, P.D.; Eriksson, L.A. Effects of calcium codoping on charge traps in LSO:Ce crystals. *IEEE Trans. Nukl. Sci.* **2009**, *56*, 2960–2965. [CrossRef]
12. Spurrier, M.A.; Szupryczynski, P.; Yang, K.; Carey, A.A.; Melcher, C.L. Effect of Ca^{2+} co-doping on the scintillation properties of LSO:Ce. *IEEE Trans. Nukl. Sci.* **2008**, *55*, 1178–1182. [CrossRef]
13. Ferrand, B.; Viana, B.; Pidol, L.; Dorenbos, P. Dense High-speed Scintillator Material of Low Afterglow. U.S. Patent 8,574,458 B2, 5 November 2013.
14. Xue, Z.; Chen, L.; Zhao, S.; Yang, F.; An, R.; Wang, L.; Sun, Y.; Feng, H.; Ding, D. Enhancement of scintillation properties of LYSO:Ce crystals by Al codoping. *Cryst. Growth Des.* **2023**, *23*, 4562–4570. [CrossRef]
15. Takasugi, T.; Yokota, Y.; Horiai, T.; Yoshino, M.; Yamaji, A.; Ohashi, Y.; Kurosawa, S.; Kamada, K.; Babin, V.; Nikl, M.; et al. Al-doping effects on mechanical, optical and scintillation properties of $Ce:(La,Gd)_2Si_2O_7$ single crystals. *Opt. Mater.* **2019**, *87*, 11–15. [CrossRef]
16. Pidol, L.; Kahn-Harari, A.; Viana, B.; Ferrand, B.; Dorenbos, P.; de Hass, J.T.M.; van Eijk, C.W.E.; Virey, E. Properties of $Lu_2Si_2O_7$:Ce^{3+}, a fast and efficient scintillator crystal. *J. Phys. Condens. Matter* **2003**, *15*, 2091–2102. [CrossRef]
17. He, F.; Guohao, R.; Yuntao, W.; Jun, X.; Qiuhong, Y.; Jianjun, X.; Mitch, C.; Chenlong, C. Optical and thermoluminescence properties of $Lu_2Si_2O_7$:Pr single crystal. *J. Rare Earths* **2012**, *30*, 775–779.
18. Laguta, V.V.; Buryi, M.; Rosa, J.; Savchenko, D.; Hybler, J.; Nikl, M.; Zazubovich, S.; Karner, T.; Stanek, C.R.; McClellan, K.J. Electron and hole traps in yttrium orthosilicate single crystals: The critical role of Si-unbound oxygen. *Phys. Rev. B* **2014**, *90*, 64104. [CrossRef]
19. Mihóková, E.; Fasoli, M.; Moretti, F.; Nikl, M.; Jary, V.; Ren, G.; Vedda, A. Defect states in Pr^{3+} doped lutetium pyrosilicate. *Opt. Mater.* **2012**, *34*, 872–877. [CrossRef]
20. Dorenbos, P.; van Eijk, C.W.E.; Bos, A.J.J.; Melcher, C.L. Afterglow and thermoluminescence properties of Lu_2SiO_5:Ce scintillation crystals. *J. Phys. Condens. Matter* **1994**, *6*, 4167. [CrossRef]

21. Laguta, V.; Buryi, M.; Wu, Y.; Ren, G.; Nikl, M. Electron and hole trapping in Ce^{3+}- and Pr^{3+}-doped lutetium pyrosilicate scintillation crystals studied by electron paramagnetic resonance. *Phys. Rev. Appl.* **2020**, *13*, 044060. [CrossRef]
22. Aykol, M.; Montoya, J.H.; Hummelshoj, J. Rational solid-state synthesis routes for inorganic materials. *J. Am. Chem. Soc.* **2021**, *143*, 9244–9259. [CrossRef]
23. Gustafsson, T.; Klintenberg, M.; Derenzo, S.E.; Weber, M.J.; Thomas, J.O. Lu_2SiO_5 by single-crystal X-ray and neutron diffraction. *Acta Crystallog. C* **2001**, *57*, 669. [CrossRef]
24. Bretheau-Raynal, F.; Lance, M.; Charpin, P. Crystal data for $Lu_2Si_2O_7$. *J. Appl. Cryst.* **1981**, *14*, 349–350. [CrossRef]
25. Pidol, L.; Guillot-Noël, O.; Kahn-Harari, A.; Viana, B.; Pelenc, D.; Gourier, D. EPR study of Ce^{3+} ions in lutetium silicate scintillators $Lu_2Si_2O_7$ and Lu_2SiO_5. *J. Phys. Chem. Solids* **2006**, *67*, 643–650. [CrossRef]
26. Pilbrow, J.R. *Transition Ion Electron Paramagnetic Resonance*; Clarendon Press: Oxford, UK, 1990.
27. Abragam, A.; Bleaney, B. *Electron Paramagnetic Resonance of Transition Ions*; Clarendon Press: Oxford, UK, 1970.
28. Buryi, M.; Laguta, V.; Babin, V.; Laguta, O.; Brik, M.G.; Nikl, M. Rare-earth ions incorporation into $Lu_2Si_2O_7$ scintillator crystals: Electron paramagnetic resonance and luminescence study. *Opt. Mater.* **2020**, *106*, 109930. [CrossRef]
29. Schirmer, O.F. O^- bond small polarons in oxides materials. *J. Phys. Condens. Matter* **2006**, *18*, R667. [CrossRef]
30. Feng, H.; Ding, D.; Li, H.; Lu, S.; Pan, S.; Chen, X.; Ren, G. Annealing effects on Czochralski grown $Lu_2Si_2O_7$:Ce^{3+} crystals under different atmospheres. *J. Appl. Phys.* **2008**, *103*, 083109. [CrossRef]
31. Pidol, L.; Viana, B.; Kahn-Harari, A.; Bessiere, A.; Dorenbos, P. Luminescence properties and scintillation mechanisms of Ce^{3+}-, Pr^{3+}- and Nd^{3+}-doped lutetium pyrosilicate. *Nukl. Instr. Methods Phys. Res. A* **2005**, *537*, 125–129. [CrossRef]

Disclaimer/Publisher's Note: The statements, opinions and data contained in all publications are solely those of the individual author(s) and contributor(s) and not of MDPI and/or the editor(s). MDPI and/or the editor(s) disclaim responsibility for any injury to people or property resulting from any ideas, methods, instructions or products referred to in the content.

Photoluminescence and Photocatalytic Properties of MWNTs Decorated with Fe-Doped ZnO Nanoparticles

Adriana Popa [1], Maria Stefan [1], Sergiu Macavei [1], Laura Elena Muresan [2], Cristian Leostean [1], Cornelia Veronica Floare-Avram [1] and Dana Toloman [1,*]

[1] National Institute for Research and Development of Isotopic and Molecular Technologies, 67-103 Donat, 400293 Cluj-Napoca, Romania; adriana.popa@itim-cj.ro (A.P.); cristian.leostean@itim-cj.ro (C.L.)
[2] Raluca Ripan Institute for Research in Chemistry, Babes-Bolyai University, 30 Fântânele, 400294 Cluj-Napoca, Romania; laura.muresan@ubbcluj.ro
* Correspondence: dana.toloman@itim-cj.ro

Abstract: The present work reports the photoluminescence (PL) and photocatalytic properties of multi-walled carbon nanotubes (MWCNTs) decorated with Fe-doped ZnO nanoparticles. MWCNT:ZnO-Fe nanocomposite samples with weight ratios of 1:3, 1:5 and 1:10 were prepared using a facile synthesis method. The obtained crystalline phases were evidenced by X-ray diffraction (XRD). X-ray Photoelectron spectroscopy (XPS) revealed the presence of both 2+ and 3+ valence states of Fe ions in a ratio of approximately 0.5. The electron paramagnetic resonance EPR spectroscopy sustained the presence of Fe^{3+} ions in the ZnO lattice and evidenced oxygen vacancies. Transmission electron microscopy (TEM) images showed the attachment and distribution of Fe-doped ZnO nanoparticles along the nanotubes with a star-like shape. All of the samples exhibited absorption in the UV region, and the absorption edge was shifted toward a higher wavelength after the addition of MWCNT component. The photoluminescence emission spectra showed peaks in the UV and visible region. Visible emissions are a result of the presence of defects or impurity states in the material. All of the samples showed photocatalytic activity against the Rhodamine B (RhB) synthetic solution under UV irradiation. The best performance was obtained using the MWCNT:ZnO-Fe(1:5) nanocomposite samples, which exhibited a 96% degradation efficiency. The mechanism of photocatalytic activity was explained based on the reactive oxygen species generated by the nanocomposites under UV irradiation in correlation with the structural and optical information obtained in this study.

Keywords: nanocomposites; photoluminescence; photocatalysis

1. Introduction

Water pollution caused by chemicals or microorganisms is a major threat to human health and the aquatic environment. Industrialization and agricultural activities are the main causes of water contamination, releasing various solvents or organic and inorganic compounds into the water. Among the pollutants discharged into water, organic dyes used in activities performed by the textile, paper or cosmetic industries are a class of major environmental pollutants due to their high toxicity [1]. Annually, approximately 7×10^7 tons of dyes, such as azo, sulfide, direct, reactive, basic and acid dyes, are fabricated [2]. For example, Rhodamine B is an azo dye widely used in the textile, plastic and leather industries, which, in high concentrations, can constitute a health risk due to its carcinogenic and neurotoxic effect [3]. Another synthetic azo dye is methyl orange (MO), which is generally used as a coloring agent in the textile and leather industries [4]. It is widely used in printing, paper manufacturing, pharmaceutical, food processing industries and in research laboratories [5,6]. MO is soluble in water, and it can cause vomiting and diarrhea. High concentrations of MO can cause death [7]. It is stable and has low biodegradability; hence, it is difficult to remove from aqueous solutions using common water purification or treatment methods [6]. Among the reported recalcitrant dyes, Remazol brilliant blue (RBB) was classified as toxic, carcinogenic and extremely harmful to aquatic and vegetative lives [8]. RBB dye is used in the textile industry to fabric nylon, wool and silk. There are many studies on the toxicity of dyes and their impact on the ecosystem [9]. The challenge remains to successfully degrade this toxic dye from the aqueous system with minimum energy and without harming the ecosystem. Over time, researchers have attempted to find effective solutions to surmount these problems. Conventional methods such as adsorption, coagulation, ion flotation [10], sedimentation [11] and ozone electrolysis [12] have led to satisfactory results, but present some shortcomings, such as [13] the large consumption of chemical reagents, generation of secondary waste products and high costs [14]. The advanced oxidation process (AOP) provides an efficient alternative to the efficient elimination of dyes from water through reactive oxygen species (ROS) generation [15]. Among the AOP methods, photocatalysis mediated by metal oxide semiconductors has been intensively studied in recent years to make the process more efficient and to avoid the appearance of by-products. This process involves exciting the semiconductor with light, resulting in electrons transfer from the valence band (VB) to the conduction band (CB), leaving a hole in the VB. If the photogenerated charges do not recombine, they can generate ROS by interacting with water molecules. ROS are reactive species able to degrade a wide variety of pollutant molecules.

In recent years, wide interest has been attributed to composite materials based on MWCNTs and oxide semiconductors as photocatalysts for water purification applications. MWCNTs, due to their large specific surface area, play a supporting role in photocatalyst. However, it has been shown that the role of nanotubes is not only to support the photocatalytic material, but can also lead to a better separation of photogenerated charges [16]. Due to its high electron storage capacity and good conductivity, MWCNTs can receive photogenerated electrons from the semiconductor material, leading to the efficient separation of e^-/h^+ pairs. Moreover, MWCNTs can also play the role of a photosensitizer by increasing the photogenerated charge density of a semiconductor by releasing its electrons into the semiconductor's CB [17]. The synergistic effects between MWCNTs and several types of oxide semiconductors, such as TiO_2 [18], WO_3 [19], ZnO [15], TiO_2-Fe_3O_4 [20], BiOI [21], MIL-101(Fe) [22] and ferrites [23], which achieve the efficient degradation of antibiotics or dyes, were reported. Another strategy to streamline photocatalytic activity is doping semiconductors with rare earth or transition metal ions [24–28]. Doping increases the ability of semiconductors to adsorb light due to the decrease in the band gap through the introduction of new energy levels inside the band gap. Moreover, altering the relative Fermi-level density of states will enhance the excitation of photogenerated electrons [29].

In this work, we aim to develop a highly efficient composite material in RhB pollutant degradation under UV light irradiations. A facile chemical method was developed to obtain nanoparticles with a controllable morphology. Additionally, using PAH to attach ZnO:Fe nanoparticles along the MWCNT represents a new strategy to obtain MWCNT-ZnO nanocomposites without other chemical modifications that could destroy the structure of the MWCNT. We studied the synergistic role of MWCNT addition and Fe doping on the ZnO photocatalytic properties. In addition, Fe ion doping generates defect states that are able to delay the photogenerated charge recombination, favoring the photocatalytic degradation process. Correlating the XPS, PL and EPR trapping technique results, a photodegradation mechanism was proposed.

2. Materials and Methods
2.1. Materials

For the preparation of MWCNT-ZnO:Fe0.5%, we used the following materials and reagents: multi-walled carbon nanotubes (MWCNTs) with a 99% purity were purchased commercially (Sigma-Aldrich, Merck, KGaA, Darmstadt, Germany), poly-allylamine hydrochloride (PAH) (Alfa–Aesar, Thermo Fisher, Kandel, Germany), sodium chloride (NaCl)-(Alpha Aesar, Bio Aqua Group, Targu Mures, Romania), zinc nitrate hexahydrate ($Zn(NO_3)_2 \cdot 6H_2O$) (Alpha Aesar, Bio

Aqua Group, Targu Mures, Romania), iron nitrate nonahydrate (Fe(NO$_3$)$_3$·9H$_2$O) (Alpha Aesar, Bio Aqua Group, Targu Mures, Romania), absolute ethanol (C$_2$H$_5$OH-EtOH) (Alpha Aesar, Bio Aqua Group, Targu Mures, Romania). All of the chemicals are of analytical grade and used without further purification. The aqueous solutions were prepared with Milli-Q water from the Direct-Q 3UV system (Millipore, Bedford, MA, USA). Dimethyl sulfoxide (DMSO; >99.9%) was purchased from VWR Chemicals, and 5,5-dimethyl-1-pyrroline N-oxide (DMPO; >97%), dimethylformamide (DMF) and Rhodamine B were purchased from Sigma-Aldrich, Merck, KGaA, Darmstadt, Germany.

2.2. Sample Preparation

2.2.1. Synthesis of ZnO:Fe Nanoparticles

The 0.5 mol% Fe-doped ZnO NPs—ZF0.5% were synthesized using the chemical precipitation method. The experimental procedure was performed according to our previous paper [30], with modifications to the reagent's concentration and temperature. Thus, stoichiometric amounts, consisting of (3−x) g of Zn(NO$_3$)$_2$·6H$_2$O (98%) and x g Fe(NO$_3$)$_2$·9H$_2$O (x represents 0.5mol% from Zn(NO$_3$)$_2$·6H$_2$O), were dissolved in 100 mL of ultrapure water and mixed under vigorous stirring to form a homogeneous solution. Subsequently, a solution of 3 M NaOH was added dropwise at a constant stirring rate until a white precipitate of zinc hydroxide was obtained. After pH = 12 was reached, the mixture was continuously stirred for 4 h at room temperature. The obtained ZF0.5% was washed with ultrapure water and dried at 65 °C for 24 h. A doping of 0.5 mol% was chosen because it presented the best photocatalytic activity in the tests performed on the Zn$_{1-x}$Fe$_x$O (x = 0, 0.1, 0.3, 0.5, 0.7) nanoparticles (Figure S1).

2.2.2. Decoration of MWCNT with ZnO:Fe

The decoration of MWCNTs with ZF0.5% nanoparticles was achieved through polymer wrapping using poly(allylamine hydrochloride) (PAH). The role of PAH is to provide functional groups (amine, OH) on the MWCNT surface, as well as stability. The MWCNTs (10 mg) were dispersed in a 0.5 wt% PAH salt solution (0.5 M NaCl, 500 mL), sonicated for 3 h, then stirred overnight at 80 °C. The separation of the MWCNT-PAH was performed by centrifugation, followed by re-dispersion in water. Before the decorating process, ZF0.5% nanoparticles were dispersed separately through sonication in ethanol for 30 min and then mixed together for another 4 h. Three different samples based on the MWCNTs and ZF0.5% were prepared, in which the molar ratio between the MWCNTs and ZF0.5% was 1:3, 1:5 and 1:10, correspondingly denoted as CZF1:3, CZF1:5 and CZF1:10.

2.3. Methods

X-ray diffraction was performed using a Rigaku-SmartLab automated Multipurpose X-ray Diffractometer with Cu-Kα radiation, operating at 45 kV, 200 mA and using a D/tex Ultra 250 detector monochromator with XRF reduction. The morphology of the composite nanoparticles was determined through scanning transmission electron microscopy (STEM). The STEM analysis was achieved using a Hitachi SU8230 microscope (Tokyo, Japan), provided with a cold field emission gun that accelerates the electron at 200 kV. The UV–Vis characterization was conducted using a JASCO V570 UV–Vis–NIR Spectrophotometer equipped with an absolute reflectivity measurement JASCO ARN-475 accessory (Easton, MD, USA). The obtained reflectance spectra were transformed in absorbance using the intern soft of the spectrophotometer. The luminescent characteristics of the samples were evaluated based on the emission spectra registered with a JASCO FP-6500 spectrofluorometer Wavell equipped with a PMT R928 photomultiplier (glass filter WG 320-ReichmannFeinoptik). X-Ray Photoelectrons Spectroscopy (XPS) was used to determine the Fe chemical state with a SPECS custom-built system using the Mg anode (1253.64 eV). The CASA software 2.3.21 was used for spectrum analysis. Electron paramagnetic resonance (EPR) spectroscopy measurements of powder samples were carried out using a Bruker E-500 ELEXSYS spectrometer (Karlsruhe, Germany.) at room temperature under identical conditions: The measurements were conducted in both X-band (9.52 GHz) and Q band (33.9 GHz), microwave power 2 mW, modulation frequency of 100 kHz. For the EPR spectra simulation, an Anisotropic SpinFit Bruker was used. To monitor the reactive oxygen species (ROS) generation, the EPR Bruker E-500 ELEXSYS X-band (9.52 GHz) spectrometer coupled with the spin trapping probe technique was used. DMPO was utilized as a spin trapping reagent. The nanoparticles (10 mg) were dispersed in DMSO (1 mL) and homogenized in an ultrasound bath (30 min) before use. DMPO of 0.2 mol/L concentration was added to the suspension. The samples were prepared immediately before measurement and transferred into the quartz flat cell, optimized for liquid measurements. A high-performance liquid chromatography-Mass spectrometry (HPLC/MS) equipped with an electrospray ionization (ESI) positive ion mode was used to determine the reaction products. The HPLC-UV system was used to separate the intermediary compounds, which were identified by soft ionization mass spectrometry (ESI–MS). The mobile phase used was methanol (solvent A) and HPLC water containing 0.1% formic acid (solvent B) (60–90% methanol over 30 min).

2.4. Evaluation of Photocatalytic Activity

The photocatalytic activity was tested by evaluating the degradation of Rhodamine B (RhB) under UV light irradiation. A laboratory reactor consisting of two UV lamps (15 W) emitting at 365 nm and a quartz reaction vessel was used for this purpose. The catalyst (2 mg) was suspended in an aqueous solution of RhB (1.0×10^{-5} mol/l, 10 mL), and then the mixture was magnetically stirred in the dark to achieve the adsorption equilibrium. Each degradation experiment was continuously conducted for 150 min. The mixture (3.5 mL) was withdrawn for analysis every 60 min and separated from the suspensions through centrifugation. The analysis was performed using a UV–Vis spectrophotometer by recording the maximum absorbance of RhB at 554 nm. The photocatalytic activity was calculated using the equation:

$$\text{Photocatalytic activity (\%)} = (1 - A_t/A_0) \times 100 \tag{1}$$

where A_t and A_0 represent the RhB absorbance located at 554 nm at time t and at time t = 0, respectively.

3. Results and Discussions

The structural characterization of the samples was achieved using XRD and EPR spectroscopy. The XRD diffraction patterns of the MWCNT, ZnO, ZF0.5% nanoparticles and CZF composite samples are illustrated in Figure 1. The diffraction spectrum of the MWCNT shows a (002) peak specific to standard graphite carbon (ICDD card 01-075-0444).

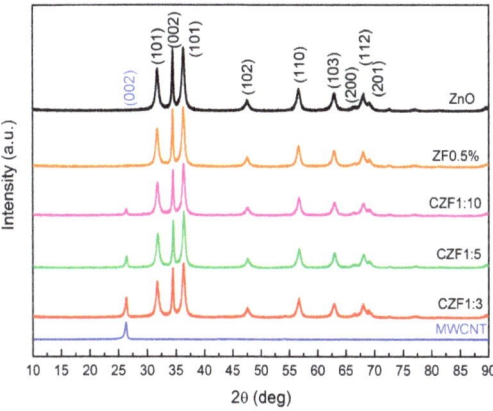

Figure 1. XRD diffraction patterns of the samples.

The diffraction patterns of ZF0.5% fit very well with those of ZnO, with a hexagonal structure (ICDD 01-080-0075). In the case of the nanocomposite samples, only the peak (002) specific to the MWCNT is observed, in addition to the patterns specific to ZnO. This peak decreases with the increase in the ZF0.5% content. The ratio between the peak intensity I_{002}(MWCNT) and I_{002}(ZnO) decreases from 0.36, in the case of the 1:3 sample, to 0.12, in the case of the 1:10 sample. Using Rietveld refinement, the lattice parameters, unit cell volume and the mean crystallites size were evaluated for the ZF0.5% and ZnO nanoparticles. The obtained parameters are presented in Table 1.

Table 1. The lattice parameters, unit cell volume, mean crystallites size corresponding to ZnO crystalline phase and the quality parameters for the Rietveld refinement.

Sample	a = b (Å)	c (Å)	V (Å^3)	D_{cryst} (nm)	R_{wp} (%)	R_p (%)	R_{exp} (%)	χ^2
ZnO	3.2472	5.2023	47.508	13.4	4.74	3.55	3.15	1.5
ZF0.5%	3.2480	5.2060	47.565	11.5	7.31	5.42	5.14	1.6
CZF1:3	3.2478	5.2059	47.560	11.5	10.84	8.39	7.41	1.7
CZF1:5	3.2479	5.2059	47.562	11.5	9.83	7.38	6.42	1.6
CZF1:10	3.2479	5.2060	47.563	11.5	8.21	6.83	5.98	1.6

The lattice parameters of the ZF0.5% nanoparticles (a = 3.248 Å, and c = 5.206 Å) are close to those of the ZnO nanoparticles (a = 3.247 Å, and c = 5.202 Å), resulting in a slight increase in the unit cell volume, up to 47.56 Å^3 from 47.50 Å^3. Comparing the ionic radius of Fe^{2+} (0.76 Å), Fe^{3+} (0.64 Å) and Zn^{2+} (0.74 Å), it is probable that both Fe ions enter into the lattice of ZnO in a substitutional position. Through doping, a slight broadening of the peaks can be observed, leading to a smaller mean crystallites size of 11.5 nm for the ZF0.5% nanoparticles compared with 13.4 nm for the ZnO nanoparticles. Similar behavior was reported in the literature, and was attributed to the reduction in the ZnO nucleation rate caused by the dopant ions [31].

EPR characterization was carried out to determine the dopant ions' location and valence state. Figure 2a shows the room temperature EPR spectra measured in the X band of the MWCNT and ZF0.5%, which were compared with the spectra corresponding to the composite samples. The EPR spectrum of the MWCNT is composed of an intense and narrow line (ΔH = 17G) at g~2.008, usually attributed to carbon radicals. The formation of these radicals is a consequence of the MWCNT backbone breaking caused by ultrasonic treatment [32]. The ZF0.5% spectrum comprises a low intense line at g~4.28 and a broad line at g~2.11. Due to their large zero-field splitting and low spin-lattice relaxation time, the Fe^{2+} ions did not provide an EPR spectrum at room temperature [33]. Consequently, the obtained EPR line could be assigned to the Fe^{3+} ions. The line at g~4.2 is usually attributed to the presence of a Fe^{3+} ion (S = 5/2) located near the nanoparticle surface in a rhombically distorted octahedral position. Similar behavior was reported in other Fe-doped ZnO nanoparticles [33].

A closer analysis of the EPR signal from g~2.11 indicates that it is composed of several overlapping lines. This sample was also measured in the Q band (33.9 GHz) to better separate these contributions. The obtained spectrum is presented in Figure 2b. After performing the simulations, three contributions were identified: (i) isolated Fe^{3+} (S = 5/2, g~2.006, ΔH = 90G, D = 620 G–C1) situated in the nanoparticles core, substituting the Zn ions [34,35]; (ii) oxygen vacancies (g~2.004, ΔH = 6 G–C2) located mainly on the nanoparticles surface; (iii) an intense and large line (g~2.1, ΔH = 940G–C3), which is due to ferromagnetically coupled Fe^{3+} ions [36].

The EPR spectra of the composite material are composed by overlapping the MWCNT-specific spectrum and that of the Fe-doped ZnO. As the concentration of the Fe-doped ZnO increases (1–3 to 1–10), its contribution to the EPR spectrum becomes predominant, and the MWCNT-specific line decreases in intensity.

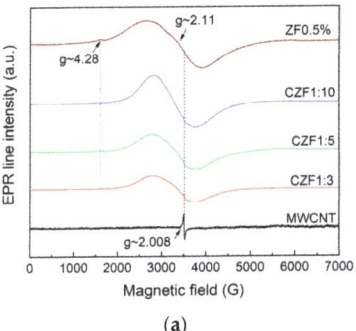
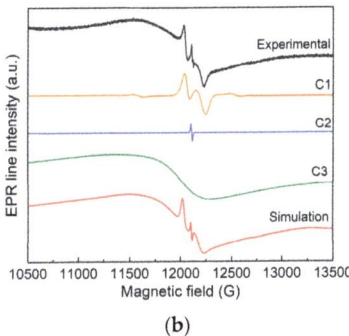

Figure 2. (a) X band spectra of MWCNT, ZF0.5% and CZF1:x (x = 3, 5, 10); (b) Q band spectra experimental and simulated spectrum of ZF0.5%. The simulated spectrum resulted from the sum of C1, C2 and C3 components.

For better identification of the Fe chemical states, XPS spectroscopy was performed. The XPS spectra of the Fe *2p* (3/2) of the CZF1:3 sample are shown in Figure 3. The broad peak suggests that both Fe^{2+} and Fe^{3+} chemical states are present. The main peaks at 710.35 eV and 711.55 eV are attributed to the Fe^{2+} and Fe^{3+} chemical states, respectively. The corresponding shake-up satellites are positioned at 714.52 eV and 718.19 eV. In addition, small quantities of surface states with lower binding energy are present at 706.98 eV (Fe^{2+}) and 710.74 eV (Fe^{3+}) [33]. The Fe^{3+}/Fe^{2+} ratio is 0.5.

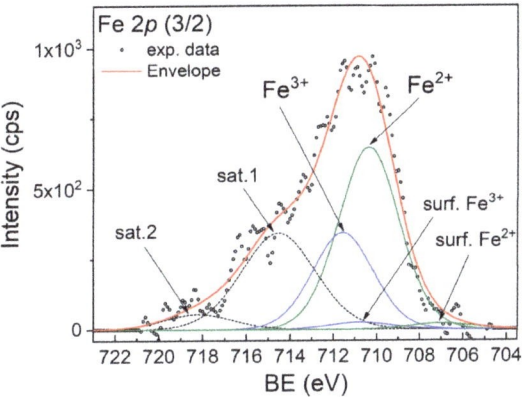

Figure 3. XPS spectrum of Fe *2p* (3/2) line corresponding to sample CZF1:3.

The morphology of the CZF1:x (x = 3, 5, 10) nanocomposites was evaluated using TEM, as presented in Figure 4. The representative images show the tubular shape of the MWCNTs decorated with ZF0.5% nanoparticles, which possess a star-like morphology, with a pronounced orientation depending on the nucleation and crystal growth. Figure S2 shows the size distribution of the ZF0.5% nanoparticles considering the length of the star petals. The mean size is approximately 210 nm. Considering the crystallite size determined by the XRD analysis (11.5 nm), the polycrystalline nature of the nanoparticles is evidenced. It is assumed that after the link between the PAH and the ZnO nanoparticle is formed, the attached nanoparticle becomes a crystallization center for the other nanoparticles from the solution. In our specific case, the experimental parameters favor this oriented growth. As expected, the grafting of the ZF0.5% nanoparticles along the MWCNT is dependent on the quantity of nanocomposites in the composite samples.

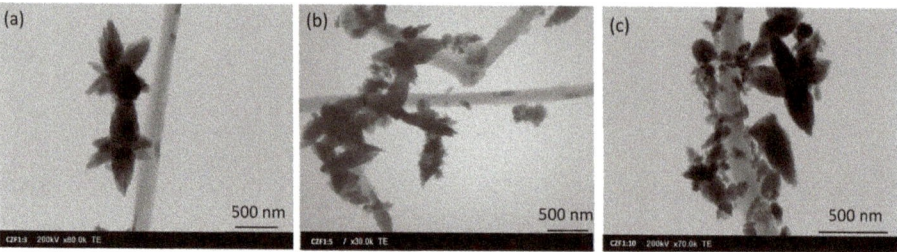

Figure 4. TEM images of CZF1:x ((**a**) x = 3, (**b**) x = 5, (**c**) x = 10) samples.

Optical absorption spectroscopy provides information about the energy gap (E_g) and bond structure based on the analysis of optically induced transitions. In the absorption process, through the absorption of a photon with a specified energy, an electron transition from a lower to a higher energy state occurs. Both reflectance and UV/Vis spectroscopy use UV-visible light to excite valence electrons to empty orbitals. The difference is that, in diffuse reflectance, we measure the relative change in the amount of light reflected by a surface, whereas in UV/Vis spectroscopy, we measure the relative change in the transmittance of light as it passes through a solution. UV–Vis diffuse reflectance spectroscopy (DRS) was used to determine the optical absorption properties of the samples. Figure 5a illustrates the UV-Vis absorption of the prepared samples. The MWCNT shows a broad absorption in all of the investigated ranges, with a maximum in the visible domain, while the ZF0.5% nanoparticles show strong absorption in the UV region, with a maximum absorbance at the 330 nm wavelength corresponding to the inter-band transition.

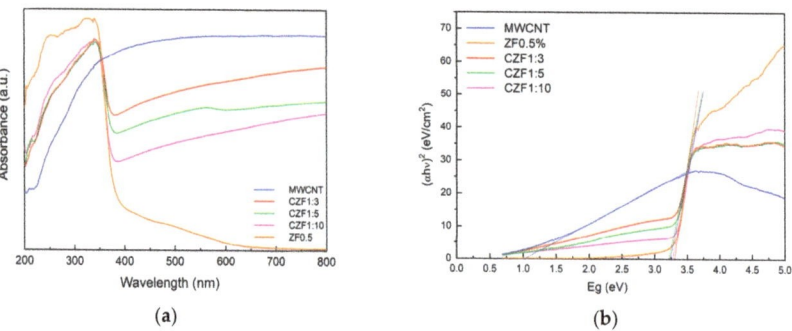

Figure 5. (**a**) UV-Vis absorption spectra of the sample. (**b**) Tauc's plots.

The low and broad absorption in the 400–600 nm wavelength range is due to the interfacial charge transfer and d–d transition between the multiplets of the $3d^5$ configuration of the high spin Fe^{3+} substituting Zn^{2+} under the influence of the tetrahedral ZnO crystal field. The observed transitions were assigned to the $^6A_{1g}$ ground state to $^4T_1$, $^4T_2$, ^4E and $^4A_1$ excited states [37,38].

In the CZF1:x composite samples, the contribution of both components can be observed and is correlated with the ratio between them. The increase in the ZnO content leads to a decrease in the intensity of the visible light absorption. Based on the absorption spectra and using the Tauc's equation [39], the band gap energy of the samples was evaluated. The Tauc's plots are presented in Figure 5b. The band gap energy of the ZF0.5% nanoparticles is 3.30 eV. For the composite samples, due to the MWCNT contribution, the band gap energy decreases from 3.29 eV, in the case of the CZF1:10 sample, to 3.20 eV in the case of the CZF1:3 sample.

The photoluminescence properties provide information about the presence of defects or impurities and the efficiency of the charge carrier recombination. The emission spectra and the Gaussian deconvolution of the composite samples obtained under an excitation wavelength of 270 nm are shown in Figure 6. Table 2 summarizes the results related to the area of the peaks obtained after the spectra deconvolution.

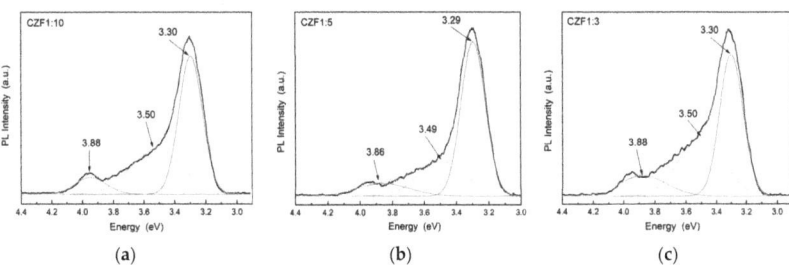

Figure 6. PL emission spectra of the samples (**a**) CFZ1:10, (**b**) CFZ1:5 and (**c**) CFZ1:3 under an excitation wavelength of 270 nm.

Table 2. The area (A_{peak}) of the deconvoluted PL peaks of Figure 5.

Sample	$A_{3.86}$	$A_{3.50}$	$A_{3.30}$
CZF1:10	0.73	3.15	4.37
CZF1:5	0.46	1.13	2.55
CZF1:3	0.82	2.28	2.88

In the range 300–400 nm (4.13–3.10 eV), there are three emission bands, known as a near band edge (NBE), due to the recombination of electrons from the CB with holes from the VB of ZnO [40]. For all three peaks, a quench of the emission can be observed by increasing the MWCNT content, probably due to a transfer of photoexcited charges from the ZnO to the MWCNT empty states [41,42].

The lowest UV emissions were observed for CZF1:5, which means that this sample shows a longer delay in the recombination process. Moreover, a Stokes shift of about 0.35 eV can be observed between the maximum of the PL emission spectra and that of the absorption spectra, which can be a result of different effects, such as interface and point defects, which give rise to a red shift of the emission bands from the absorption edge [43]. To evidence the emission bands from the visible domain, the emissions spectra were obtained under an excitation wavelength of 360 nm. The obtained PL spectra of the composite samples and their Gaussian deconvolution are shown in Figure 7. Table 3 summarizes the results related to the area of the peaks obtained after the spectra deconvolution.

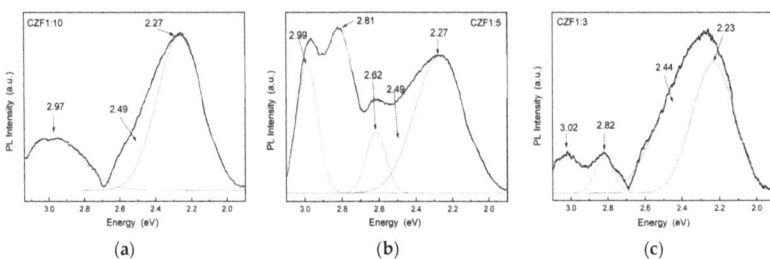

Figure 7. PL emission spectra of the samples (**a**) CFZ1:10, (**b**) CFZ1:15 and (**c**) CFZ1:3 under an excitation wavelength of 360 nm.

Table 3. The area (A_{peak}) of the deconvoluted PL peaks of Figure 6.

Sample	$A_{3.02}$	$A_{2.82}$	$A_{2.62}$	$A_{2.50}$	$A_{2.27}$
CZF1:10	3.63	-		1.86	8.39
CZF1:5	2.56	1.94	1.18	1.58	6.56
CZF1:3	0.40	0.20		1.32	1.66

In the violet range, the composite samples exhibit one emission at 2.99 eV (415 nm). In the literature, there are two scenarios in the assignment of this band. Some authors attribute this band to: (i) Zn atoms in interstitial sites (Zn_i), specific to an electronic transition between the Zn_i level and oxygen vacancies [42,44,45], and others to: (ii) unintentional H dopants, which can appear in ZnO samples prepared using chemical precipitation methods [46,47]. Indeed, in the synthesis process used in our samples, the PAH binding polymer was used to attach Fe-doped ZnO nanoparticles onto the surface of the MWCNT. The protonated functional groups from the PAH structure should contribute to these states. The blue emissions centered at 2.82 eV (439 nm), or 2.62 eV (473 nm), were attributed to the transition between the Zn_i level and V_{Zn} or to surface deep trap states [48]. The green emission bands centered at 2.5 eV (496 nm) and 2.21 eV (560 nm) were assigned to the transitions from Zn_i to the oxygen vacancies. The two green emissions can arise from different shallow energy levels in direct proximity to the band gap associated with the interface trap in the grain boundaries and dislocations [49]. From the analysis of the data obtained after the spectra deconvolution (Table 2) results, as in the case of the UV emissions, those in the visible range are quenched with the increase in the MWCNT content due to a delay In the recombination process induced by the charge transfer between defects levels of the ZnO nanoparticles and MWCNT.

The photocatalytic activity was tested on a synthetic solution of RhB under UV light irradiation. Before starting the photodegradation process, the solutions with a given concentration of photocatalysts were kept in the dark for 60 min until adsorption-desorption equilibrium was achieved. The adsorption increases with the increase in the MWCNT content in the composite samples. This may be due to the electrostatic interaction between the positively charged MWCNT and positively charged pollutant molecules [20]. After adsorption, the samples were irradiated with 365 nm UV light for 3 h. The photodegradation efficiency of the composite samples compared to the ZF0.5% is shown in Figure 8a. In addition, this figure contains the photocatalytic activity of ZnO, MWCNT and RhB photolysis. The best photocatalytic activity, 96%, was shown by the CZF1:5 sample.

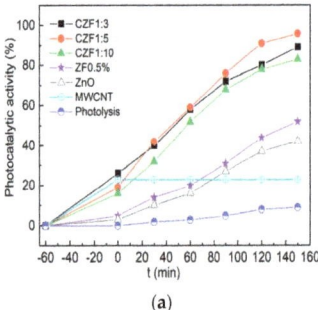
(a)

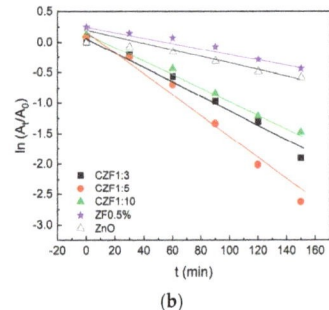
(b)

Figure 8. (a) Photocatalytic activity of the samples under UV irradiation against RhB. (b) Photodegradation kinetics.

The photodegradation kinetics of RhB were analyzed by applying the first kinetic order model. The equation describing this model is:

$$\ln(A_t/A_0) = -k_i \times t \qquad (2)$$

where A_t and A_0 represent the absorbance of RhB at time t and in the dark, respectively, k_i is the apparent rate constant. The plots obtained are shown in Figure 8b, with a linear dependence in t being observed. The apparent rate constant, k_i, and the correlation coefficients R^2 were obtained from the linear fitting of the dates and are presented in Table 4. These results support the conclusion that the CZF1:5 sample has the best photocatalytic activity.

Table 4. The apparent rate constant, k_i, and the correlation coefficients R^2.

Sample	$k_i \times 10^{-3}$ (min^{-1})	R^2
ZnO	4.1	0.98
ZF0.5%	4.6	0.98
CZF1:3	12.6	0.96
CZF1:5	20.16	0.99
CZF1:10	11.2	0.96

Some works about organic pollutant degradation using MWCNT-based composite nanomaterials are presented in Table 5. Previously, similar results were obtained, but direct comparison is difficult due to the differences in the photocatalyst concentration, the concentration of initial pollutant, the UV lamp power and the irradiation time.

Table 5. Some works related to pollutant photodegradation using MWCNT-based photocatalysts.

Photocatalyst	Pollutant Type	Pollutant Concentr.	Light Irrad./Power Source or Illuminance	Irrad. Time (min)	Catalyst Concentr(mg/mL)	Removal Rate (%)	Ref
MWCNT-ZnO:Fe	RhB	1×10^{-5} M	UV/30 W	150	0.2	95	This work
g-C$_3$N$_4$/ZnO/MWCNT	Alprazolam	0.1 mmol/L	UV/125 W	15	1.0	100	[50]
ZnO-Ag/MWCNT	Phenol	10 mg/l	UV/not specified	240	1.0	81	[51]
NiFe$_2$O$_4$/MWCNT	Methylene blue	20 mg/l	UV/233 lux	150	0.025	88	[52]
Bi$_{0.5}$Na$_{0.5}$TiO$_3$/MWCNT	RhB	5 mg/L	UV/300 W	40	1.0	100	[53]

One of the parameters influencing the degradation efficiency of pollutants is the solution pH. Thus, the degradation efficiency of RhB using the CZF1:5 sample was tested by varying the pH between three and eight and maintaining a constant photocatalyst concentration (0.2 mg/mL). The results are presented in Figure 9. It can be observed that a pH value of six assures the best photocatalytic degradation. For higher pH, the degradation efficiency decreases, probably due to the repulsion between negatively charged pollutant molecules and nanoparticles [9]. In contrast, by utilizing an acidic solution (pH of about 3), the lowest photocatalytic activity was obtained, probably caused by the slight dissolution of the catalyst [54].

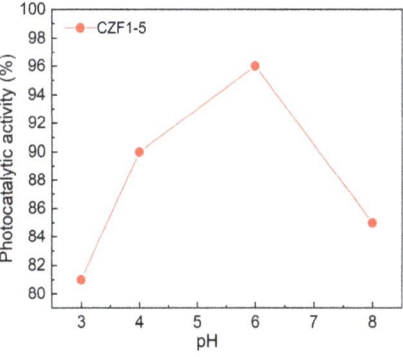

Figure 9. Effect of initial solution pH on pollutant photodegradation.

The effect of the RhB concentration on the CZF1:5 sample's photocatalytic activity was studied using three concentrations of RhB (2 mg/mL, 5 mg/mL and 15 mg/mL), and is shown in Figure 10. It can be observed that the adsorption capacity of the sample increases with the increase in the RhB concentration solution. This behavior can be explained based on the increased number of dye molecules from a diluted solution to a concentrated one. After 150 min irradiation, the 5 mg/ml solution was degraded the most, followed by the 2 mg/mL solution, and the lowest degradation was obtained for the 15 mg/L solution.

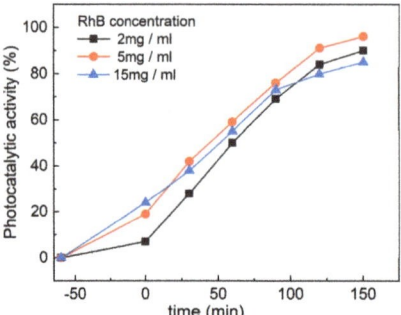

Figure 10. Effect of RhB concentration on the CZF1:5 sample photocatalytic activity.

To evaluate the stability of the photocatalyst, several consecutive RhB photodegradation cycles were performed. Between two photodegradation cycles, the pollutant solution was removed, and the sample was washed with ethanol/water and dried overnight. Figure 11 shows the photocatalytic activity after five runs using the CZF1:5 sample as a photocatalyst. It can be observed that the sample efficiency remains practically unmodified after five photodegradation cycles and sustained the good stability of the sample. In addition, the CZF1-5 sample stability was verified through FT-IR (Figure S3, Table S1). No modification of the spectrum was observed after 150 min of UV irradiation.

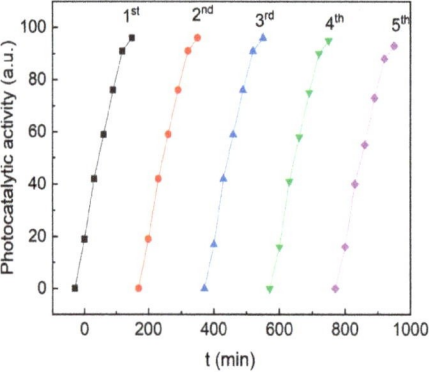

Figure 11. The reusability of CZF1:5 sample for degradation of RhB dye for five cycles under UV irradiation.

To elucidate the mechanism of RhB pollutant photodegradation, we investigated the generation of reactive oxygen species by the CZF1:5 sample under the action of UV light. We used the ESR, coupled with the spin trapping technique and DMPO (5,5-Dimethyl-1-pyrroline N-oxide), as a spin trapper. No ESR signal was obtained for the unirradiated sample, but a complex spectrum was obtained after 25 min of irradiation, as seen in Figure 12. The simulation was performed to identify each component of the spectrum.

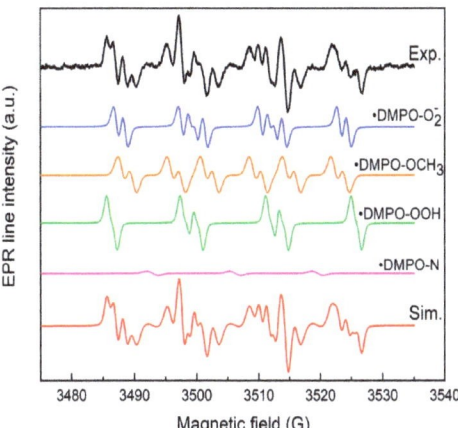

Figure 12. Experimental and simulated spectra of DMPO spin adducts generated by CZF1:5 sample after 25 min irradiation.

The simulation results reveal that the spectrum is composed of the superposition of the following components: •DMPO-OCH$_3$ (a_N = 13.2 G, a_H = 7.8 G, a_H = 1.6 G, relative concentration 38%), •DMPO-OOH (a_N = 13.8 G, a_H = 11.8, a_H = 0.9 G, g = 2.0098, relative concentration 29%), •DMPO-O$_2^-$ (a_N = 12.8 G, a_H = 10.4 G, a_H = 1.4 G, relative concentration 31%) and •DMPO-N (a_N = 13.9 G, relative concentration 2%). The •DMPO-OCH$_3$ adduct spin is produced via the interaction of hydroxyl radicals (•OH) with DMSO solvent, and •DMPO-OOH is a result of •O$_2^-$ radicals protonation. The •DMPO-N appears through the cleavage of the N–C bond in DMPO [55]. These results indicate that the hydroxyl and superoxide radicals are generated under irradiation.

To explain the photocatalytic mechanism, a band alignment corresponding to the nanocomposite was proposed, considering the Eg values estimated by the UV-VIS and ZnO VB energy position of −7.44 eV [56]. For the MWCNT, an ionization energy of 5.01 eV was considered [57,58]. Analyzing the band edge energies, it can be observed that a type-II heterostructure was formed [56]. The proposed photocatalytic mechanism is illustrated in Figure 13. Following the UV light excitation of the composites, the electrons from the ZnO VB are promoted into the CB, generating holes in the VB. The photogenerated electrons from the LUMO band of the MWCNT can be transferred towards ZnO CB [56]. The electrons from ZnO CB can reach the photocatalyst surface, where interacting with adsorbed O$_2$ will create •O$_2^-$ species. However, if the electrons have enough energy, they can attend the Fe^{3+} ions from the nanoparticle surface, generating Fe^{2+} ions. As Fe^{2+} ions are unstable, they will interact with the O$_2$ molecules, forming Fe^{3+} and O$_2^-$ reactive species. In addition, Fe^{2+} ions can receive and electron form unstable Fe^{1+} ions, which will further interact with the O$_2$ molecules, generating O$_2^-$ and Fe^{2+} ions [33]. In addition, the photogenerated electrons can move to the Zn$_i$ levels, followed by transitions towards the V$_O$ and V$_{Zn}$ levels, leading to radiative emissions [42,59]. The remaining holes in the ZnO VB will pass to the MWCNT via impurity levels in the ZnO bandgap and can interact with the H$_2$O molecules, generating •OOH reactive species [33]. Consequently, the MWCNT's role is to efficiently separate the photogenerated charges, allowing electrons to generate ROS species and increasing the photocatalytic efficiency [60]. The photocatalytic investigation reveals the existence of an optimum amount of ZF0.5%, which assures the best photocatalytic efficiency. A higher concentration of ZF0.5% nanoparticles on the MWCNT surface could prevent incident light from reaching the photocatalyst's active surface, leading to a decrease in the photocatalyst efficiency [61].

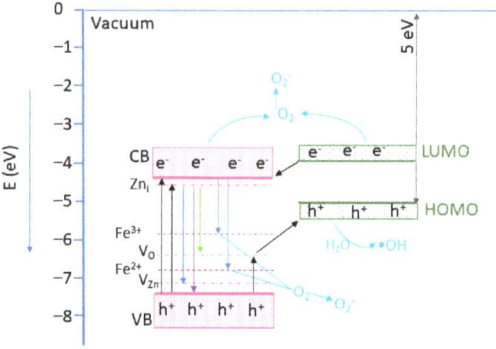

Figure 13. The proposed photocatalytic mechanism based on the energy band structure. The energy level positions were draw according to [33,56–58,62].

To better understand the photodegradation mechanism, the intermediates produced by irradiating the RhB solution for 150 min in the presence of the sample with the best photocatalytic activity (CZF1-5) were identified through HPLC/MS (Figures S4 and S5). The fragmentation of RhB with a m/z value of 443 generates five major N-de-ethylated intermediates, including N, N-diethyl-N′-ethylrhodamine (m/z = 415), N, N-diethylrhodamine/N′-ethyl-N-ethylrhodamine (m/z = 387), N-ethylrhodamine (m/z = 359) and rhodamine (m/z = 331). Consequently, the photodegradation of RhB occurs through the N-de-ethylation process, which is schematically shown in Figure 14.

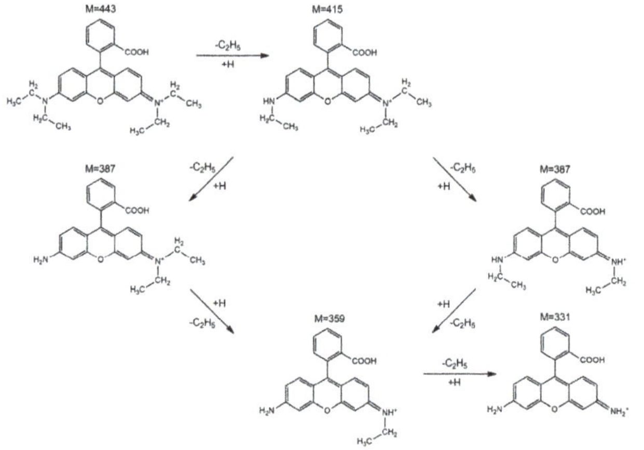

Figure 14. The proposed photodegradation pathways of RhB based on the intermediates identified by the HPLC–MS method.

The sample with the best photocatalytic efficiency (CZF1-5) was tested on an antibiotic (oxytetracycline) degradation to demonstrate the sample's ability to degrade other types of pollutants (Figure S5). The obtained degradation efficiencies were 85% (by monitoring the 360nm absorption peak) and 30% (by monitoring the 270 nm absorption peak). A longer irradiation time would likely have been necessary to improve the photocatalytic performance.

4. Conclusions

In this work, MWCNTs decorated with 0.5%Fe-doped ZnO nanoparticles were prepared through polymer wrapping using poly(allylamine hydrochloride). Three samples, in which the molar ratio between the MWCNT and ZF0.5% was 1:3, 1:5 and 1:10, were prepared. XRD diffraction certified the formation of composite samples, evidencing the patterns specific to ZnO, with a hexagonal structure, and that specific to MWCNTs. From the EPR spectra analysis results, ZF0.5% presents three types of paramagnetic species: isolated Fe^{3+} ions, ferromagnetically coupled Fe^{3+} ions and oxygen vacancies. The TEM images show a tubular shape of the MWCNT decorated with ZF0.5% nanoparticles with a star-like morphology. The band gap energies of the samples decrease with the increase in the MWCNT content. The samples show emissions in both the UV and visible ranges. The emissions from UV are known as the near band edge, and that from the visible range are associated with the emission caused by different defect states, such as Zn in interstitial sites or oxygen vacancy. Both types of emissions are quenched with the increase in the MWCNT content due to a delay in the recombination process induced by the charge transfer between the defect levels of the ZnO nanoparticles and MWCNT. All of the samples show photocatalytic activity against RhB under UV irradiation. The best photocatalytic activity was obtained for the CZF1:5 sample, and this sample's stability was proven by performing five consecutive cycles. Through HPLC characterization, it was evidenced that the photodegradation of RhB occurs through the N-de-ethylation process. Analyzing the ROS generated by the CZF1:5 sample under UV irradiation shows that hydroxyl and superoxide radicals are involved in the photocatalytic reactions. The photocatalytic mechanism was elucidated based on the ROS generated under UV irradiation and in correlation with the structural and optical properties.

Supplementary Materials: The following supporting information can be downloaded at: https://www.mdpi.com/article/10.3390/ma16072858/s1.

Author Contributions: Conceptualization, A.P. and D.T.; investigation, A.P., M.S., S.M., L.E.M., C.L., C.V.F.-A. and D.T.; writing—original draft preparation, A.P., M.S. and D.T.; writing—review and editing, A.P., D.T, C.L. and C.V.F.-A. All authors have read and agreed to the published version of the manuscript.

Funding: This research was funded by the MCID through the "Nucleu" Programe within the National Plan for Research, Development and Innovation 2022–2027, project PN 23 24 01 03 and through Program 1—Development of the national research and development system, Subprogram 1.2—Institutional performance—Projects that finance the RDI excellence, Contracts no. 37PFE/30.12.2021.

Institutional Review Board Statement: Not applicable.

Informed Consent Statement: Not applicable.

Data Availability Statement: The data presented in this study are available on reasonable request.

Acknowledgments: The authors acknowledge financial support from the MCID through the "Nucleu" Programe within the National Plan for Research, Development and Innovation 2022–2027, project PN 23 24 01 03 and through Program 1—Development of the national research and development system, Subprogram 1.2—Institutional performance—Projects that finance the RDI excellence, Contracts no. 37PFE/30.12.2021. In addition, the authors wish to thank L. Barbu-Tudoran from the Integrated Laboratory of Electron Microscopy at INCDTIM for acquiring the TEM images.

Conflicts of Interest: The authors declare no conflict of interest.

References

1. Nazir, M.A.; Najam, T.; Shahzad, K.; Wattoo, M.A.; Hussain, T.; Tufail, M.K.; Shah, S.S.A.; Rehman, A.U. Heterointerface engineering of water stable ZIF-8@ZIF-67: Adsorption of rhodamine B from water. *Surf. Interfaces* **2022**, *34*, 102324. [CrossRef]
2. Tóth, A.J.; Fózer, D.; Mizsey, P.; Varbanov, P.S.; Klemeš, J.J. Physicochemical methods for process wastewater treatment: Powerful tools for circular economy in the chemical industry. *Rev. Chem. Eng.* **2022**. [CrossRef]
3. Al-Tohamy, R.; Ali, S.S.; Li, F.; Okasha, K.M.; Mahmoud, Y.A.-G.; Elsamahy, T.; Jiao, H.; Fu, Y.; Sun, J. A critical review on the treatment of dye-containing wastewater: Ecotoxicological and health concerns of textile dyes and possible remediation approaches for environmental safety. *Ecotoxicol. Environ. Saf.* **2022**, *231*, 113160–113172. [CrossRef] [PubMed]
4. Nazir, M.A.; Najam, T.; Jabeen, S.; Wattoo, M.A.; Bashir, M.S.; Shah, S.S.A.; Rehman, A. Facile synthesis of Tri-metallic layered double hydroxides (NiZnAl-LDHs): Adsorption of Rhodamine-B and methyl orange from water. *Inorg. Chem. Commun.* **2022**, *145*, 110008. [CrossRef]

5. Znad, H.; Abbas, K.; Hena, S.; Awual, M.R. Synthesis a novel multilamellar mesoporous TiO_2/ZSM-5 for photo-catalytic degradation of methyl orange dye in aqueous media. *J. Environ. Chem. Eng.* **2018**, *6*, 218–227. [CrossRef]
6. Bhowmik, M.; Deb, K.; Debnath, A.; Saha, B. Mixed phase Fe_2O_3/Mn_3O_4 magnetic nanocomposite for enhanced adsorption of methyl orange dye: Neural network modeling and response surface methodology optimization. *Appl. Organomet. Chem.* **2018**, *32*, 4186. [CrossRef]
7. Sejie, F.P.; Tabbiruka, M.S.N. Removal of methyl orange (MO) from water by adsorption onto modified local clay (kaolinite). *Phys. Chem.* **2016**, *6*, 39–48.
8. Eljiedi, A.A.A.; Kamari, A. Removal of methyl orange and methylene blue dyes from aqueous solution using lala clam (*Orbicularia orbiculata*) shell. *AIP Conf. Proc.* **2017**, *1847*, 040003. [CrossRef]
9. Mate, C.J.; Mishra, S. Synthesis of borax cross-linked Jhingan gum hydrogel for remediation of Remazol Brilliant Blue R (RBBR) dye from water: Adsorption isotherm, kinetic, thermodynamic and biodegradation studies. *Int. J. Biol. Macromol.* **2020**, *151*, 677–690. [CrossRef]
10. Ahmad, M.A.; Puad, N.A.A.; Bello, O.S. Kinetic, equilibrium and thermodynamic studies of synthetic dye removal using pomegranate peel activated carbon prepared by microwave-induced KOH activation. *Water Resour. Ind.* **2014**, *6*, 18–35. [CrossRef]
11. Nafi, A.W.; Taseidifar, M. Removal of hazardous ions from aqueous solutions: Current methods, with a focus on green ion flotation. *J. Environ. Manag.* **2022**, *319*, 115666. [CrossRef] [PubMed]
12. Opitz, J.; Bauer, M.; Alte, M.; Peiffer, S. Development of a novel sizing approach for passive mine water treatment systems based on ferric iron sedimentation kinetics. *Water Res.* **2023**, *233*, 119770. [CrossRef] [PubMed]
13. Heebner, A.; Abbassi, B. Electrolysis catalyzed ozonation for advanced wastewater treatment. *J. Water Process. Eng.* **2022**, *46*, 102638. [CrossRef]
14. Ramu, A.G.; Choi, D. Highly efficient and simultaneous catalytic reduction of multiple toxic dyes and nitrophenols waste water using highly active bimetallic PdO–NiO nanocomposite. *Sci. Rep.* **2021**, *11*, 22699. [CrossRef] [PubMed]
15. Shariati, M.; Babaei, A.; Azizi, A. Synthesis of Ag_2CrO_4/Ag/Fe_3O_4/RGO nanocomposite as a suitable photocatalyst for degradation of methylene blue in aqueous media: RSM modeling, kinetic and energy consumption studies. *Inorg. Chem. Commun.* **2022**, *145*, 110004. [CrossRef]
16. Wahba, M.A.; Yakout, S.M.; Khaled, R. Interface engineered efficient visible light photocatalytic activity of MWCNTs/Co doped ZnO nanocomposites: Morphological, optical, electrical and magnetic properties. *Opt. Mater.* **2021**, *115*, 111039. [CrossRef]
17. Irani, E.; Amoli-Diva, M. Hybrid adsorption–photocatalysis properties of quaternary magnetoplasmonic ZnO/MWCNTs nanocomposite for applying synergistic photocatalytic removal and membrane filtration in industrial wastewater treatment. *J. Photochem. Photobiol. A* **2020**, *391*, 112359–112360. [CrossRef]
18. Guo, J.; Li, Y.; Zhu, S.; Chen, Z.; Liu, Q.; Zhang, D.; Moon, W.-J.; Song, D.-M. Synthesis of WO_3@Graphene composite for enhanced photocatalytic oxygen evolution from water. *RSC Adv.* **2011**, *2*, 1356–1363. [CrossRef]
19. Bárdos, E.; Kovács, G.; Gyulavári, T.; Németh, K.; Kecsenovity, E.; Berki, P.; Baia, L.; Pap, Z.; Hernádi, K. Novel synthesis approaches for WO_3-TiO_2/MWCNT composite photocatalysts-problematic issues of photoactivity enhancement factors. *Catal. Today* **2018**, *300*, 28–38. [CrossRef]
20. Jeevitha, G.; Sivaselvam, S.; Keerthana, S.; Mangalaraj, D.; Ponpandian, N. Highly effective and stable MWCNT/WO3 nanocatalyst for ammonia gas sensing, photodegradation of ciprofloxacin and peroxidase mimic activity. *Chemosphere* **2022**, *297*, 134023. [CrossRef]
21. Dobaradaran, S.; Nabizadeh Nodehi, R.; Yaghmaeian, K.; Jaafari, J.; Hazrati Niari, M.; Bharti, A.K.; Agarwal, S.; Gupta, V.K.; Azari, A.; Shariatifar, N. Catalytic decomposition of 2-chlorophenol using an ultrasonic-assisted Fe_3O_4–TiO_2@MWCNT system: Influence factors, pathway and mechanism study. *J. Colloid. Interface. Sci.* **2018**, *512*, 172–189. [CrossRef] [PubMed]
22. Gao, P.; Huang, S.; Tao, K.; Li, Z.; Feng, L.; Liu, Y.; Zhan, L. Synthesis of adjustable {312}/{004} facet heterojunction MWCNTs/Bi_5O_7I photocatalyst for ofloxacin degradation: Novel insights into the charge carriers transport. *J. Hazard. Mater.* **2022**, *437*, 129374. [CrossRef] [PubMed]
23. Yan, D.; Hu, H.; Gao, N.; Ye, J.; Ou, H. Fabrication of carbon nanotube functionalized MIL-101(Fe) for enhanced visible-light photocatalysis of ciprofloxacin in aqueous solution. *Appl. Surf. Sci.* **2019**, *498*, 143836. [CrossRef]
24. Stefan, M.; Leostean, C.; Popa, A.; Toloman, D.; Perhaita, I.; Cadis, A.; Macavei, S.; Pana, O. Highly stable MWCNT-$CoFe_2O_4$ photocatalyst. EGA-FTIR coupling as efficient tool to illustrate the formation mechanism. *J. Alloy. Compd.* **2022**, *928*, 167188. [CrossRef]
25. Tang, X.; Xue, Q.; Qi, X.; Cheng, C.; Yang, M.; Yang, T.; Chen, F.; Qiu, F.; Quan, X. DFT and experimental study on visible-light driven photocatalysis of rare-earth-doped TiO_2. *Vacuum* **2022**, *200*, 110972. [CrossRef]
26. Toloman, D.; Stefan, M.; Pana, O.; Rostas, A.M.; Silipas, T.D.; Pogacean, F.; Pruneanu, S.; Leostean, C.; Barbu-Tudoran, L.; Popa, A. Transition metal ions as a tool for controlling the photocatalytic activity of MWCNT-TiO_2 nanocomposites. *J. Alloy Compd.* **2022**, *921*, 166095–166111. [CrossRef]
27. Lemos, S.C.S.; de Lima Rezende, T.K.; Assis, M.; da Costa Romeiro, F.; Peixoto, D.A.; de Oliveira Gomes, E.; Jacobsen, G.M.; Teodoro, M.D.; Gracia, L.; Ferrari, J.L.; et al. Efficient Ni and Fe doping process in ZnO with enhanced photocatalytic activity: A theoretical and experimental investigation. *Mater. Res. Bull.* **2022**, *152*, 111849. [CrossRef]
28. Lee, H.; Jang, H.S.; Kim, N.Y.; Joo, J.B. Cu-doped TiO_2 hollow nanostructures for the enhanced photocatalysis under visible light conditions. *J. Ind. Eng. Chem.* **2021**, *99*, 352–363. [CrossRef]

29. Feng, C.; Chen, Z.; Li, W.; Zhang, F.; Li, X.; Xu, L.; Sun, M. First-principle calculation of the electronic structures and optical properties of the metallic and nonmetallic elements-doped ZnO on the basis of photocatalysis. *Phys. B: Condens. Matter* **2019**, *555*, 53–60. [CrossRef]
30. Râpă, M.; Stefan, M.; Popa, P.; Toloman, D.; Leostean, C.; Borodi, G.; Vodnar, D.; Wrona, M.; Salafranca, J.; Nerín, C.; et al. Electrospun Nanosystems Based on PHBV and ZnO for Ecological Food Packaging. *Polymers* **2021**, *13*, 2123. [CrossRef]
31. Rocha, M.; Araujo, F.P.; Castro-Lopes, S.; de Lima, I.S.; Silva-Filho, E.C.; Anteveli Osajima, J.; Oliveira, C.S.; Viana, B.C.; Almeida, L.C.; Guerra, Y.; et al. Synthesis of Fe–Pr co-doped ZnO nanoparticles: Structural, optical and antibacterial properties. *Ceram. Intl.* **2023**, *49*, 2282–2295. [CrossRef]
32. Dubey, P.; Sonkar, S.K.; Majumder, S.; Tripathi, K.M.; Sarkar, S. Isolation of water soluble carbon nanotubes with network structure possessing multipodal junctions and its magnetic property. *RSC Adv.* **2013**, *3*, 7306–7312. [CrossRef]
33. Popa, A.; Pana, O.; Stefan, M.; Toloman, D.; Stan, M.; Leostean, C.; Suciu, R.C.; Vlad, G.; Ulinici, S.; Baisan, G.; et al. Interplay between ferromagnetism and photocatalytic activity generated by Fe^{3+} ions in iron doped ZnO nanoparticles grown on MWCNTs. *Phys. E: Low-Dimens. Syst. Nanostructures* **2020**, *129*, 114581–114590. [CrossRef]
34. Buz, E.; Zhou, D.; Kittilstved, K.R. Air-stable n-type Fe-doped ZnO colloidal nanocrystals. *J. Chem. Phys.* **2019**, *151*, 134702–134708. [CrossRef]
35. Misra, S.K.; Andronenko, S.I.; Thurber, A.; Punnoose, A.; Nalepa, A. An X-and Q-band Fe^{3+} EPR study of nanoparticles of magnetic semiconductor $Zn_{1-x}Fe_xO$. *J. Magn. Magn. Mater.* **2014**, *363*, 82–87. [CrossRef]
36. Raita, O.; Popa, A.; Stan, M.; Suciu, R.C.; Biris, A.; Giurgiu, L.M. Effect of Fe Concentration in ZnO Powders on Ferromagnetic Resonance Spectra. *Appl. Magn. Reson.* **2012**, *42*, 499–509. [CrossRef]
37. Rajan, C.P.; Abharana, N.S.; Jha, N.; Bhattacharyya, D.; John, T.T. Local Structural Studies Through EXAFS and Effect of Fe^{2+} or Fe^{3+} Existence in ZnO Nanoparticles. *J. Phys. Chem. C* **2021**, *125*, 13523–13533. [CrossRef]
38. Kim, K.J.; Park, Y.R. Optical investigation of $Zn_{1-x}Fe_xO$ films grown on $Al_2O_3(0001)$ by radio-frequency sputtering. *J. Appl. Phys.* **2004**, *96*, 4150–4153. [CrossRef]
39. Toloman, D.; Mesaros, A.; Popa, A.; Silipas, T.D.; Neamtu, S.; Katona, G. V-doped ZnO particles: Synthesis, structural, optical and photocatalytic properties. *J. Mater. Sci: Mater Electron* **2016**, *27*, 5691–5698. [CrossRef]
40. Bai, S.; Guo, T.; Zhao, Y.; Sun, J.; Li, D.; Chen, A.; Liu, C.C. Sensing performance and mechanism of Fe-doped ZnO microflowers. *Sens. Actuators B: Chem.* **2014**, *195*, 657–666. [CrossRef]
41. El Mel, A.-A.; Buffière, M.; Ewels, C.P.; Molina-Luna, L.; Faulques, E.; Colomer, J.-F.; Kleebe, H.-J.; Konstantinidis, S.; Snyders, R.; Bittencourt, C. Zn based nanoparticle–carbon nanotube hybrid materials: Interaction and charge transfer. *Carbon* **2013**, *66*, 442–449. [CrossRef]
42. Jangir, L.K.; Kumari, Y.; Kumar, A.; Kumar, M.; Awasthi, K. Investigation of luminescence and structural properties of ZnO nanoparticles, synthesized with different precursors. *Mater. Chem. Front.* **2017**, *1*, 1413–1421. [CrossRef]
43. Sagar, P.; Shishodia, P.; Mehra, R.; Okada, H.; Wakahara, A.; Yoshida, A. Photoluminescence and absorption in sol–gel-derived ZnO films. *J. Lumin.* **2007**, *126*, 800–806. [CrossRef]
44. Zeng, H.; Duan, G.; Li, Y.; Yang, S.; Xu, X.; Cai, W. Blue Luminescence of ZnO Nanoparticles Based on Non-Equilibrium Processes: Defect Origins and Emission Controls. *Adv. Funct. Mater.* **2010**, *20*, 561–572. [CrossRef]
45. Ghoderao, K.P.; Jamble, S.N.; Kale, R.B. Hydrothermally synthesized Cd-doped ZnO nanostructures with efficient sunlight-driven photocatalytic and antibacterial activity. *J. Mater. Sci. Mater. Electron.* **2019**, *30*, 11208–11219. [CrossRef]
46. Van de Walle, C.G. Hydrogen as a Cause of Doping in Zinc Oxide. *Phys. Rev. Lett.* **2000**, *85*, 1012. [CrossRef] [PubMed]
47. Chen, C.; Lu, Y.; He, H.; Xiao, M.; Wang, Z.; Chen, L.; Ye, Z. Violet Emission in ZnO Nanorods Treated with High-Energy Hydrogen Plasma. *ACS Appl. Mater. Interfaces* **2013**, *5*, 10274–10279. [CrossRef]
48. Debnath, B.; Halder, G.; Bhattacharyya, S. One-Step Synthesis, Structural and Optical Characterization of Self-Assembled ZnO Nanoparticle Clusters with Quench-Induced Defects. *Sci. Adv. Mater.* **2014**, *6*, 1160–1169.
49. Ahn, C.H.; Kim, Y.Y.; Kim, D.C.; Mohanta, S.K.; Cho, H.K. A comparative analysis of deep level emission in ZnO layers deposited by various methods. *J. Appl. Phys.* **2009**, *105*, 013502–013510. [CrossRef]
50. Shi, X.; Liu, J.B.; Hosseini, M.; Shemshadi, R.; Razavi, R.; Parsaee, Z. Ultrasound-asisted photodegradation of Alprazolam in aqueous media using a novel high performance nanocomosite hybridization g-C3N4/MWCNT/ZnO. *Catal. Today* **2019**, *335*, 582–590. [CrossRef]
51. Hosseini, F.; Kasaeian, A.; Pourfayaz, F.; Sheikhpour, M.; Wen, D. Novel ZnO-Ag/MWCNT nanocomposite for the photocatalytic degradation of phenol. *Mater. Sci. Semicond. Procc.* **2018**, *83*, 175–185. [CrossRef]
52. Hazarika, M.; Chinnamuthu, P.; Borah, J. Enhanced photocatalytic efficiency of MWCNT/NiFe2O4 nanocomposites. *Phys. E Low-Dimens. Syst. Nanostructures* **2022**, *139*, 115177. [CrossRef]
53. Wang, P.; Zhong, S.; Lin, M.; Lin, C.; Lin, T.; Gao, M.; Zhao, C.; Li, X.; Wu, X. Signally enhanced piezo-photocatalysis of $Bi_{0.5}Na_{0.5}TiO_3$/MWCNTs composite for degradation of rhodamine B. *Chemosphere* **2022**, *308*, 136596. [CrossRef] [PubMed]
54. Kakarndee, S.; Nanan, S. SDS capped and PVA capped ZnO nanostructures with high photocatalytic performance toward photodegradation of reactive red (RR141) azo dye. *J. Environ. Chem. Eng.* **2018**, *6*, 74–94. [CrossRef]
55. Martínez, J.M.L.; Denis, M.F.L.; Piehl, L.L.; de Celis, E.R.; Buldain, G.Y.; Orto, V.C.D. Studies on the activation of hydrogen peroxide for color removal in the presence of a new Cu(II)-polyampholyte heterogeneous catalyst. *Appl. Catal. B: Environ.* **2008**, *82*, 273–283. [CrossRef]

56. Sharma, M.D.; Mahala, C.; Basu, M. Sensitization of vertically grown ZnO 2D thin sheets by MoSx for efficient charge separation process towards photoelectrochemical water splitting reaction. *Int. J. Hydrog. Energy* **2020**, *45*, 12272–12282. [CrossRef]
57. Baldoni, M.; Sgamellotti, A.; Mercuri, F. Finite-Length Models of Carbon Nanotubes Based on Clar Sextet Theory. *Org. Lett.* **2007**, *9*, 4267–4270. [CrossRef]
58. Riaz, A.; Alam, A.; Selvasundaram, P.B.; Dehm, S.; Hennrich, F.; Kappes, M.M.; Krupke, R. Near-Infrared Photoresponse of Waveguide-Integrated Carbon Nanotube–Silicon Junctions. *Adv. Electron. Mater.* **2018**, *5*, 1800265. [CrossRef]
59. Elhamdi, I.; Souissi, H.; Taktak, O.; Elghoul, J.; Kammoun, S.; Dhahri, E.; Benilde, F.; Costa, O. Experimental and modeling study of ZnO:Ni nanoparticles for near-infrared light emitting diodes. *RSC Adv.* **2022**, *12*, 13074–13086. [CrossRef]
60. Mancuso, A.; Sacco, O.; Mottola, S.; Pragliola, S.; Moretta, A.; Vaiano, V.; De Marco, I. Synthesis of Fe-doped ZnO by supercritical antisolvent precipitation for the degradation of azo dyes under visible light. *Inorg. Chim. Acta* **2023**, *549*, 121407–121415. [CrossRef]
61. Farhadian, M.; Sangpour, P.; Hosseinzadeh, G. Preparation and photocatalytic activity of WO_3–MWCNT nanocomposite for degradation of naphthalene under visible light irradiation. *RSC Adv.* **2016**, *6*, 39063–39073. [CrossRef]
62. Vempati, S.; Mitra, J.; Dawson, P. One-step synthesis of ZnO nanosheets: A blue-white fluorophore. *Nanoscale Res. Lett.* **2012**, *7*, 470. [CrossRef] [PubMed]

Disclaimer/Publisher's Note: The statements, opinions and data contained in all publications are solely those of the individual author(s) and contributor(s) and not of MDPI and/or the editor(s). MDPI and/or the editor(s) disclaim responsibility for any injury to people or property resulting from any ideas, methods, instructions or products referred to in the content.

Article

Neodymium-Doped Gadolinium Compounds as Infrared Emitters for Multimodal Imaging

Maxime Delaey, Seppe Van Bogaert, Ewoud Cosaert, Wout Mommen and Dirk Poelman *

LumiLab, Department of Solid State Sciences, Ghent University, 9000 Ghent, Belgium; maxime.delaey@ugent.be (M.D.); ewoud.cosaert@ugent.be (E.C.)
* Correspondence: dirk.poelman@ugent.be

Abstract: This study aims to investigate the optical properties of multiple neodymium-doped gadolinium compounds as a means to examine their eligibility as optical probes for fluorescence imaging. $GdVO_4$, $GdPO_4$, $GdAlO_3$, Gd_2SiO_5 and $Gd_3Ga_5O_{12}$ (GGG) samples were synthesized through solid-state reactions with varying neodymium doping levels to compare their optical properties in great detail. The optimal doping concentration was generally found to be approximately 2%. Furthermore, the luminescence lifetime, which is a valuable parameter for time-gated imaging, was determined to range from 276 down to 14 μs for the highest doping concentrations, resulting from energy transfer and migration assisted decay.

Keywords: luminescence; near infrared; spectroscopy; neodymium; gadolinium compounds; bioimaging; luminescence decay

1. Introduction

Fluorescence imaging is a biomedical technique in which optical probes are utilized to attain clinically relevant knowledge [1]. Fluorescent dyes and proteins have predominantly been investigated as optical probes; however, inorganic nanoparticles have attracted interest due to their much higher stability, tunable pharmacokinetic properties and resistance to photobleaching [2]. In this work, the possibility of using neodymium-doped gadolinium compounds as optical probes is explored. The presence of gadolinium presents additional advantages: due to its paramagnetic nature and its strong X-ray absorption cross section, the material could also serve as a contrast agent for both magnetic resonance imaging (MRI) and X-ray computed tomography (CT) scans, allowing for true multimodal imaging using a single type of material [3,4].

In order to be an eligible candidate, the material in question should meet various requirements. For the purpose of allowing both in vivo excitation and detection of the emitted light, both the excitation and emission wavelengths of the nanoparticles should lie withing the biological windows, which are wavelength ranges for which the absorption coefficient of biological tissue is minimal. As Nd^{3+} may be excited at 808 nm and exhibits intense emission around 1060 nm, the relevant biological windows are 650–950 nm and 1000–1350 nm [5]. However the autofluorescence of tissues poses another challenge: removing this background signal from the luminescence of the tissues themselves requires the emission wavelength to be larger than 1100 nm [2] or thee use of time-gated imaging [6]—detecting the particle luminescence after the autofluorescence has decayed—which is made possible through the long decay times of Nd^{3+} of the order of 100 μs. Lastly, the nanoparticles should be biocompatible and have an appropriate size, as this dictates the pharmacokinetic properties [7].

While in most work, the properties of Nd^{3+} in only a single host are presented, here, we explore the the luminescence properties of the Nd ions in relation to the structural properties of a series of different hosts. This allows us to pinpoint a number of guidelines in selecting a suitable host for Nd^{3+}-doped near-infrared-emitting phosphors. The optical properties of the bulk materials were investigated in order to examine the influence of the doping concentration and local environment of Nd^{3+} on the excitation spectra, emission intensities and luminescence lifetimes. This study allows for a founded selection of an optimum material and subsequent development of a suitable nanoparticle synthesis method.

2. Materials and Methods

2.1. Synthesis

All of the samples were synthesized through solid state reactions with Gd_2O_3 (99.99%, Alfa Aesar, Ward Hill, MA, USA), Nd_2O_3 (99.99%, Sigma-Aldrich, Burlington, MA, USA), Al_2O_3 (99.99%, Alfa Aesar), SiO_2 (99.95%, Alfa Aesar), Ga_2O_3 (99.99%, Sigma-Aldrich), $(NH_4)_2HPO_4$ (99%, Acros Organics, Geel, Belgium) and V_2O_5 (99.99%, Thermo Scientific, Waltham, MA, USA) as precursors, which were weighed in a stoichiometric manner such that the chemical reactions listed below are valid. The doping level (x) was defined as $\frac{[Nd]}{[Nd]+[Gd]}$. The precursors were mixed using a mortar and pestle, after which they were heated to the temperatures mentioned below, with a heating rate of 300 °C/h. The ovens used for synthesis operated in an air atmosphere.

2.1.1. $GdAlO_3$

Gadolinium aluminate was prepared by heating the precursors at 1500 °C for 6 h, followed by dry grinding and further heating at 1550 °C for 2 h in a tube furnace (ETF 30-50/18-S, Entech, Ängelholm, Sweden).

$$(1-x)Gd_2O_3 + Al_2O_3 + x\,Nd_2O_3 \longrightarrow 2\,Gd_{1-x}Nd_xAlO_3$$

2.1.2. Gd_2SiO_5

Gd_2SiO_5 was also prepared at 1500 °C for 6 h but with the addition of 2 wt% of BaF_2 as a flux to facilitate the reaction, as performed in [8]. Afterwards, the samples were heated at 1550 °C for 2 h. The same tube furnace was employed to synthesize the batch of Gd_2SiO_5 samples.

$$(1-x)Gd_2O_3 + SiO_2 + x\,Nd_2O_3 \longrightarrow Gd_{2(1-x)}Nd_{2x}SiO_5$$

2.1.3. GGG ($Gd_3Ga_5O_{12}$)

GGG was prepared at 1450 °C for 6 h, followed by dry grinding and further heating at 1500 °C for 2 h. Similarly to the last two samples, GGG was also heated in a tube furnace.

$$3(1-x)Gd_2O_3 + 5Ga_2O_3 + 3x\,Nd_2O_3 \longrightarrow 2\,Gd_{3(1-x)}Nd_{3x}Ga_5O_{12}$$

2.1.4. $GdPO_4$

Gadolinium phosphate was synthesized using Li_2CO_3 (99.998%, Alfa Aesar) as a flux [9]. The precursors and flux were heated to 900 °C for 4 h in a muffle furnace (Nabertherm LT 5/13).

$$(1-x)Gd_2O_3 + 2(NH_4)_2HPO_4 + x\,Nd_2O_3 \longrightarrow 2\,Gd_{(1-x)}Nd_xPO_4 + 3H_2O + 4NH_3$$

2.1.5. $GdVO_4$

The precursor mixture was heated up to 800 °C for 1 h. Secondly, the precursors were ground and placed an oven at 1100 °C for 3 h [4], after which they were heated once more at 1250 °C for 2 h. The $GdVO_4$ samples were synthesized in the same muffle furnace that was employed for $GdPO_4$.

$$(1-x)Gd_2O_3 + V_2O_5 + x\,Nd_2O_3 \longrightarrow 2\,Gd_{(1-x)}Nd_xVO_4$$

2.2. X-ray Diffraction

In order to evaluate the structure of the samples and verify the synthesis process, X-ray powder diffraction (XRD) was performed. The patterns were measured from 5° to 80° with a step size of 0.02° and an integration time of 1.2 s per step for the GGG, GdPO$_4$ and GdVO$_4$ samples, while an integration time of 4.8 s per step of 0.04° for the Gd$_2$SiO$_5$ and GdAlO$_3$ samples was utilized. A θ–2θ diffractometer (Siemens D5000) with Cu Kα radiation (λ = 0.15406 nm) and generator settings of 40 kV and 40 mA was employed.

2.3. Scanning Electron Microscopy

Scanning electron microscopy (SEM) was conducted using an FEI Quanta 200 FEG SEM, which operates at high vacuum. Before the morphology of the samples was probed using secondary electrons, the samples were coated with a thin gold layer to prevent them from collecting charges.

2.4. Optical Absorption

As the optical absorbance of a powder cannot directly be measured, Kubelka–Munk approximation was employed [10] to convert diffuse reflectance measurements ($R(\lambda)$) into absorption:

$$\frac{k(\lambda)}{s(\lambda)} = \frac{(1 - R(\lambda))^2}{2R(\lambda)}, \tag{1}$$

where $k(\lambda)$ is the absorption coefficient, $s(\lambda)$ is the back-scattering coefficient and $R(\lambda)$ is the reflectance. The diffuse reflectance measurements were performed on powders that were pressed onto a sample holder and kept into place using a quartz slide. A spectrophotometer (LAMBDA 1050 S UV/Vis/NIR, PerkinElmer, Waltham, MA, USA) equipped with a Spectralon 150 mm integrating sphere with a photomultiplier (PMT) for UV and visible detection and an InGaAs diode for the near-infrared range was then used to obtain the diffuse reflectance spectra of the powders.

2.5. Optical Emission

In order to prepare the samples, the powders were pressed onto aluminum sample holders using glass slides to obtain a smooth surface. The slides were removed during the measurements. The emission spectra of the samples were examined with both an Edinburgh FS920 photoluminescence spectrometer using a 450 W Xe arc lamp and a double monochromator as excitation source and a liquid-nitrogen-cooled germanium detector for detection. Spectra were also measured using an InGaAs array spectrometer (AvaSpec-NIR512-1.7-HSC-EVO, Avantes, Apeldoorn, the Netherlands). In the latter setup, the samples were excited with an 808 nm diode laser.

2.6. Luminescence Lifetime Measurements

The powders were fixed with carbon tape on an aluminum sheet. The samples were then excited using a Nd:YAG laser-pumped optical parametric oscillator (OPO) tuned to a wavelength of 808 nm and with a pulse repetition rate of 10 Hz (Ekspla NT342, Ekspla, Vilnius, Lithuania). Upon excitation, the samples emitted light, which passed through a 1002 nm long pass filter to remove scattered light produced by the laser from the signal. Similarly, the light resulting from the $^4F_{3/2} \longrightarrow\ ^4I_{11/2}$ transition was separated from the rest with the use of filters such that only the luminescence lifetime of that transition was measured. Subsequently, the signal was transferred to an InGaAs amplified photodetector (PDA20C/M, Thorlabs, Newton, NJ, USA) employing an optical fiber (P400-2-VIS-NIR, Ocean Insight, Orlando, FL, USA). Then, the light intensity was transformed into a potential difference in a linear manner. Both the photodetector and trigger of the laser, which served as a start indicator, were connected to a USB oscilloscope (Picoscope 5244D, Pico Technology, St Neots, UK). Given the 3–5 ns pulse duration of the OPO laser and the bandwidth of

5 MHz of the photodetector (rise time of 70 ns), a response time below 100 ns was achieved, which is orders of magnitude lower than the decay time of the Nd^{3+} luminescence.

3. Results

Table 1 encompasses several properties of the researched hosts, namely $GdVO_4$, $GdPO_4$, $GdAlO_3$, Gd_2SiO_5 and GGG. Upon doping these materials with Nd^{3+}, neodymium ions were substituted on the Gd^{3+} sites, considering that their ionic radii match closely, as seen in Table 2. Since local symmetry and the environment of these sites vary among the hosts, the host–dopant interactions may differ as well, thus resulting in diverse spectra and luminescence lifetimes. In particular, it is important that the local site symmetry does not contain an inversion center, which causes the transitions to be forced electric dipole transitions instead of electric dipole-forbidden by the mixing of d and f orbitals [11]. Furthermore, the effect of concentration quenching is influenced by the distance between the Nd^{3+} ions and, as a consequence, the intersite distances in the different hosts.

Table 1. Structural properties of the hosts.

	$GdAlO_3$	Gd_2SiO_5	GGG	$GdPO_4$	$GdVO_4$
Melting point (°C) [12–16]	~2070	~1950	~1720	1899–1920	~1800
Space group [17]	Pnma	$P2_1/c$	$Ia\bar{3}d$	$P2_1/c$	$I4_1/amd$
Coordination number [19–23]	8	site A: 7 site B: 9	8	9	8
Local site symmetry [19–23]	C_s	site A: C_s site B: C_{3v}	D_2	C_1	D_{2d}
Distance between Gd^{3+} sites [Å] [17,24]	2 × 3.67 2 × 3.75 2 × 3.79	A { 1 × 3.51 2 × 3.57 1 × 3.73 B { 1 × 3.36 2 × 3.67 1 × 3.73	4 × 3.84 8 × 5.87 2 × 6.28	2 × 4.00 1 × 4.01 2 × 4.19	4 × 3.95 4 × 5.98 8 × 6.02

Table 2. Ionic radii of Gd^{3+} and Nd^{3+} with varying coordination numbers [25]

Coordination Number:	VI	VII	VIII	IX
Ionic radii of Gd^{3+} [Å]	0.938	1	1.053	1.107
Ionic radii of Nd^{3+} [Å]	0.983	no data	1.109	1.163

The relevant energy levels and transitions of Nd^{3+} are shown in Figure 1. Upon excitation by photons with a wavelength of around 808 nm, the system transitions from the ground level into an excited state: $^4I_{9/2} \longrightarrow {}^2H_{9/2}, {}^4F_{5/2}$. These excited states then decay non-radiatively into $^4F_{3/2}$. Afterwards, various radiative decay channels towards lower energy states, such as $^4I_{9/2}$, $^4I_{11/2}$ and $^4I_{13/2}$, are possible. These three transitions lead to the main emission features of Nd^{3+} in the NIR at around 900 nm, 1060 nm and 1350 nm, respectively. Furthermore, in the case of another nearby Nd^{3+} ion that is in the ground state, cross relaxation may occur as well, in which the excited ion decays to $^4I_{15/2}$ while exciting the nearby atom to $^4I_{15/2}$ [21]. Evidently, this phenomenon becomes more prominent with increasing doping levels and contributes to luminescence quenching, as well as migration-assisted decay.

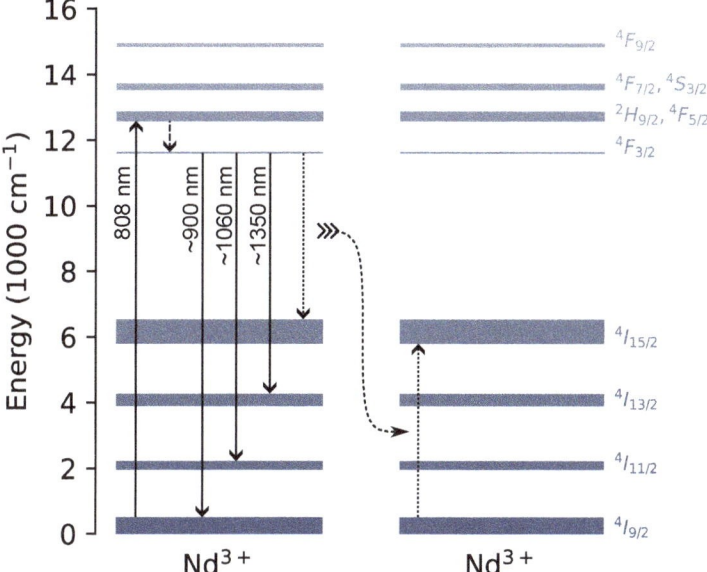

Figure 1. Partial energy diagram of Nd^{3+} with ground state absorption (upward solid arrow), non-radiative decay (dashed downward arrow), radiative decay (solid downward arrows) and cross relaxation (dotted arrows). Data were taken from [26].

3.1. X-ray Diffraction

$GdVO_4$, GGG and $GdPO_4$ were found to be phase-pure, with the intended stoichiometry and phase and without any traces of impurity phases, while some $GdAlO_3$ samples exhibited peaks that can be attributed to trace amounts of Gd_2O_3 and some Gd_2SiO_5 samples that also showed trace amounts of $Gd_{9.33}(SiO_4)_6O_2$. The XRD patterns did not show any appreciable peak broadening due to finite cystallite size or lattice strain, as seen in Figure 2.

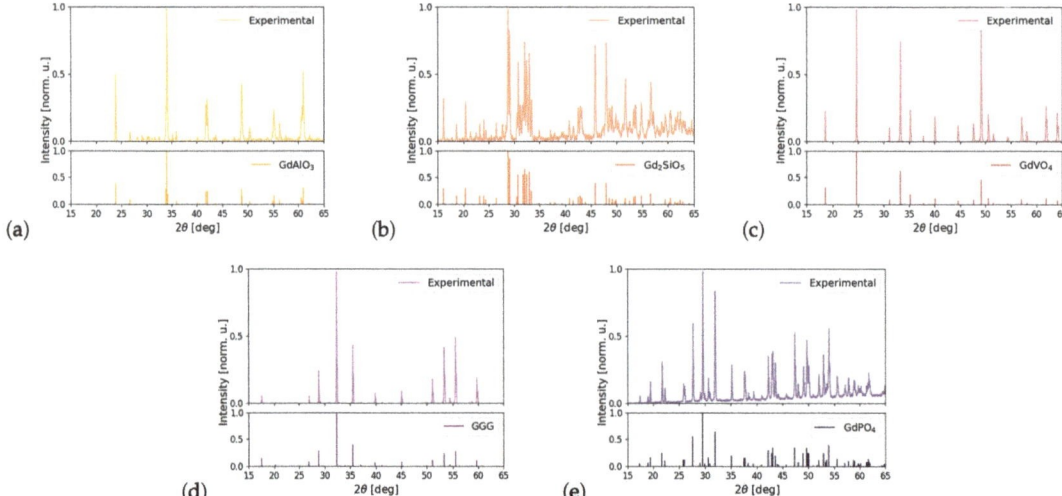

Figure 2. (a–e) Measured XRD patterns of GdAlO$_3$, Gd$_2$SiO$_5$, GdVO$_4$, GGG and GdPO$_4$ with reference patterns of 00-046-0395, 01-074-1795, 01-086-0996, 01-088-0574 and 00-032-0386, respectively.

3.2. Scanning Electron Microscopy

Figure 3 compares the morphology of doped and undoped GdPO$_4$. As expected, the morphology of the two samples is similar, since the ionic radii of Gd^{3+} and Nd^{3+} are comparable. Thus, substituting some gadolinium ions with neodymium ions has little affect on the structure.

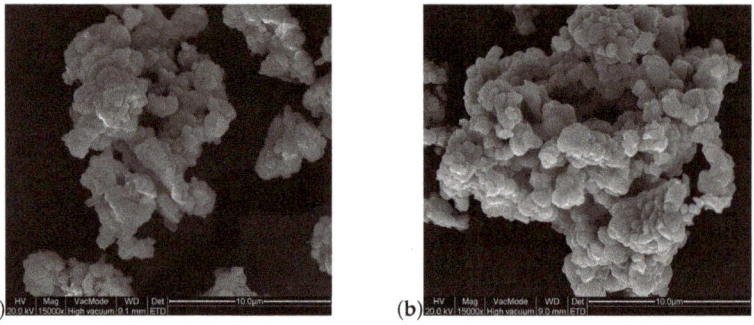

Figure 3. SEM images of (a) undoped GdPO$_4$ and (b) GdPO$_4$ doped with 2% Nd^{3+} observed with a magnification of 15,000×.

The SEM images shown in Figure 4 reveal strong agglomeration, as well as similar grain sizes of the order of µm across all samples, with the exception of GdPO$_4$. The latter exhibits smaller structures than the other samples, possibly due to the low heating temperature required for its synthesis.

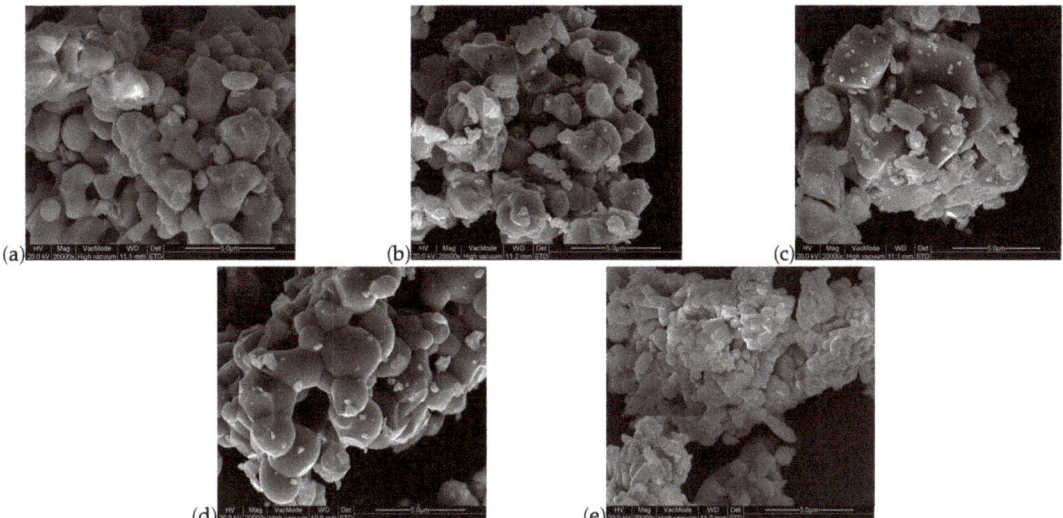

Figure 4. SEM images of (**a**) GdAlO$_3$, (**b**) Gd$_2$SiO$_5$, (**c**) GdVO$_4$, (**d**) GGG and (**e**) GdPO$_4$, all doped with 2% Nd^{3+} and observed with a magnification of 20,000×.

3.3. Optical Absorption

As can be seen in Figure 5, neodymium-doped GGG and GdVO$_4$ may be appropriately excited within the 730–760 nm and 790–820 nm regions. It has been suggested that excitation of Nd^{3+} by an 808 nm diode laser could be optimized by tuning its emission wavelength by heating or cooling of the laser [2]. Peculiarly, the absorption cross section of the transitions from $^4I_{9/2}$ towards $^4S_{3/2}$ and $^4F_{7/2}$, roughly around 740 nm, is relatively high in the case of GdPO$_4$ and especially in the case of GdAlO$_3$. These transitions lie within the biological windows as well. One should consider that even within the biological windows, the attenuation coefficient varies. For instance, light with a wavelength of 740 nm is less attenuated by oxygenated whole blood than 800 nm light. However, it is attenuated more by deoxygenated whole blood [5].

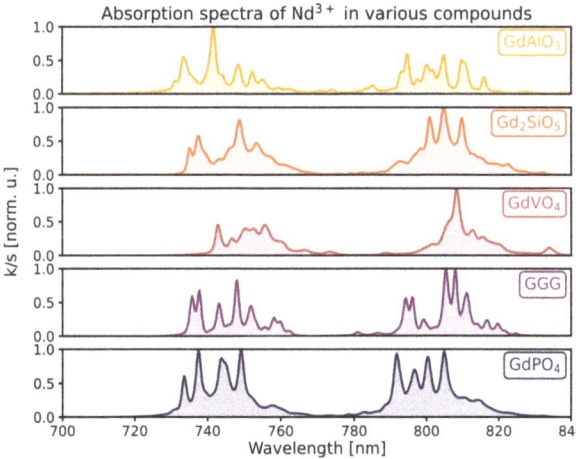

Figure 5. Absorption spectra of the samples doped with 2% neodymium in the region of interest as derived from diffuse reflection spectra.

The total integrated absorbance from the transitions seen in Figure 5 obtained by Kubelka–Munk transform is shown in Figure 6 for various doping levels. The trend is similar for all materials; as more neodymium ions occupy gadolinium sites within the host, the absorbance increases. The fact that the absorption increases linearly with dopant concentration for all hosts is a good indication that all Nd is properly incorporated into the host lattice without the occurrence of precipitation or secondary phases.

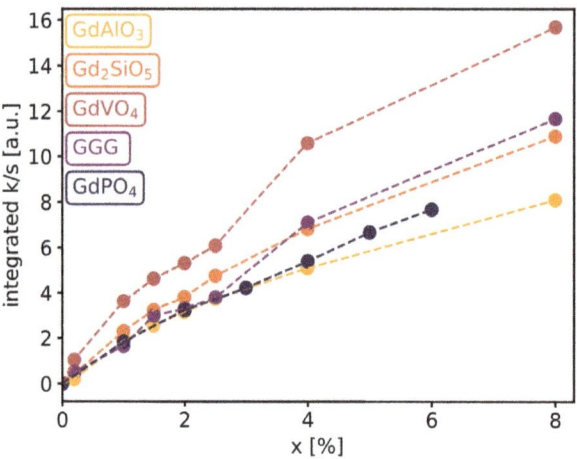

Figure 6. Integrated absorption of the different phosphors as a function of dopant concentration from 720 to 840 nm.

3.4. Optical Emission

Figure 7 shows the emission spectra of all samples doped with 2% neodymium. The three transitions discussed in Figure 1 that effectuate radiative decay ($^4F_{3/2} \longrightarrow {}^4I_{13/2}$, $^4I_{11/2}$ and $^4I_{9/2}$) can be observed among all samples. The branching ratio defines the ratio of the emission intensities of the three main emission peaks.. It is evident that this ratio is most favorable for the 1060 nm peak, corresponding to $^4F_{3/2} \longrightarrow {}^4I_{11/2}$, which is responsible for more than 63% of the emitted intensity in all samples. Remarkably, Nd^{3+} in $GdAlO_3$ features a maximum shift towards longer wavelengths in contrast to other samples, at 1075 nm. Advantageously, the absorption spectra of many of the most prominent constituents of biological tissue reveal a dip at roughly 1100 nm, and the attenuation decreases from 1060 nm to larger wavelengths until the minimum [5]. As expected from the energy level structure of Gd^{3+}, we did not observe any absorption or emission features related to gadolinium [27]. It is also anticipated that the spectra are similar, as all transitions are between 4f orbitals, which are shielded well from the crystal field by other orbitals. Nonetheless, the local symmetry of the site determines the further splitting of the energy levels. Emission spectra of samples with hosts that have low-symmetry Gd^{3+} sites appear broader due to the additional transitions. Furthermore, the neodymium ions may be located on two sites with different symmetry in the case of Gd_2SiO_5.

However, as seen in Figure 8, showing the integrated emission intensity as a function of dopant concentration, Nd^{3+} in $GdAlO_3$ is, at most, only half as intense as the brightest sample, which is $GdVO_4$, with a doping level of around 1.5–2.5%; followed by GGG, with 2% Nd^{3+}; Gd_2SiO_5, with 1.5% Nd^{3+}; $GdAlO_3$, with 1% Nd^{3+}; and, lastly, $GdPO_4$, with 2% Nd^{3+}. These samples exhibit maxima, as an increase in ions results in more available luminescent centers but also amplifies the effect of concentration quenching due to cross relaxation and migration-assisted decay. The maxima of $GdAlO_3$ and Gd_2SiO_5 appear at lower doping concentrations, which is in line with the distances between Gd^{3+} sites, as seen in Table 1. Shorter distances between neodymium ions facilitate the transfer of excitation energy from one ion to another, resulting in stronger concentration quenching.

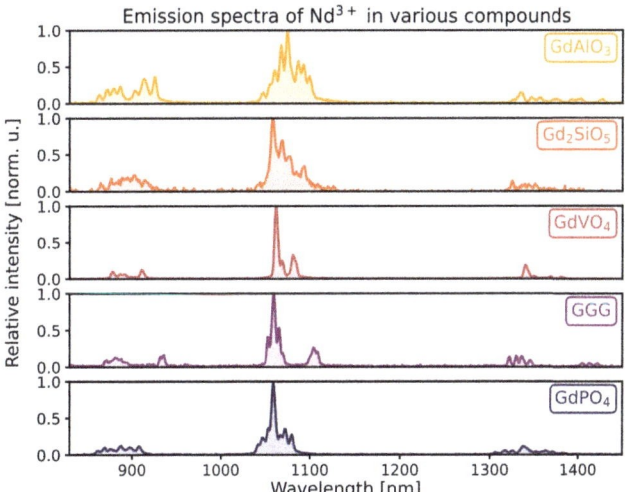

Figure 7. Emission spectra of the samples consisting of various hosts doped with 2% Nd^{3+}.

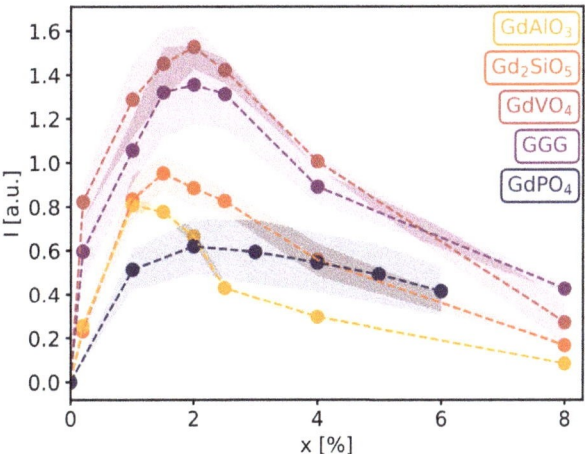

Figure 8. Integrated emission intensity from 1000 to 1200 nm for all samples shown as a function of the doping concentration, with shaded areas indicating the standard error.

3.5. Luminescence Decay

The luminescence lifetime (τ) is an optical property of great significance, especially when considering the application of time-gated imaging. The cross relaxation shown in Figure 1 is a contributor to the concentration quenching of luminescence in neodymium-doped compounds, wherein an ion with an excited state of $^4F_{3/2}$ partially transfers its energy to a nearby neodymium ion originally in the ground state. Not only does it result in a reduction in emission intensity beyond a certain concentration of Nd^{3+} ions; it also decreases the luminescence lifetimes. At low doping concentrations, this effect on the decay profiles is well described by [28]:

$$I(t) = I(0)e^{-\frac{t}{\tau_0} - \Gamma(1-\frac{3}{s})\frac{c}{c_0}\left(\frac{t}{\tau_0}\right)^{\frac{3}{s}}}, \tag{2}$$

where τ_0 is the lifetime for radiative decay in the absence of other nearby Nd^{3+} ions; $s = 6, 8$ and 10 if the interaction between the ions is dipole–dipole, dipole–quadrupole and quadrupole–quadrupole, respectively; c is the dopant concentration; c_0 a parameter called the critical transfer concentration, defined as $\frac{3}{4\pi R_0^3}$; and R_0 is the distance between two ions for which the radiative decay rate is equal to the energy transfer rate. Lastly, Γ is the gamma function. In the case of dipole–dipole interactions, Equation (2) simplifies to:

$$I(t) = I(0)e^{-\frac{t}{\tau_0} - \sqrt{\pi}\frac{c}{c_0}\sqrt{\frac{t}{\tau_0}}}. \tag{3}$$

However, at higher concentrations, energy migration effects are inevitable and should be taken into account as well. The excitation energy of a neodymium ion with an energy level of $^4F_{3/2}$ may migrate to another nearby neodymium ion, resulting in that ion being excited to $^4F_{3/2}$ [29]. This process may repeat itself, resulting in migration-assisted decay. The rate of migration-assisted decay (W) has been described using various method, such as a hopping model or a diffusion model under different assumptions. What most of the results of these methods have in common is a rate that is proportional to the square of the ion concentration in the case of self-quenching, which has also been experimentally observed [30,31]. In the dipole–dipole approximation, the hopping model yields:

$$W = \pi\left(\frac{2\pi}{3}\right)^{5/2} c^2 \sqrt{C_{da}C_{dd}}, \tag{4}$$

where C_{da} is a microparameter of the cross relaxation and C_{dd} is the microparameter of the donor–donor interaction, which corresponds to migration.

Taking both mechanisms into account, the luminescence decay of most neodymium-doped materials in the dipole–dipole approximation is then described by equation [29,30,32]:

$$I(t) = I(0)e^{-\frac{t}{\tau_0} - \gamma\sqrt{t} - Wt}, \tag{5}$$

where both γ and W are macroparameters. γ depends linearly on the ion concentration, while W is proportional to the square of the ion concentration [29–33]. Therefore, it is possible to write both as:

$$\begin{cases} \gamma = \frac{c}{c_0}\sqrt{\frac{\pi}{\tau_0}} = c\gamma', \\ W = \pi\left(\frac{2\pi}{3}\right)^{5/2} c^2 \sqrt{C_{da}C_{dd}} = c^2 W', \end{cases} \tag{6}$$

allowing a sequence of decay profiles derived from samples prepared at a multitude of doping levels to be fit with only three fixed parameters: τ_0, γ' and W' (in addition to the amplitude ($I(0)$) of each sample). The mean lifetime (τ_m) can be generally defined according to Equation (7) [28]. By using this definition in conjunction with Equation (5), the mean lifetime can be presented as a function of the doping concentration.

$$\tau_m = \frac{\int_0^\infty t I(t) dt}{\int_0^\infty I(t) dt} \tag{7}$$

The decay profiles are shown in Figure 9. Equation (5) was fit to the ensemble of decay profiles for all dopant concentrations simultaneously by minimizing the sum of all the squared errors of each fit using the Levenberg–Marquardt algorithm, while keeping τ_0, γ' and W' from Equation (6) fixed across every profile and c equal to the nominal neodymium concentration of the synthesis.

It can be seen that this approach properly describes the luminescence decay of a multitude of samples with a limited number of parameters. Small deviations from the experimental data can be attributed to slight deviations of the actual neodymium concentration from the nominal concentration. This process was not conducted for Gd_2SiO_5, as it has two non-equivalent gadolinium sites in which the Nd^{3+} may be located, as seen in Table 1. This may also contribute to the non-exponential trend of its luminescence decay, since there are two different environments in which the neodymium ion could reside. Instead, a biexponential fit was performed on each Gd_2SiO_5 sample separately.

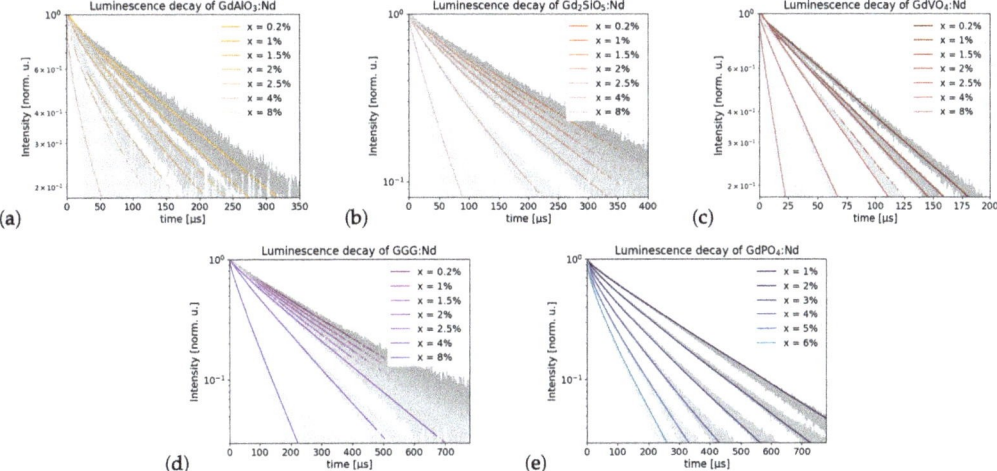

Figure 9. (a–e) Luminescence decay of $GdVO_4$, Gd_2SiO_5, $GdAlO_3$, GGG and $GdPO_4$ in semi-log scale respectively, each fit using Equation (5), except (**b**), which was fit using Equation (8).

As the used fitting parameters are independent of the dopant concentration, three parameters can be used to construct luminescence decay curves at various doping levels. From these decay curves, the mean lifetime (τ_m) can be derived using Equation (7), where the numerator and denominator were numerically calculated. This allows for a continuous visualization of the mean lifetime as a function of the doping level, as seen in Figure 10. To corroborate the validity of this approach, the biexponential function below (8) was fit on each sample separately, as denoted by dots in Figure 10:

$$I(t) = A_1 e^{-\frac{t}{\tau_1}} + A_2 e^{-\frac{t}{\tau_2}}, \tag{8}$$

with A_1, A_2, τ_1 and τ_2 as parameters. While this results in a lot more parameters with little physical meaning, it does describe the individual decay profiles well. The mean lifetime can once again be calculated using Equation (7). τ_m as a function of doping concentration may be observed in Figure 10, the lifetimes obtained by fitting a biexponential function separately correspond closely to the lifetimes obtained through Equation (5). A comparison with other works was conducted in the case of $GdVO_4$, as shown in Table 3.

Table 3. Luminescent lifetime of Nd^{3+} in $GdVO_4$.

Doping Level (%)	This Work (µs)	Other Work (µs)	Reference
0.5	104	107, 88	[4,34]
0.9	98	97	[32]
1	96	95, 88, 84	[4,35]
1.2	92	90	[36]
2	75	81–63, 44	[4,37]
5	31	34, 26	[4,37]
10	10	9	[4]

From Figure 10, it is also clear that there is a substantial difference in the luminescence lifetime of neodymium doped in a variety of hosts. For instance, the τ_0 of neodymium-doped GGG and GdPO$_4$ is nearly three times that of GdVO$_4$. The mean lifetimes of Nd^{3+} in GdAlO$_3$ and Gd$_2$SiO$_5$ are also similar, with the former having a τ_0 of 200 µs.

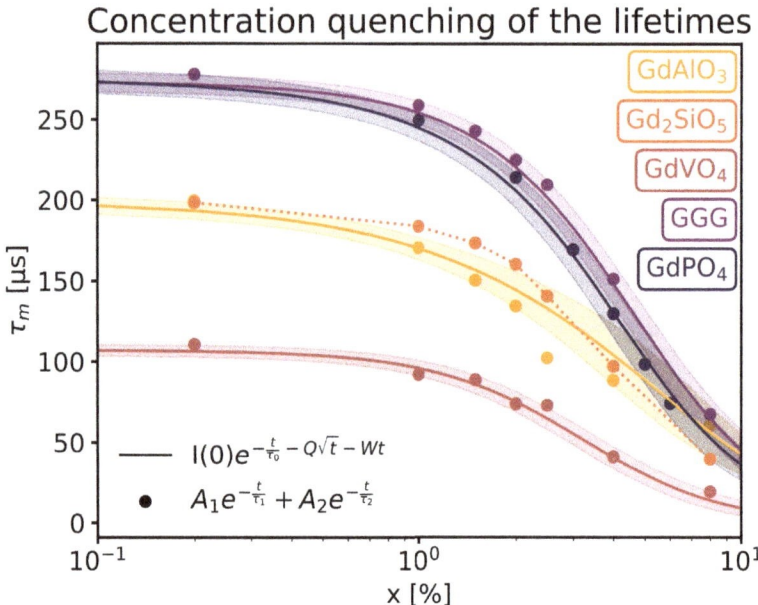

Figure 10. Mean luminescence lifetime as a function of the doping concentration on a semi-log scale, with shaded areas indicating standard error resulting from the errors in the fitting parameters propagated to the lifetimes.

4. Discussion

A total of 34 samples with 5 different hosts and varying doping levels were synthesized using solid-state reactions, as shown in Figure 11, which also presents brightness in the NIR.

As seen in Figure 10, the lifetimes vary substantially among the samples, ranging from 276 to 14 µs. Autofluorescence of biological tissues in the near-infrared range is a common source of background signals in bioimaging. Since this autofluorescence has a much shorter decay time—between 0.1 ns and 7 ns [38]—than that of the Nd-doped particles, it is easy to separate the two signals using time-gated imaging. In this study, the autofluorescence signal was suppressed using pulsed excitation, incorporating a delay of at least 10 ns before measuring the signal from the luminescent particles.

The emission intensity is another significant factor. Figure 8 reveals GdVO$_4$ doped with 1.5–2.5% Nd^{3+} to be the brightest sample. The spectra in hosts with low symmetry gadolinium sites appear broader due to the low symmetry, in the case of Gd$_2$SiO$_5$ further broadening is caused by a second inequivalent site in which the neodymium ions may be located. The width of the peaks separately is a result of the limited resolution of the detectors. This work indicates that the different Nd-doped Gd-based compounds show excellent optical characteristics and are potentially suitable for bioimaging. Ultimately, the choice between the different hosts for bioimaging will come down to the ease of preparation of high-performance nanoparticles with good biocompatibility and a narrow size distribution.

Figure 11. All samples in an acrylic sample holder. Each row corresponds to the host denoted on the right, and the doping concentration (x, in %) is denoted by numbers. Host names comprise the sample with 2% Nd^{3+}, and "LumiLab" consists of $GdVO_4$ doped with 2% Nd^{3+}. (**a**) Sample in visible light and (**b**) illuminated by ambient light photographed in the NIR region with a Xeva 1.7 320 TE3 USB 100 InGaAs camera.

Author Contributions: Conceptualization, D.P.; methodology, all authors; software, validation, formal analysis and investigation, M.D., S.V.B. and W.M.; original draft preparation, M.D. and D.P.; review and editing, all authors; supervision, project administration and funding acquisition, D.P. All authors have read and agreed to the published version of the manuscript.

Funding: This research was funded by FWO (Fund for Scientific Research-Flanders) projects I002418N and G025322N.

Data Availability Statement: Data available upon reasonable request from the authors.

Acknowledgments: Olivier Janssens is acknowledged for performing XRD measurements and providing SEM images.

Conflicts of Interest: The authors declare no conflict of interest.

References

1. Schneider, A.; Feussner, H. *Biomedical Engineering in Gastrointestinal Surgery*; Rutgers University Press: New Brunswick, NJ, USA, 2017.
2. del Rosal, B.; Perez, A.; Misiak, M.; Bednarkiewicz, A.; Vanetsev, A.; Orlovskii, Y.; Jovanovic, D.; Dramicanin, M.; Rocha, U.; Kumar, K.; et al. Neodymium-doped nanoparticles for infrared fluorescence bioimaging: The role of the host. *J. Appl. Phys.* **2015**, *118*, 143104. [CrossRef]
3. Carbonati, T.; Cionti, C.; Cosaert, E.; Nimmegeers, B.; Meroni, D.; Poelman, D. NIR emitting $GdVO_4$:Nd nanoparticles for bioimaging: The role of the synthetic pathway. *J. Alloys Compd.* **2021**, *862*, 158413. [CrossRef]
4. Nimmegeers, B.; Cosaert, E.; Carbonati, T.; Meroni, D.; Poelman, D. Synthesis and Characterization of $GdVO_4$:Nd Near-Infrared Phosphors for Optical Time-Gated In Vivo Imaging. *Materials* **2020**, *13*, 3564. [CrossRef] [PubMed]
5. Smith, A.; Mancini, M.; Nie, S. Bioimaging: Second window for in vivo imaging. *Nat. Nanotechnol.* **2009**, *4*, 710–711. [CrossRef] [PubMed]

6. Del Rosal, B.; Ortgies, D.H.; Fernández, N.; Sanz-Rodríguez, F.; Jaque, D.; Rodríguez, E.M. Overcoming Autofluorescence: Long-Lifetime Infrared Nanoparticles for Time-Gated In Vivo Imaging. *Adv. Mater.* **2016**, *28*, 10188–10193. [CrossRef] [PubMed]
7. Hoshyar, N.; Gray, S.; Han, H.; Bao, G. The effect of nanoparticle size on in vivo pharmacokinetics and cellular interaction. *Nanomedicine* **2016**, *11*, 673–92. [CrossRef] [PubMed]
8. Zhang, X.; Chen, Y.; Zhou, L.; Pang, Q.; Gong, M.L. Synthesis of a Broad-Band Excited and Multicolor Tunable Phosphor Gd_2SiO_5:Ce^{3+},Tb^{3+},Eu^{3+} for Near-Ultraviolet Light-Emitting Diodes. *Ind. Eng. Chem. Res.* **2014**, *53*, 6694–6698. [CrossRef]
9. Cao, C.; Yang, H.; Moon, B.; Choi, B.; Jeong, J. Host Sensitized White Luminescence of Dy^{3+} Activated $GdPO_4$ Phosphors. *J. Electrochem. Soc.* **2011**, *158*, J6. [CrossRef]
10. Kubelka, P.; Munk, F. An article on optics of paint layers. *Tech. Phys.* **1931**, *53*, 259–274.
11. Tanner, P. *Lanthanide Luminescence in Solids*; Springer: Berlin/Heidelberg, Germany, 2010; Volume 7, pp. 183–233. [CrossRef]
12. Mazelsky, R.; Kramer, W.; Hopkins, R. Crystal growth of $GdAlO_3$. *J. Cryst. Growth* **1968**, *2*, 209–214. [CrossRef]
13. Utsu, T.; Akiyama, S. Growth and applications of Gd_2SiO_5: Ce scintillators. *J. Cryst. Growth* **1991**, *109*, 385–391. [CrossRef]
14. Garem, H.F.E.; Rabier, J.; Veyssiére, P. Slip systems in gadolinium gallium garnet single crystals. *J. Mater. Sci.* **1982**, *17*, 878–884. [CrossRef]
15. Lessing, P.A.; Erickson, A.W. Synthesis and characterization of gadolinium phosphate neutron absorber. *J. Eur. Ceram. Soc.* **2003**, *23*, 3049–3057. [CrossRef]
16. Antić, Ž.; Dramićanin, M.D.; Prashanthi, K.; Jovanović, D.; Kuzman, S.; Thundat, T. Pulsed Laser Deposited Dysprosium-Doped Gadolinium–Vanadate Thin Films for Noncontact, Self-Referencing Luminescence Thermometry. *Adv. Mater.* **2016**, *28*, 7745–7752. [CrossRef] [PubMed]
17. Jain, A.; Ong, S.P.; Hautier, G.; Chen, W.; Richards, W.D.; Dacek, S.; Cholia, S.; Gunter, D.; Skinner, D.; Ceder, G.; et al. Commentary: The Materials Project: A materials genome approach to accelerating materials innovation. *APL Mater.* **2013**, *1*, 011002. [CrossRef]
18. Momma, K.; Izumi, F. VESTA3 for three-dimensional visualization of crystal, volumetric and morphology data. *J. Appl. Crystallogr.* **2011**, *44*, 1272–1276. [CrossRef]
19. Bagnato, V.; Nunes, L.; Zilio, S.; Scheel, H.; Castro, J. Infrared electronic transitions of Eu^{+3} in $GdAlO_3$. *Solid State Commun.* **1984**, *49*, 27–30. [CrossRef]
20. De Camargo, A.S.S.; Davolos, M.R.; Nunes, L.A.O. Spectroscopic characteristics of Er^{3+} in the two crystallographic sites of Gd_2SiO_5. *J. Phys. Condens. Matter* **2002**, *14*, 3353. [CrossRef]
21. Monteseguro, V.; Rathaiah, M.; Linganna, K.; Lozano-Gorrín, A.; Hernández-Rodríguez, M.; Martin, I.; Babu, P.; Rodríguez-Mendoza, U.; Manjon, F.J.; Munoz, A.; et al. Chemical pressure effects on the spectroscopic properties of Nd^{3+}-doped gallium nano-garnets. *Opt. Mater. Express* **2015**, *8*, 1661–1673. [CrossRef]
22. Huittinen, N.; Arinicheva, Y.; Schmidt, M.; Neumeier, S.; Stumpf, T. Using Eu^{3+} as an atomic probe to investigate the local environment in $LaPO_4$–$GdPO_4$ monazite end-members. *J. Colloid Interface Sci.* **2016**, *483*, 139–145. [CrossRef]
23. Gavrilovic, T.; Marinkovic, D.; Lojpur, V.; Dramicanin, M. Multifunctional Eu- and Er/Yb-doped $GdVO_4$ nanoparticles synthesized by reverse micelle method. *Sci. Rep.* **2014**, *4*, 4209. [CrossRef] [PubMed]
24. Larsen, A.; Mortensen, J.; Blomqvist, J.; Castelli, I.; Christensen, R.; Dulak, M.; Friis, J.; Groves, M.; Hammer, B.; Hargus, C.; et al. The Atomic Simulation Environment—A Python library for working with atoms. *J. Phys. Condens. Matter* **2017**, *29*, 273002. [CrossRef]
25. Shannon, R.D. Revised effective ionic radii and systematic studies of interatomic distances in halides and chalcogenides. *Acta Crystallogr. Sect. A* **1976**, *32*, 751–767. [CrossRef]
26. Carnall, W.T.; Goodman, G.L.; Rajnak, K.; Rana, R.S. A systematic analysis of the spectra of the lanthanides doped into single crystal LaF_3. *J. Chem. Phys.* **1989**, *90*, 3443–3457. [CrossRef]
27. Wegh, R.T.; Donker, H.; Meijerink, A.; Lamminmäki, R.J.; Hölsä, J. Vacuum-ultraviolet spectroscopy and quantum cutting for Gd^{3+} in $LiYF_4$. *Phys. Rev. B* **1997**, *56*, 13841–13848. [CrossRef]
28. Inokuti, M.; Hirayama, F. Influence of Energy Transfer by the Exchange Mechanism on Donor Luminescence. *J. Chem. Phys.* **2004**, *43*, 1978–1989. [CrossRef]
29. Caird, J.A.; Ramponi, A.J.; Staver, P.R. Quantum efficiency and excited-state relaxation dynamics in neodymium-doped phosphate laser glasses. *J. Opt. Soc. Am. B* **1991**, *8*, 1391–1403. [CrossRef]
30. Voronko, I.; Mammadov, T.; Osiko, V.; Prokhorov, A.; Sakun, V.; Shcherbakov, I. Investigation of the nature of nonradiative relaxation of energy of excitation in condensed media with high activator concentration. *Zh. Eksp. Teor. Fiz.* **1976**, *71*, 478–496.
31. Burshtein, A. Concentration self-quenching. *Zh. Eksp. Teor. Fiz.* **1983**, *84*, 2001–2013.
32. Ostroumov, V.; Jensen, T.; Meyn, J.P.; Huber, G.; Noginov, M.A. Study of luminescence concentration quenching and energy transfer upconversion in Nd-doped $LaSc_3(BO_3)_4$ and $GdVO_4$ laser crystals. *J. Opt. Soc. Am. B* **1998**, *15*, 1052–1060. [CrossRef]
33. Tkachuk, A.; Ivanova, S.; Joubert, M.F.; Guyot, Y.; Guy, S. Luminescence self-quenching from $^4F_{3/2}$, $^2P_{3/2}$ and $^4D_{3/2}$ neodymium levels in double sodium–yttrium fluoride crystals. *J. Lumin.* **2001**, *94–95*, 343–347. [CrossRef]
34. Xia, H.R.; Jiang, H.D.; Zheng, W.Q.; Lu, G.W.; Meng, X.L.; Zhang, H.J.; Liu, X.S.; Zhu, L.; Wang, J.Y. Optical parameters and luminescent properties of Nd:$GdVO_4$ crystals. *J. Appl. Phys.* **2001**, *90*, 4433–4436. [CrossRef]
35. CASTECH Web Site. Available online: http://www.castech-us.com/ndgdvo4.htm (accessed on 20 June 2023).
36. Jensen, T.; Ostroumov, V.; Meyn, J.P.; Huber, G.; Zagumennyi, A.; Shcherbakov, I. Spectroscopic characterization and laser performance of diode-laser-pumped Nd: $GdVO_4$. *Appl. Phys. B* **1994**, *58*, 373–379. [CrossRef]

37. Ogawa, T.; Urata, Y.; Wada, S.; Onodera, K.; Machida, H.; Sagae, H.; Higuchi, M.; Kodaira, K. Efficient laser performance of N:dGdVO$_4$ crystals grown by the floating zone method. *Opt. Lett.* **2003**, *28*, 2333–2335. [CrossRef]
38. Berezin, M.Y.; Achilefu, S. Fluorescence Lifetime Measurements and Biological Imaging. *Chem. Rev.* **2010**, *110*, 2641–2684. [CrossRef] [PubMed]

Disclaimer/Publisher's Note: The statements, opinions and data contained in all publications are solely those of the individual author(s) and contributor(s) and not of MDPI and/or the editor(s). MDPI and/or the editor(s) disclaim responsibility for any injury to people or property resulting from any ideas, methods, instructions or products referred to in the content.

Article

ZGSO Spinel Nanoparticles with Dual Emission of NIR Persistent Luminescence for Anti-Counterfeiting Applications

Guanyu Cai [1,2], Teresa Delgado [1], Cyrille Richard [2,*] and Bruno Viana [1,*]

1 Université PSL, Chimie ParisTech, CNRS, IRCP, Institut de Recherche de Chimie Paris, 75005 Paris, France
2 Université Paris Cité, CNRS, INSERM, UTCBS, Unité de Technologies Chimiques et Biologiques pour la Santé, Faculté de Pharmacie, 75006 Paris, France
* Correspondence: cyrille.richard@u-paris.fr (C.R.); bruno.viana@chimieparistech.psl.eu (B.V.)

Abstract: The property of persistent luminescence shows great potential for anti-counterfeiting technology and imaging by taking advantage of a background-free signal. Current anti-counterfeiting technologies face the challenge of low security and the inconvenience of being limited to visible light emission, as emitters in the NIR optical windows are required for such applications. Here, we report the preparation of a series of $Zn_{1+x}Ga_{2-2x}Sn_xO_4$ nanoparticles (ZGSO NPs) with persistent luminescence in the first and second near-infrared window to overcome these challenges. ZGSO NPs, doped with transition-metal (Cr^{3+} and/or Ni^{2+}) and in some cases co-doped with rare-earth (Er^{3+}) ions, were successfully prepared using an improved solid-state method with a subsequent milling process to reach sub-200 nm size particles. X-ray diffraction and absorption spectroscopy were used for the analysis of the structure and local crystal field around the dopant ions at different Sn^{4+}/Ga^{3+} ratios. The size of the NPs was ~150 nm, measured by DLS. Doped ZGSO NPs exhibited intense photoluminescence in the range from red, NIR-I to NIR-II, and even NIR-III, under UV radiation, and showed persistent luminescence at 700 nm (NIR-I) and 1300 nm (NIR-II) after excitation removal. Hence, these NPs were evaluated for multi-level anti-counterfeiting technology.

Keywords: spinel; photoluminescence; persistent luminescence; nanoparticles; anti-counterfeiting; NIR

Citation: Cai, G.; Delgado, T.; Richard, C.; Viana, B. ZGSO Spinel Nanoparticles with Dual Emission of NIR Persistent Luminescence for Anti-Counterfeiting Applications. *Materials* 2023, 16, 1132. https://doi.org/10.3390/ma16031132

Academic Editor: Toma Stoica

Received: 19 December 2022
Revised: 13 January 2023
Accepted: 17 January 2023
Published: 28 January 2023

Copyright: © 2023 by the authors. Licensee MDPI, Basel, Switzerland. This article is an open access article distributed under the terms and conditions of the Creative Commons Attribution (CC BY) license (https://creativecommons.org/licenses/by/4.0/).

1. Introduction

Photoluminescence has become an increasingly important imaging technology, with popular applications in many fields including lasers, sensors, lighting, and bio-probe imaging [1–3]. Anti-counterfeiting also employs photoluminescent materials due to their high throughput, low cost, flexibility, and stability [4,5]. Current optical-based anti-counterfeiting technologies are mainly based on down-conversion or up-conversion luminescence under excitation [5–12]. However, real-time fluorescence of lanthanides for anti-counterfeiting purposes is susceptible to producing interference with the background and excitation light [13]. In this context, the development of a background-free anti-counterfeiting technology is highly desirable [14,15]. To avoid background noise, near-infrared (NIR) or short-wavelength infrared (SWIR) luminescence imaging has begun to be developed as a result of InGaAs-based NIR detectors that can be employed not only for anti-counterfeiting applications but also to provide images with better clarity and spatial resolution [16–20].

The optical window in NIR (or SWIR) ranges can be artificially divided into deep-red NIR-I (700–950 nm), for which a Si-based detector could be used, and NIR-II (1000–1400 nm) and NIR-III (1500–1700 nm) subregions, for which an InGaAs-based detector is required [21–23]. However, current anti-counterfeiting technologies face the challenge of low security and the inconvenience of being limited to visible light emission as emitters in the second and even third optical windows are required [24,25]. Ni^{2+} and Er^{3+} ions, for example, could play an excellent role as emitters due to their luminescence in NIR-II and NIR-III [24–28]. Indeed, previous studies in the field have proposed the use of RE^{3+} ions in Sn-free ZGO NPs as sensitizers to enhance the NIR-I persistent luminescence of Cr^{3+} at 700 nm [29–33]. We have reported NIR-I persistent luminescence of Cr^{3+} ions in bulk ZGSO ceramics corresponding to the $^2E \rightarrow {}^4A_2$ (^4F) transition [34], and Qiu et al. were able to tune the optical properties of ZGSO by precisely controlling the crystal field by use of Sn in these bulk materials [35].

In this study, we report the synthesis of a series of nanomaterials based on an Ni^{2+}-doped $Zn_{1+x}Ga_{2-2x}Sn_xO_4$ (ZGSO) matrix with varying amounts of Sn (from x = 0 to x = 0.6). After selecting the most favorable composition in terms of duration and intensity of persistent luminescence, the effect of co-dopant Cr^{3+} and/or rare-earth (Er^{3+}) ions was evaluated, in order to provide tunable multi-emission from deep red to NIR-III. At the nanoscale, these materials (ZGSO) can be highly stable in an aqueous medium for several days, which allows their use in anti-counterfeiting, as the patterns can be easily machine printed or hand painted with the aid of an aqueous solvent. By introducing a "multi-level" anti-counterfeiting method based on ZGSO persistent luminescence NPs with emission in multiple ranges, we could effectively cover the deep-red to NIR-III ranges. New anti-counterfeiting systems with persistent luminescence nanoparticles could be developed with various duration times. The detection tools being employed are Si-based cameras/detectors for detection in deep-red to NIR-I ranges and InGaAs-based cameras/detectors for NIR-II and III ranges. Indeed, ZGSO NPs not only expand the area of anti-counterfeiting technology but also provide higher security compared to traditional phosphors emitting visible light and detectable with the naked eye.

2. Materials and Methods

2.1. Materials

Materials were ZnO (99.99%, Strem Chemicals,), Ga_2O_3 (99.99%, Strem Chemicals), SnO_2 (99.99%, Strem Chemicals), and dopant element sources were NiO (99.99% Aldrich Chemistry, St. Louis, MO, USA), Er_2O_3 (99%, Alfa Aesar, Haverhill, MA, USA), and Cr_2O_3 (99.99%, Aldrich Chemistry). All chemicals were used as received without further purification.

2.2. Synthesis of Persistent Luminescence Phosphors

The powder samples were synthesized by a high-temperature solid-state method starting from binary oxides. The raw material mixture was prepared according to the stoichiometric ratio of the compounds. For the example of ZGSO-3:Ni^{2+} ($Zn_{1.3}Ga_{1.39}Ni_{0.01}Sn_{0.3}O_{3.995}$), 1.3 mmol of ZnO, 1.3 mmol of Ga_2O_3, 0.3 mmol of SnO_2, and 0.1 mmol of NiO were weighed and ground homogenously in an agate mortar, and the mixtures were introduced into alumina crucibles. The alumina crucible was subsequently placed in a high-temperature furnace at 1300 °C for 6 h in the air to produce the final sample [36]. After cooling to room temperature, the phosphors were ground to fine powders. The other samples in this paper were prepared using the same method.

2.3. Recovery of Persistent Luminescence NPs

The ZGSO powders (~500 mg) were crushed using a Pulverisette 7 Fritsch Planetary Ball Mills, with dropped 1 mL of 5 mM HCl solution, at a speed of 1000 rpm for 2–4 h, to reduce their size. The different materials were then transferred into a flask and stirred vigorously for 24 h at room temperature. The final ZGSO NPs were obtained from polydisperse colloidal mixtures by centrifugation with a SANYO MSE Mistral 1000 at 3500 rpm for 5 min. ZGSO NPs were recovered from the supernatant with a size of ~150 nm measured by Dynamic Light Scattering (DLS), as shown in Figure S8. The supernatants were gathered and concentrated to a final ~5 mg/mL suspension.

2.4. Characterization

2.4.1. X-ray Diffraction (XRD)

X-ray diffraction (XRD) patterns of the materials showed crystalline phases with cubic spinel structures. XRD was performed with an X-ray diffractometer (XPert PRO,

PANalytical, Malvern Panalytical Ltd., Malvern, UK) equipped with a Ge111 single-crystal monochromator and by selecting the $K_{\alpha 1}$ radiation wavelength of the Cu X-ray tube (0.15405 nm).

2.4.2. Dynamic Light Scattering (DLS)

The hydrodynamic diameter of the ZGSO NPs was characterized by dynamic light scattering (DLS) performed with a Zetasizer Nano ZS (Malvern Instruments, Southborough, MA, USA) equipped with a 632.8 nm helium–neon laser and 5 mW power, with a detection angle of 173° (non-invasive backscattering).

2.4.3. Absorption Spectroscopy

Absorption measurements of the dry ZGSO NPs were carried out in a UV/Vis/NIR spectrophotometer (Varian Cary 6000 i, Agilent). The resolution was 0.1 nm for the Cr bands.

2.4.4. Photoluminescence

NIR photoluminescence (PL) measurements of dry ZGSO NPs were performed with an NIR camera (PyLoNIR, Princeton Instruments, Trenton, NJ, USA) for the NIR-II range cooled at −100 °C and coupled to a monochromator (Acton Spectra Pro, Princeton Instruments), with 300 grooves per mm and centered at 1200 nm.

Visible and deep-red (or NIR-I) photoluminescence (PL) measurements of dry ZGSO NPs were performed using a CCD camera (Roper Pixis 100, Princeton Instruments) cooled at −65 °C and coupled to a monochromator, with 300 grooves per mm and centered at 500 nm.

2.4.5. Persistent Luminescence

The dry ZGSO NPs samples were thermally detrapped before each experiment and then kept in the dark. The samples were loaded with a 365 nm lamp for 5 min at 290 K, and after excitation removal, the afterglow was recorded for 15 min at the same temperature. The signal was followed with the same camera as in the PL experiment. Afterglow curves were obtained by integrating the intensity of the persistent luminescence spectra as a function of time.

2.4.6. Anti-Counterfeiting Applications

Anti-counterfeiting patterns were hand painted using a ZGSO NP suspension (10 mg of NPs into 1 mL of water), dropped onto paper to form anti-counterfeiting marks, and then dried at room temperature (RT) for 24 h. Using this method, the "UTCBS" acronym was marked using ZGSO-3:Ni^{2+}, Er^{3+} NPs. The other "MPOE" acronym was marked using ZGSO-3:Ni^{2+}, Er^{3+}, Cr^{3+} NPs. Thus, the tests became fluorescent signals able to provide an effective property of persistent luminescence. These could be employed in anti-counterfeiting applications by creating persistent luminescence images.

Imaging in the NIR-I range was recorded with a photon-counting device (Optima, Biospace Lab, Nesles-la-Vallée, France). The persistent luminescence images of the various figures were recorded for 5 min after removal of the 365 nm lamp excitation.

Imaging signals in the NIR-II range were recorded with an InGaAs camera (Princeton NIRvana camera). The persistent luminescence features were recorded after removal of the 365 nm lamp excitation. The camera collected an image every 3 min (during the 15 min period following excitation removal). In addition, the PL images of the patterns were recorded with the camera under various excitations ranging from 365 nm (lamp) to 808 nm and 980 nm (NIR laser).

3. Results and Discussion

A normal spinel (ZGO) and an inverse spinel (ZSO), in addition to a series of complex spinel nanomaterials, were designed as combinations of the normal spinel $ZnGa_2O_4$

and inverse spinel Zn_2SnO_4 with different Sn amounts, as represented by the formula $Zn_{1+x}Ga_{2-2x}Sn_xO_4$, x = 0, 0.1, 0.2, 0.3, ... , 0.6, denoted as ZGSO (from ZGSO-0 to ZGSO-6, respectively). The content of Ni^{2+}-doped ZGSO NPs can be seen in Table S1 and Figure S6 in the Supplementary Materials. The most favorable persistent luminescence properties suggested a nominal doping content for the activator Ni^{2+} at 0.5 mol % in respect to Ga^{3+} ions in the matrix; this will be discussed in detail in the following sections of the paper. Divalent nickel ions as dopants in this host are in the octahedral site, as the stabilization energy is fairly important (greater than 80 kJ.mol^{-1}) [37]. Therefore, only the octahedral sites are populated with Ni^{2+} ions as dopants. Furthermore, one should notice the very similar ionic radius of Ni^{2+} with Ga^{3+} cations in addition to the strong octahedral coordination preference [38]. In addition, other contents of doping cations (Ni^{2+}, Er^{3+}, and Cr^{3+}) in ZGSO-3 NPs are shown in Tables S2 and S3, and Figure S10. Thermoluminescence spectra of Ni^{2+} in ZGO and ZGSO matrices have also been obtained; these are shown in Figure S11. As previously observed for ZGSO:Cr [34], the TL glow presents shallower traps when Sn^{4+} is introduced with an increase in signal (see the signal-over-noise ratio) and the resulting broadening.

3.1. Crystal Structure Analysis by Powder X-ray Diffraction (PXRD)

The ionic radius of Sn^{4+} is 0.83 Å, which is slightly greater than that of Ga^{3+} (0.76 Å). Thus, the Sn^{4+} ion is well adapted to an octahedral configuration with Sn–O distances (2.05 Å), which is similar to the Ga–O distances (1.98 Å). Sn^{4+} ions can be introduced into the $ZnGa_2O_4$ host [34] without affecting its structural properties. As shown in Figure 1a, the diffraction peaks (at 2θ = 18.4°, 30.3°, 35.7°, 57.4°, 63.0°) of sample ZGSO-0 (Sn free, normal ZGO) relate well to the standard lattice planes [(111), (220), (311), (511), (440), respectively] of $ZnGa_2O_4$ (JCPDS No. 01-086-0413). For the series of $Zn_{1+x}Ga_{2-2x}Sn_xO_4$ materials with increasing x values (concentration of Sn), the PXRD patterns show a slight shift in the diffraction peaks, as seen in Figure 1. The diffraction peaks shift to lower 2θ values with an increase in Sn^{4+}/Ga^{3+} ratio. A comparison of the ZGSO-2 (x = 2) and ZGSO-4 (x = 4) patterns is shown in Figure S1.

The Ni^{2+} cation crystal field could be modified by tuning the Sn^{4+}/Ga^{3+} ratios, as seen in the following section of the paper. This is also the case for the introduction of Cr^{3+}, as previously reported [35]. In contrast, any shift in the position of the diffraction peaks is observed with a 0.5% Ni^{2+} doping ratio (Figure 1b). Ni^{2+}, Cr^{3+}, and to a lesser extent Er^{3+}, as the coordination is less favorable, are cations that can be incorporated effectively (up to 1 or 2%, as shown previously) into octahedral lattice sites, due to a similar ionic radius and valence with Ga^{3+} cations [39]. Hence, as it has been suggested that multi-emission is effectively covered in the range from visible to NIR-III, low doping ratios of Er^{3+} (1%) and Ni^{2+} (0.5%) or Cr^{3+} (0.5%) have been introduced into hosts without concern for the stability of the crystal structure and while retaining the normal spinel structure of $ZnGa_2O_4$ in all samples regardless of the tested Sn^{4+} contents, as seen in Figure 1.

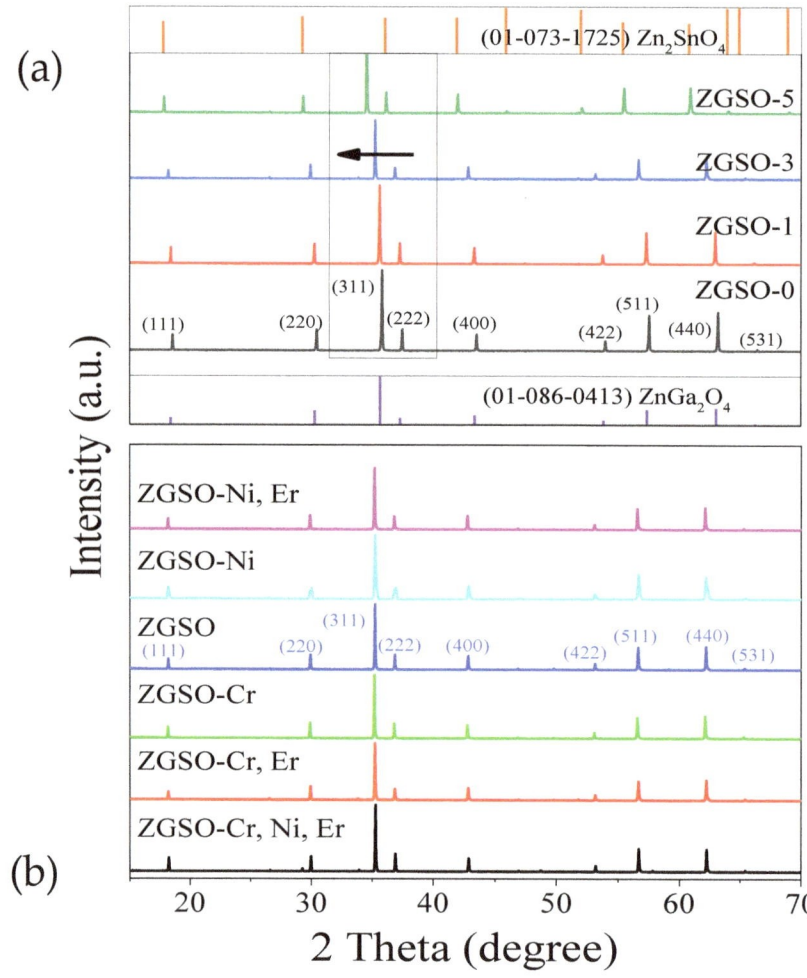

Figure 1. X-ray diffraction spectra of (**a**) ZGSO 0, 0.1, 0.3, and 0.5 concentration ratios of Sn/Ga matched to samples ZGSO-0, ZGSO-1, ZGSO-3, ZGSO-5, respectively. (**b**) Transition-metal (Ni^{2+} and/or Cr^{3+}) and rare-earth (Er^{3+}) ion-doped ZGSO-3.

3.2. Absorption Spectra Analysis

The absorption spectra of the samples were recorded to further validate the relationship between the change in crystal structure and the energy difference between electronic transitions. ZGSO-1, ZGSO-3, and ZGSO-5 are shown in Figure 2a. Figure S2 in the supplementary materials shows the normalized absorption spectra of Ni^{2+}-doped ZGSO-2 and ZGSO-4. Strong, broad excitation bands located between ~260 nm and ~320 nm are observed. For example, in the ZGSO-3: Ni sample, the observed shoulders at around 370 nm are attributed to the $^3A_2$ (^3F) → $^3T_1$ (^3P) spin-allowed transition of Ni^{2+}, almost overlapping with the bandgap edge of the ZGSO matrix. This bandgap absorption edge originates from the O–Ga charge-transfer transition in the ZGSO host, which is also seen in Figure 2b, and the bandgap values decrease with Sn^{4+} content. The bandgap energy values can be extracted and calculated using the Kubelka–Munk function [35]. With an increase in

Sn^{4+} content in the ZGSO matrix, the bandgap values varied from 3.5 to 4.6 eV using the following formula:

$$E = \frac{hc}{\lambda}$$

where E is the bandgap energy (in eV) and λ is the experimental absorption wavelength, as shown in Figure 2.

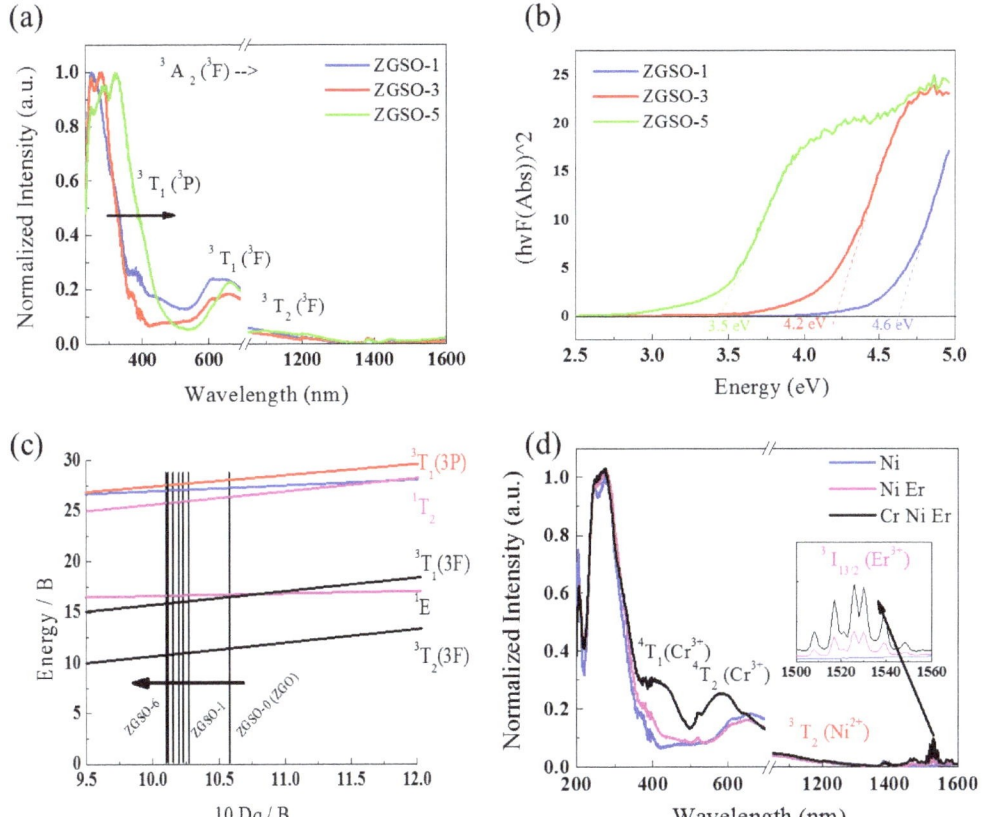

Figure 2. (a) Normalized absorption spectra of samples Ni^{2+}-doped ZGSO-1 (Zn$_{1.1}$Ga$_{1.79}$Ni$_{0.01}$Sn$_{0.1}$O$_{3.995}$), ZGSO-3 (Zn$_{1.3}$Ga$_{1.39}$Ni$_{0.01}$Sn$_{0.3}$O$_{3.995}$), and ZGSO-5 (Zn$_{1.5}$Ga$_{0.09}$Ni$_{0.01}$Sn$_{0.5}$O$_{3.995}$). (b) Bandgap calculation of samples Ni^{2+}-doped ZGSO-1, ZGSO-3, and ZGSO-5. (c) Part of the Tanabe–Sugano diagram of Ni^{2+} with d^8 electron configuration in the complex spinel samples of ZGSO-0~ZGSO-6 with different Sn concentrations (see Table S3). (d) Transition-metal (Ni^{2+} and/or Cr^{3+}) and rare-earth (Er^{3+}) doped ZGSO-3. The illustration is a partial enlargement of Er^{3+} absorption.

In addition, the d^8 Tanabe–Sugano diagrams shown in Figure 2c and Figure S3 (in the Supplementary Materials) effectively show that the shifted absorption peaks of Ni^{2+} towards a longer wavelength depend on an increase in the Sn^{4+}/Ga^{3+} ratio corresponding to a decrease in crystal field strength [40]. Two absorption peaks of ZGSO-3: Ni^{2+} observed at ~600 nm and ~1050 nm originate from the $^3A_2$ (^3F) → $^3T_1$ (^3F) and $^3A_2$ (^3F) → $^3T_2$ (^3F) transitions of Ni^{2+}, respectively. For a more quantitative analysis, the crystal field strength value D_q and the Racah parameter B can be estimated using the energies of various absorption peaks as follows [29]:

$$10\,D_q = v_2 \tag{1}$$

$$\frac{v_1^2 + 2 v_3^2 - 3 v_1 v_3}{15 v_1 - 27 v_3} \tag{2}$$

where v_1 and v_3 are the energies corresponding to Ni^{2+} ions [$^3A_2$ (^3F) → $^3T_1$(^3P)] and [$^3A_2$ (^3F) → $^3T_2$ (^3F)] transitions, respectively.

As a result, variation in the crystal field strength of ZGSO was observed. As Sn^{4+} content varied from 0.1 to 0.3, and even up to 0.5, the energy level of the divalent nickel corresponded to lower crystal field strength values (10 D_q decreased from 10.58 to 10.12). Crystal field values, v_3 (10 D_q/B), in addition to other related parameters, are shown in Table 1.

Table 1. Peak position of absorption and estimated crystal field parameters of Ni^{2+} in the spinel samples ZGSO.

No.	Sample/Material	$^3A_2$→$^3T_1$ [nm]	v_1 [cm^{-1}]	$^3A_2$→$^3T_1$ [nm]	v_2 [cm^{-1}]	$^3A_2$→$^3T_2$ [nm]	v_3 [cm^{-1}]	10 D_q [cm^{-1}]	B [cm^{-1}]	10 D_q/B
ZGSO-0	ZnGa$_2$O$_4$:0.5%Ni	370	27,027	622	16,077	1029	9718	9718	919	10.58
ZGSO-1	Zn$_{1.1}$Ga$_{1.8}$Sn$_{0.1}$O$_4$:0.5%Ni	371	26,954	625	16,000	1046	9560	9560	932	10.26
ZGSO-2	Zn$_{1.2}$Ga$_{1.6}$Sn$_{0.2}$O$_4$:0.5%Ni	372	26,882	628	15,924	1050	9524	9524	931	10.23
ZGSO-3	Zn$_{1.3}$Ga$_{1.4}$Sn$_{0.3}$O$_4$:0.5%Ni	374	26,738	632	15,823	1057	9461	9461	927	10.20
ZGSO-4	Zn$_{1.4}$Ga$_{1.2}$Sn$_{0.4}$O$_4$:0.5%Ni	376	26,596	636	15,723	1065	9390	9390	925	10.15
ZGSO-5	Zn$_{1.5}$Ga$_1$Sn$_{0.5}$O$_4$:0.5%Ni	379	26,385	641	15,601	1075	9302	9302	919	10.12
ZGSO-6	Zn$_{1.6}$Ga$_{0.8}$Sn$_{0.6}$O$_4$:0.5%Ni	382	26,178	646	15,480	1084	9225	9225	912	10.11

Furthermore, Er^{3+} and Cr^{3+} ions were used as dopants to obtain other emission wavelengths. The absorption spectra of Er^{3+} and Er^{3+}/Cr^{3+} co-doped ZGSO: Ni^{2+} NPs are shown in Figure 2d. The absorption peak at 1530 nm is attributed to the Er^{3+} $^4I_{15/2}$ → $^4I_{13/2}$ transition. The broad absorption bands at 410 nm and 570 nm are attributed to the $^4A_2$ (^4F) → $^4T_1$ (^4F) and $^4A_2$ (^4F) → $^4T_2$ (^4F) transitions of Cr^{3+}, respectively.

3.3. Photoluminescence (PL) Spectral Analysis

Using a control of tin composition, a tunable $^3T_2$ (^3F) → $^3A_2$ (^3F) transition of Ni^{2+} PL extends from 1270 to 1340 nm, as shown in Figure 3a. The variation in wavelength is also effectively explained in the Tanabe–Sugano diagram d^8 in Figure 2d.

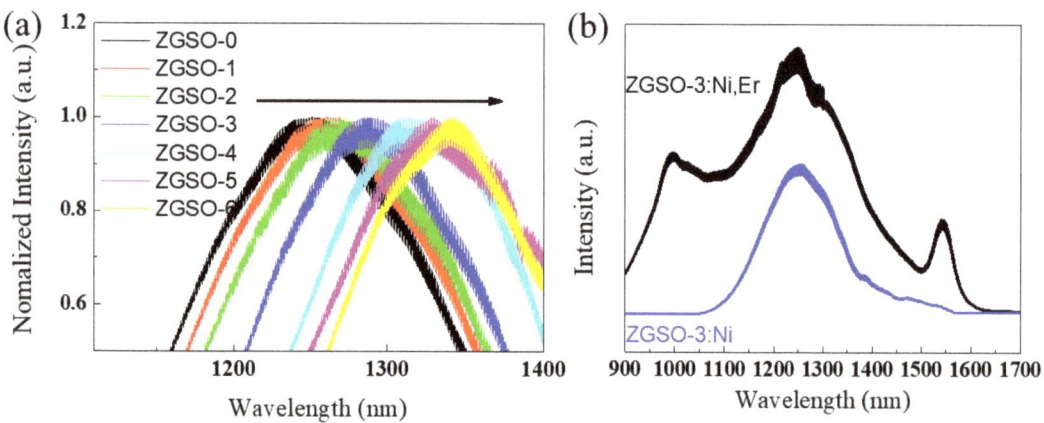

Figure 3. (a) PL spectra of 0.5% Ni^{2+}-doped ZGSO-0 to ZGSO-6 samples. (b) PL of ZGSO-3:Ni^{2+} (blue) and ZGSO-3:Ni^{2+} Er^{3+} (black). All samples were excited using a 365 nm UV lamp.

As co-dopants, Er^{3+} cations with several emissions in the visible, NIR-I, and even SWIR range may be very interesting, and we have focused our attention on the ZGSO-3: Ni^{2+}, Er^{3+} sample, which presents the most favorable persistent luminescence properties, as shown in the following section. As shown in Figure 3b, ZGSO-3: Ni^{2+}, Er^{3+} NPs show PL peaks at 980 nm (NIR-I range) and 1300 nm and 1533 nm (NIR-II range) under 365 nm irradiation. The peak at 1300 nm is attributed to the $^3T_2$ (^3F) → $^3A_2$ (^3F) transition of octahedral Ni^{2+} ions [41].

On the other hand, Er^{3+} is responsible for the emission at 980 nm (NIR-I) and 1533 nm (NIR-II), attributed to the $^4I_{11/2}$ → $^4I_{15/2}$ and $^4I_{13/2}$ → $^4I_{15/2}$ transitions, respectively, as shown in Figure S4c. These optical properties—with emission in the visible, deep-red, and NIR ranges—have been obtained due to knowledge within our laboratories, which have been working on the zinc gallate matrix for several years [42–47]. The PL spectra of (a) ZGSO:Cr^{3+}, (b) ZGSO:Er^{3+}, and (c) ZGSO:Cr^{3+}, Er^{3+} in the visible range are shown in Figure 4. ZGSO-3: Ni^{2+}, Er^{3+}, Cr^{3+} NPs were also considered within this work because Cr^{3+} provides a 700 nm persistent luminescence in the deep-red (or NIR-I) range [46,47]. Under 365 nm UV excitation, the multi-emission on the PL spectrum of ZGSO-3: Ni^{2+}, Er^{3+}, Cr^{3+} NPs covers visible, NIR-I, NIR-II, and NIR-III ranges; the PL features will be compared to the persistent luminescence properties in the following section of the paper.

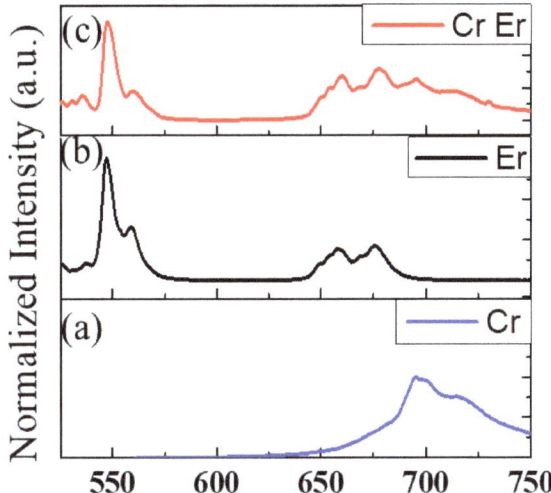

Figure 4. PL spectra of (**a**) ZGSO-3:Cr, (**b**) ZGSO-3:Er, (**c**) ZGSO-3:Cr, Er. All PL measurements are obtained from the samples under 365 nm UV excitation.

3.4. Persistent Luminescence Spectral Analysis

After UV excitation removal, a persistent luminescence signal is observed for all Ni^{2+}-doped ZGSO NPs. This persistent luminescence is assigned to the $^3T_2$ (^3F) → $^3A_2$ (^3F) transition of octahedral Ni^{2+} cations [42]. The position of the broad emission peak of Ni^{2+} shifts from 1270 to 1340 nm by increasing the Sn (x value, Table S1) content in the matrix (see Figure 5a), due to a decrease in the strength of the crystal field. Figure S4 shows a possible mechanism to explain the persistent luminescence signal for ZGSO-3: Ni^{2+} NPs. Under UV excitation, charges are formed and trapped in the ZGSO host defects. Based on this system, the release with temperature of the trapped charges after excitation removal and recombination toward the Ni^{2+} centers lead to the persistent luminescence signal in the SWIR range, attributed to the $^3T_2$ (^3F) → $^3A_2$ (^3F) transition (Figure S4b). An increase in Zn^{2+} and Sn^{4+} content leads to a decrease in crystal field around Ni^{2+}, similar to that seen with photoluminescence, and a shift in the persistent luminescence signal towards a longer wavelength (i.e., lower energy) is observed, as seen in Figure 5a.

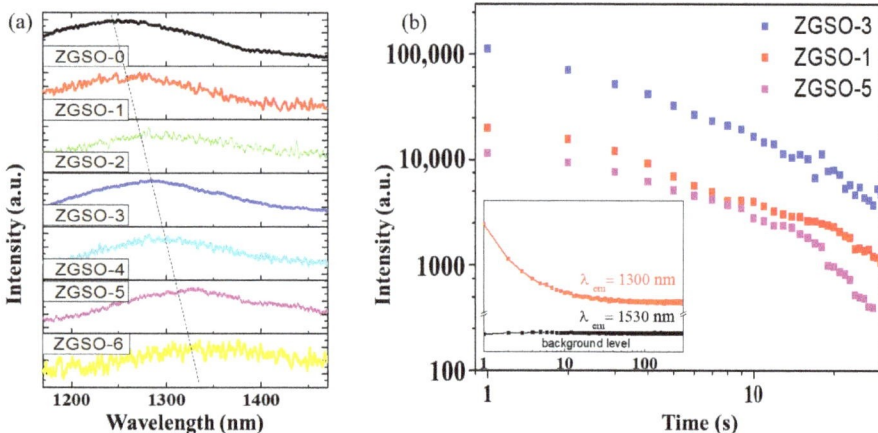

Figure 5. (a) Persistent luminescence spectra of the ZGSO-0–ZGSO-6 samples (1 min after ceasing UV excitation). (b) Corresponding decay curves of the ZGSO-1, ZGSO-3, ZGSO-5 samples after removing excitation. The inset presents the time dependence of the NIR persistent luminescence intensity of emission wavelength at 1300 nm and 1530 nm for Er^{3+}, Ni^{2+} co-doped ZGSO-3. The emission at 1530 nm is at around the background level. All samples were excited using a 365 nm UV lamp for 5 min.

Although all samples show a rapid decrease in persistent luminescence, the initial intensity depends on the sample. For example, as shown in Figure 5b, of ZGSO-1:Ni^{2+}, ZGSO-3:Ni^{2+}, and ZGSO-5:Ni^{2+}, ZGSO-3:Ni^{2+} (namely, $Zn_{1.3}Ga_{1.4}Sn_{0.3}O_4$:0.5%$Ni^{2+}$) shows the highest initial intensity value of persistent luminescence at 1300 nm, as shown in Figure S5. Furthermore, the Ni^{2+} doping concentration was optimized (Figure S6) and was revealed to be 0.5%; in comparison, the persistent luminescence intensity of the Ni^{2+}-doped ZGSO-3 samples with different nickel concentrations was 0.25%, 0.5%, and 1%, as shown in Figure S6. As a result, the persistent luminescence at 1300 nm increased as the crystal field value (10 D_q/B) decreased to 10.20 from an initial value of 10.58. $Zn_{1.3}Ga_{1.4}Sn_{0.3}O_4$:0.5%$Ni^{2+}$ is the optimal material as it has the strongest persistent luminescence intensity of all the samples prepared and reported in Table S1.

The effect of the amount of Er^{3+} in the ZGSO matrix was also evaluated. As shown in Figure S10, of the different concentrations tested, the optimal amount of Er^{3+} is 1% atomic. On the other hand, the inset in Figure 5b shows the time dependence of the NIR persistent luminescence with wavelengths at 1300 nm and 1530 nm from Er^{3+}, Ni^{2+} co-doped ZGSO-3 NPs. While the persistent luminescence intensity at 1300 nm reduced slowly, the emission at 1530 nm (see Figure 3b) fully disappeared when the excitation was removed, and no persistent emission could be recorded. For the ZGSO:Cr^{3+}, Er^{3+} samples, the visible persistent emissions in the red and deep-red ranges are shown in Figure S7. Persistent luminescence spectra of ZGSO:Cr^{3+} and ZGSO:Cr^{3+}, Er^{3+} are similar, and no red emission is observed (at 650 nm for Er^{3+} cations); the persistent luminescence under UV excitation is governed solely by Cr^{3+} decay. Meanwhile, Figure S4 shows schematic diagrams of the absorption/emission bands; trapping/detrapping and energy-transfer mechanisms lead to deep-red emissions of (a) ZGSO-3:Cr^{3+} NPs, NIR emission of (b) ZGSO-3:Ni^{2+} NPs, and 4f-4f transition of (c) ZGSO-3:Er^{3+} NPs.

The spectra in Figure 6 indicate the PL (Figure 6a) and persistent luminescence (Figure 6b) of ZGSO:Ni^{3+}, Cr^{3+}, Er^{3+}, detected with an Si-based detector. In the SWIR (from NIR-II to NIR-III) range; Figure 6c,d also provide the spectrum of PL (Figure 6c) and persistent luminescence (Figure 6c) of the same sample ZGSO:Ni^{2+}, Cr^{3+}, Er^{3+}, detected with an InGaAs-based detector. Peaks of Er^{3+} vanished at ~550 nm and ~660 nm in the visible range and at ~980 nm and ~1530 nm in the NIR, corresponding to a lack of Er^{3+} remaining in the samples. With these features of PL and persistent luminescence, anti-counterfeiting applications have been proposed.

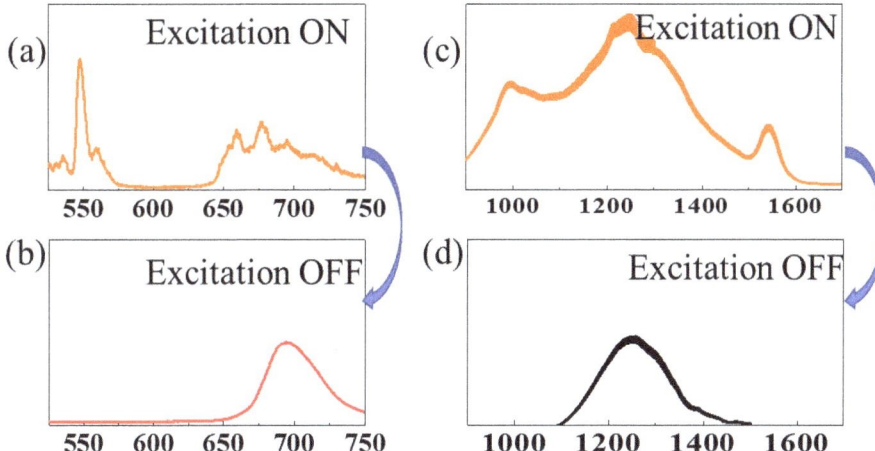

Figure 6. (**a**) Spectrum of PL (excitation ON) and (**b**) persistent luminescence (excitation OFF) detected with Si-based detector; (**c**) spectrum of PL (excitation ON) and (**d**) persistent luminescence (excitation OFF) detected with an InGaAs detector. All the PL or persistent luminescence spectra were from ZGSO-3:Ni, Cr, Er NPs under excitation or after excitation by the 365 nm UV lamp for 5 min, respectively.

3.5. Towards Anti-Counterfeiting Applications

3.5.1. A Two-Step Anti-Counterfeiting Method in the NIR-II Range

A new two-step anti-counterfeiting method in the NIR range based on ZGSO:Ni^{2+}, Er^{3+} NPs is introduced. As described above, NIR multi-emissions of PL at ~980 nm, ~1300 nm, and ~1530 nm are obtained from NPs under UV irradiation as a first security step (for bright emissions, UV excitation wavelengths should be in the range 275–400 nm). As shown by the example image, the "UTCBS" acronym noted by ZGSO:Ni^{2+}, Er^{3+} NPs can be easily detected due to their strong PL (Figure 7a), corresponding to the PL in the NIR range.

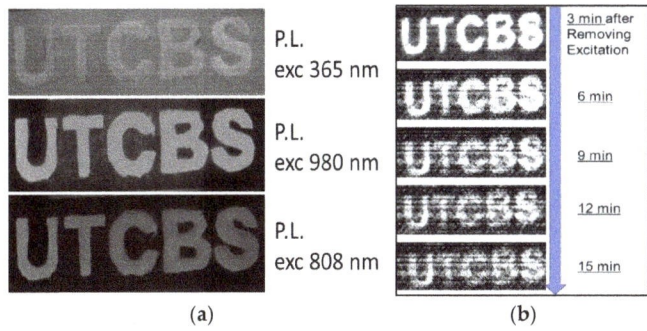

Figure 7. (**a**) PL of the "UTCBS" pattern noted by ZGSO-3:Ni^{2+}, Er^{3+} NPs under excitation at 365 nm, 980 nm, and 808 nm, respectively. (**b**) Persistent luminescence images of the "UTCBS" pattern noted by ZGSO-3:Ni^{2+}, Er^{3+} NPs after removal of 365 nm excitation for 5 min. All images were collected with an InGaAs camera.

For a second security step, after removing UV excitation, the persistent luminescence pattern "UTCBS" is obtained with an InGaAs-based camera for at least 15 min (Figure 7b). Additionally, another security step could be observation of the change in signal shape between PL and persistent luminescence (see Figure 3b). The Er^{3+} emission in the SWIR vanishes in the persistent luminescence spectrum, which provides a new step in the field of anti-counterfeiting, NIR (or SWIR) emission giving higher accuracy and security. In addition, the PL of the pattern "UTCBS" noted by ZGSO:Ni^{2+}, Er^{3+} NPs excited by the 980 nm and 808 nm NIR lasers, respectively, in the Er^{3+} absorption bands is shown in Figure 7. No persistent luminescence was observed under these NIR excitation wavelengths.

3.5.2. Multi-Level Anti-Counterfeiting Method in Visible, NIR-I, and NIR-II Ranges

To advance this proof of concept using $Zn_{1.3}Ga_{1.4}Sn_{0.3}O_4$ material, we have shown that we can introduce several dopants to propose a multi-level anti-counterfeiting compound in the visible and NIR ranges. (i) First, this work benefits from stable NIR-II photoluminescence from Er^{3+} and NIR-II persistent luminescence from Ni^{2+}, in addition to NIR-I persistent emission when doping with Cr^{3+} (Figures 4 and 6). (ii) Second, for further certification requirements, information and images could be read in both the visible and NIR ranges. To further address the advantages of anti-counterfeiting by this type of novel NP, patterns based on ZGSO:Ni^{2+}, Er^{3+}, Cr^{3+} NPs were evaluated by both Si-based and InGaAs-based cameras. A multi-level anti-counterfeiting system could then be obtained, as follows:

The changes in spectra obtained by optical spectroscopy (shown in Figure 6) are also observed in the images under both Si-based and InGaAs-based cameras. For example, the "MPOE" pattern was hand painted using ZGSO:Ni^{2+}, Er^{3+}, Cr^{3+} NPs. Figure 8 shows (a) an NIR-I persistent luminescence image from an "MPOE" pattern detected with a Si-based visible camera and (b) an NIR-II persistent luminescence image from an "MPOE" pattern detected with an InGaAs-based NIR camera after UV excitation removal. Figure S9 shows (a) the "MPOE" pattern template for hand painting, using ZGSO-3:Ni^{2+}, Er^{3+}, Cr^{3+} NPs and (b) an NIR-II PL image from the "MPOE" pattern under UV excitation, detected with an InGaAs-based NIR camera with minimum contrast to avoid saturation. In comparison with the NIR-II PL image, the persistent luminescence signal takes advantage of avoidance of an autofluorescence background and provision of an enhanced anti-counterfeiting effect.

Figure 8. (a) Persistent luminescence image from the "MPOE" pattern, hand painted using ZGSO-3: Ni^{2+}, Er^{3+}, Cr^{3+} NPs detected with Si-based visible camera; (b) persistent luminescence image from the "MPOE" pattern, hand printed using ZGSO-3: Ni^{2+}, Er^{3+}, Cr^{3+} NPs, detected with InGaAs-based NIR camera.

4. Conclusions

We successfully prepared several transition-metal (Ni^{2+} and/or Cr^{3+}) cations and rare-earth (Er^{3+}) co-doped ZGSO spinel nanocrystals with controllable photoluminescence wavelengths by tuning the concentration of Sn^{4+} and intensity emission and persistent luminescence for chromium and nickel transition-metal cations. Of these, multi-emission of PL in the NIR range benefits from the Er^{3+} ions working as emitters in nanoparticles of $Zn_{1.3}Ga_{1.4}Sn_{0.3}O_4$ co-doped with Ni^{2+}, Er^{3+}, and Cr^{3+} cations. Furthermore, $Zn_{1.3}Ga_{1.4}Sn_{0.3}O_4$ NPs exhibit persistent luminescence at 700 nm in the NIR-I range and 1300 nm NIR-II range, and their decay can last for more than 5 min after excitation removal.

A multi-level anti-counterfeiting pattern was designed by recording PL and persistent luminescence in the red, deep-red, and SWIR ranges. To further improve security, alternative excitation sources could be envisioned in the future due to the very wide excitation ranges of persistent luminescence materials [48–52].

Supplementary Materials: The following supporting information can be downloaded at https://www.mdpi.com/article/10.3390/ma16031132/s1: Table S1: Names and formulae of the Ni^{2+}-doped ZGSO samples; Table S2: Names and formulae of the co-doped ZGSO samples; Table S3: Names and formulae of the Cr^{3+}-doped ZGSO samples; Figure S1: XRD patterns of 0.5% Ni^{2+}-doped ZGSO-2, ZGSO-4, and ZSO samples; Figure S2: Normalized absorption spectra of 0.5% Ni^{2+}-doped ZGSO-2 and ZGSO-4 samples; Figure S3: Tanabe–Sugano diagram of Ni^{2+} (d^8 configuration) in the complex spinel samples of ZGSO-0 ($ZnGa_2O_4$:0.5%Ni^{2+}) and ZGSO-6 ($Zn_{1.6}Ga_{0.8}Sn_{0.6}O_4$:0.5%$Ni^{2+}$) corresponding to different Sn^{4+} concentrations; Figure S4: Schematic diagram of the absorption/emission bands; trapping/detrapping and energy-transfer mechanisms leading to deep-red emission of (a) ZGSO-3:Cr^{3+} NPs, and NIR emission of (b) ZGSO-3:Ni^{2+} NPs, and 4f-4f transition of (c) ZGSO-3:Er^{3+} NPs; Figure S5: Persistent luminescence intensity of Ni^{2+}-doped ZGSO-0–ZGSO-6 samples (detection 1 min after cutting off the 365 nm UV lamp excitation). Same quantity of powder and same excitation time is used for all samples. Detection with InGaAs camera; Figure S6: Persistent luminescence intensity of the Ni^{2+}-doped ZGSO-3 for three nominal different doping concentrations, namely 0.25%, 0.5% and 1% (detection with the InGaAs camera, 1 min after cutting off the 365 nm UV lamp excitation); Figure S7: Persistent luminescence spectra of (a) ZGSO-3:Cr^{3+} and (b) ZGSO-3:Er^{3+}, Cr^{3+}, respectively (detection with the Si camera 1 min after removal of UV lamp excitation (365 nm, 5 min)); Figure S8: NP size measurement of ZGSO-3:Ni^{2+} NPs and ZGSO-3:Ni^{2+}, Er^{3+}, Cr^{3+} NPs obtained by DLS; Figure S9: (a) Image of the "MPOE" pattern template for hand painting with ZGSO-3:Ni^{2+}, Er^{3+}, Cr^{3+} NPs. (b) PL image came from the "MPOE" pattern hand printed using ZGSO-3: Ni^{2+}, Er^{3+}, Cr^{3+} NPs under UV excitation, detected with InGaAs-based NIR camera with minimum contrast to avoid saturation.

Author Contributions: Conceptualization, G.C.; methodology, B.V.; software, C.R.; validation, T.D.; data curation and writing—original draft preparation, G.C.; writing—review and editing, T.D., C.R. and B.V.; supervision, C.R. and B.V. All authors have read and agreed to the published version of the manuscript.

Funding: This research was funded by Agence Nationale de la Recherche (ANR-18-CE08-0012 PERSIST) and the CSC Grant program.

Institutional Review Board Statement: Not applicable.

Informed Consent Statement: Not applicable.

Data Availability Statement: The data presented in this study are available on request from the corresponding author.

Acknowledgments: The authors would like to thank the Agence Nationale de la Recherche (ANR-18-CE08-0012 PERSIST) and the CSC Grant program.

Conflicts of Interest: The funders had no role in the design of the study; in the collection, analyses, or interpretation of data; in the writing of the manuscript; or in the decision to publish the results.

References

1. De Guzman, G.N.A.; Fang, M.H.; Liang, C.H.; Bao, Z.; Hu, S.F.; Liu, R.S. Near-infrared phosphors and their full potential: A review on practical applications and future perspectives. *J. Lumin.* **2020**, *219*. [CrossRef]
2. Liu, J.; Lecuyer, T.; Seguin, J.; Mignet, N.; Scherman, D.; Viana, B.; Richard, C. Imaging and therapeutic applications of persistent luminescence nanomaterials. *Adv. Drug Deliv. Rev.* **2019**, *138*, 193–210. [CrossRef] [PubMed]

3. Bessière, A.; Lecointre, A.; Benhamou, R.A.; Suard, E.; Wallez, G.; Viana, B. How to induce red persistent luminescence in biocompatible Ca$_3$(PO$_4$)$_2$. *J. Mater. Chem. C* **2013**, *1*, 1252–1259. [CrossRef]
4. Sun, Z.Y.; Yang, J.X.; Huai, L.W.; Wang, W.X.; Ma, Z.D.; Sang, J.K.; Zhang, J.C.; Li, H.H.; Ci, Z.P.; Wang, Y.H. Spy Must Be Spotted: A Multistimuli-Responsive Luminescent Material for Dynamic Multimodal Anticounterfeiting and Encryption. *Acs Appl. Mater. Interfaces* **2018**, *10*, 21451–21457. [CrossRef] [PubMed]
5. Lei, L.; Chen, D.Q.; Li, C.; Huang, F.; Zhang, J.J.; Xu, S.Q. Inverse thermal quenching effect in lanthanide-doped upconversion nanocrystals for anti-counterfeiting. *J. Mater. Chem. C* **2018**, *6*, 5427–5433. [CrossRef]
6. Liu, J.; Rijckaert, H.; Zeng, M.; Haustraete, K.; Laforce, B.; Vincze, L.; Van Driessche, I.; Kaczmarek, A.M.; Van Deun, R. Simultaneously excited downshifting/upconversion luminescence from lanthanide-doped core/shell fluoride nanoparticles for multimode anticounterfeiting. *Adv. Funct. Mater.* **2018**, *28*, 1707365. [CrossRef]
7. Suo, H.; Zhu, Q.; Zhang, X.; Chen, B.; Chen, J.; Wang, F. High-security anti-counterfeiting through upconversion luminescence. *Mater. Today Phys.* **2021**, *21*, 100520. [CrossRef]
8. Zhang, Z.; Ma, N.; Kang, X.; Li, X.; Yao, S.; Han, W.; Chang, H. Switchable up and down-conversion luminescent properties of Nd (III)-nanopaper for visible and near-infrared anti-counterfeiting. *Carbohydr. Polym.* **2021**, *252*, 117134. [CrossRef]
9. Zhang, Z.; Chang, H.; Xue, B.; Zhang, S.; Li, X.; Wong, W.-K.; Li, K.; Zhu, X. Near-infrared and visible dual emissive transparent nanopaper based on Yb (III)–carbon quantum dots grafted oxidized nanofibrillated cellulose for anti-counterfeiting applications. *Cellulose* **2018**, *25*, 377–389. [CrossRef]
10. Wang, X.; Li, T.; Liang, W.; Zhu, C.; Guo, L. Triple NIR light excited up-conversion luminescence in lanthanide-doped BaTiO$_3$ phosphors for anti-counterfeiting. *J. Am. Ceram. Soc.* **2021**, *104*, 5826–5836. [CrossRef]
11. Zhou, S.; Wang, Y.; Hu, P.; Zhong, W.; Jia, H.; Qiu, J.; Fu, J. Cascaded Photon Confinement-Mediated Orthogonal RGB-Switchable NaErF$_4$-Cored Upconversion Nanoarchitectures for Logicalized Information Encryption and Multimodal Luminescent Anti-Counterfeiting. *Laser Photonics Rev.* **2022**, 2200531. [CrossRef]
12. Lecuyer, T.; Durand, M.A.; Volatron, J.; Desmau, M.; Lai-Kuen, R.; Corvis, Y.; Seguin, J.; Wang, G.; Alloyeau, D.; Scherman, D.; et al. Degradation of ZnGa$_2$O$_4$:Cr^{3+} luminescent nanoparticles in lysosomal-like medium. *Nanoscale* **2020**, *12*, 1967–1974. [CrossRef] [PubMed]
13. Liu, F.Y.; Zhao, Q.; You, H.P.; Wang, Z.X. Synthesis of stable carboxy-terminated NaYF$_4$:Yb^{3+}, Er^{3+}@SiO$_2$ nanoparticles with ultrathin shell for biolabeling applications. *Nanoscale* **2013**, *5*, 1047–1053. [CrossRef] [PubMed]
14. Ma, C.Q.; Liu, H.H.; Ren, F.; Liu, Z.; Sun, Q.; Zhao, C.J.; Li, Z. The Second Near-Infrared Window Persistent Luminescence for Anti-Counterfeiting Application. *Cryst. Growth Des.* **2020**, *20*, 1859–1867. [CrossRef]
15. Hong, G.S.; Antaris, A.L.; Dai, H.J. Near-infrared fluorophores for biomedical imaging. *Nat. Biomed. Eng.* **2017**, 1. [CrossRef]
16. Yang, J.; Zhou, Y.; Ming, H.; Song, E.; Zhang, Q. Site-Selective Occupancy of Mn^{2+} Enabling Adjustable Red/Near-Infrared Multimode Luminescence in Olivine for Dynamic Anticounterfeiting and Encryption. *ACS Appl. Electron. Mater.* **2022**, *4*, 831–841. [CrossRef]
17. Gao, G.; Busko, D.; Joseph, R.; Howard, I.A.; Turshatov, A.; Richards, B.S. Highly efficient La$_2$O$_3$: Yb^{3+}, Tm^{3+} single-band NIR-to-NIR upconverting microcrystals for anti-counterfeiting applications. *ACS Appl. Mater. Interfaces* **2018**, *10*, 39851–39859. [CrossRef]
18. Zhang, Y.; Huang, R.; Li, H.; Lin, Z.; Hou, D.; Guo, Y.; Song, J.; Song, C.; Lin, Z.; Zhang, W. Triple-Mode Emissions with Invisible Near-Infrared After-Glow from Cr^{3+}-Doped Zinc Aluminum Germanium Nanoparticles for Advanced Anti-Counterfeiting Applications. *Small* **2020**, *16*, 2003121. [CrossRef]
19. Kang, H.; Lee, J.W.; Nam, Y. Inkjet-printed multiwavelength thermoplasmonic images for anticounterfeiting applications. *ACS Appl. Mater. Interfaces* **2018**, *10*, 6764–6771. [CrossRef]
20. Yu, X.; Zhang, H.; Yu, J. Luminescence anti-counterfeiting: From elementary to advanced. *Aggregate* **2021**, *2*, 20–34. [CrossRef]
21. Zhu, S.J.; Herraiz, S.; Yue, J.Y.; Zhang, M.X.; Wan, H.; Yang, Q.L.; Ma, Z.R.; Wang, Y.; He, J.H.; Antaris, A.L.; et al. 3D NIR-II Molecular Imaging Distinguishes Targeted Organs with High-Performance NIR-II Bioconjugates. *Adv. Mater.* **2018**, *30*, 1705799. [CrossRef] [PubMed]
22. Przybylska, D.; Grzyb, T.; Erdman, A.; Olejnik, K.; Szczeszak, A. Anti-counterfeiting system based on luminescent varnish enriched by NIR-excited nanoparticles for paper security. *Sci. Rep.* **2022**, *12*, 19388. [CrossRef] [PubMed]
23. Antaris, A.L.; Chen, H.; Cheng, K.; Sun, Y.; Hong, G.S.; Qu, C.R.; Diao, S.; Deng, Z.X.; Hu, X.M.; Zhang, B.; et al. A small-molecule dye for NIR-II imaging. *Nat. Mater.* **2016**, *15*, 235–242. [CrossRef] [PubMed]
24. Chan, M.H.; Huang, W.T.; Chen, K.C.; Su, T.Y.; Chan, Y.C.; Hsiao, M.; Liu, R.S. The optical research progress of nanophosphors composed of transition elements in the fourth period of near-infrared windows I and II for deep-tissue theranostics. *Nanoscale* **2022**, *14*, 7123–7136. [CrossRef]
25. Sordillo, L.A.; Pu, Y.; Pratavieira, S.; Budansky, Y.; Alfano, R.R. Deep optical imaging of tissue using the second and third near-infrared spectral windows. *J. Biomed. Opt.* **2014**, *19*, 056004. [CrossRef]
26. Xu, J.; Tanabe, S.; Sontakke, A.D.; Ueda, J. Near-infrared multi-wavelengths long persistent luminescence of Nd^{3+} ion through persistent energy transfer in Ce^{3+}, Cr^{3+} co-doped Y$_3$Al$_2$Ga$_3$O$_{12}$ for the first and second bio-imaging windows. *Appl. Phys. Lett.* **2015**, *107*. [CrossRef]
27. Xu, J.; Murata, D.; Katayama, Y.; Ueda, J.; Tanabe, S. Cr^{3+}/Er^{3+} co-doped LaAlO$_3$ perovskite phosphor: A near-infrared persistent luminescence probe covering the first and third biological windows. *J. Mater. Chem. B* **2017**, *5*, 6385–6393. [CrossRef] [PubMed]

28. Xu, J.; Murata, D.; Ueda, J.; Tanabe, S. 1.5 um persistent luminescence of Er^{3+} in $Gd_3Al_{5-x}Ga_xO12$ (GAGG) garnets via persistent energy transfer. In Proceedings of the Conference on Optical Components and Materials XVI, San Francisco, CA, USA, 4–6 February 2019.
29. Abdukayum, A.; Chen, J.T.; Zhao, Q.; Yan, X.P. Functional Near Infrared-Emitting Cr^{3+}/Pr^{3+} Co-Doped Zinc Gallogermanate Persistent Luminescent Nanoparticles with Superlong Afterglow for in Vivo Targeted Bioimaging. *J. Am. Chem. Soc.* **2013**, *135*, 14125–14133. [CrossRef]
30. Gong, Z.; Liu, Y.X.; Yang, J.; Yan, D.T.; Zhu, H.C.; Liu, C.G.; Xu, C.S.; Zhang, H. A Pr^{3+} doping strategy for simultaneously optimizing the size and near infrared persistent luminescence of ZGGO:Cr^{3+} nanoparticles for potential bio-imaging. *Phys. Chem. Chem. Phys.* **2017**, *19*, 24513–24521. [CrossRef]
31. Castaing, V.; Sontakke, A.D.; Carrion, A.J.F.; Allix, M.; Viana, B. Deep Red and Near Infrared Persistent Luminescence in Yb^{3+},Cr^{3+} co-Doped $ZnGa_2O_4$ Nano Glass Ceramics. In Proceedings of the Conference on Lasers and Electro-Optics Europe/European Quantum Electronics Conference (CLEO/Europe-EQEC), Munich, Germany, 23–27 June 2019.
32. Glais, E.; Pellerin, M.; Castaing, V.; Alloyeau, D.; Touati, N.; Viana, B.; Chaneac, C. Luminescence properties of $ZnGa_2O_4$:Cr^{3+},Bi^{3+} nanophosphors for thermometry applications. *Rsc Adv.* **2018**, *8*, 41767–41774. [CrossRef]
33. Castaing, V.; Sontakke, A.D.; Xu, J.; Fernandez-Carrion, A.J.; Genevois, C.; Tanabe, S.; Allix, M.; Viana, B. Persistent energy transfer in ZGO:Cr^{3+},Yb^{3+}: A new strategy to design nano glass-ceramics featuring deep red and near infrared persistent luminescence. *Phys. Chem. Chem. Phys.* **2019**, *21*, 19458–19468. [CrossRef] [PubMed]
34. Pan, Z.F.; Castaing, V.; Yan, L.P.; Zhang, L.L.; Zhang, C.; Shao, K.; Zheng, Y.F.; Duan, C.K.; Liu, J.H.; Richard, C.; et al. Facilitating Low-Energy Activation in the Near-Infrared Persistent Luminescent Phosphor $Zn_{1+x}Ga_{2-2x}Sn_xO_4$:Cr^{3+} via Crystal Field Strength Modulations. *J. Phys. Chem. C* **2020**, *124*, 8347–8358. [CrossRef]
35. Nie, J.M.; Li, Y.; Liu, S.S.; Chen, Q.Q.; Xu, Q.; Qiu, J.R. Tunable long persistent luminescence in the second near-infrared window via crystal field control. *Sci. Rep.* **2017**, *7*, 12392. [CrossRef]
36. Bessiere, A.; Jacquart, S.; Priolkar, K.; Lecointre, A.; Viana, B.; Gourier, D. $ZnGa_2O_4$:Cr^{3+}: A new red long-lasting phosphor with high brightness. *Opt. Express* **2011**, *19*, 10131–10137. [CrossRef] [PubMed]
37. Pellerin, M.; Castaing, V.; Gourier, D.; Chaneac, C.; Viana, B. Persistent luminescence of transition metal (Co, Ni center dot center dot center dot)-doped $ZnGa_2O_4$ phosphors for applications in the near-infrared range. In Proceedings of the Conference on Oxide-Based Materials and Devices IX, San Francisco, CA, USA, 28 January–1 February 2018.
38. Bessiere, A.; Sharma, S.K.; Basavaraju, N.; Priolkar, K.R.; Binet, L.; Viana, B.; Bos, A.J.J.; Maldiney, T.; Richard, C.; Scherman, D.; et al. Storage of Visible Light for Long-Lasting Phosphorescence in Chromium-Doped Zinc Gallate. *Chem. Mater.* **2014**, *26*, 1365–1373. [CrossRef]
39. Maldiney, T.; Doan, B.T.; Alloyeau, D.; Bessodes, M.; Scherman, D.; Richard, C. Gadolinium-Doped Persistent Nanophosphors as Versatile Tool for Multimodal In Vivo Imaging. *Adv. Funct. Mater.* **2015**, *25*, 331–338. [CrossRef]
40. Balti, I.; Mezni, A.; Dakhlaoui-Omrani, A.; Leone, P.; Viana, B.; Brinza, O.; Jouini, N. Comparative Study of Ni- and Co-Substituted ZnO Nanoparticles: Synthesis, Optical, and Magnetic Properties. *J. Phys. Chem. C* **2011**, *115*, 15758–15766. [CrossRef]
41. Zhou, S.F.; Feng, G.F.; Wu, B.T.; Jiang, N.; Xu, S.Q.; Qiu, J.R. Intense infrared luminescence in transparent glass-ceramics containing beta-Ga_2O_3: Ni^{2+} nanocrystals. *J. Phys. Chem. C* **2007**, *111*, 7335–7338. [CrossRef]
42. Maldiney, T.; Bessiere, A.; Seguin, J.; Teston, E.; Sharma, S.K.; Viana, B.; Bos, A.J.; Dorenbos, P.; Bessodes, M.; Gourier, D.; et al. The in vivo activation of persistent nanophosphors for optical imaging of vascularization, tumours and grafted cells. *Nat. Mater.* **2014**, *13*, 418–426. [CrossRef]
43. Katayama, Y.; Viana, B.; Gourier, D.; Xu, J.; Tanabe, S. Photostimulation induced persistent luminescence in $Y_3Al_2Ga_3O_{12}$:Cr^{3+}. *Opt. Mater. Express* **2016**, *6*, 1405–1413. [CrossRef]
44. Sharma, S.K.; Bessière, A.; Basavaraju, N.; Priolkar, K.R.; Binet, L.; Viana, B.; Gourier, D. Interplay between chromium content and lattice disorder on persistent luminescence of $ZnGa_2O_4$:Cr^{3+} for in vivo imaging. *J. Lumin.* **2014**, *155*, 251–256. [CrossRef]
45. Lecuyer, T.; Teston, E.; Ramirez-Garcia, G.; Maldiney, T.; Viana, B.; Seguin, J.; Mignet, N.; Scherman, D.; Richard, C. Chemically engineered persistent luminescence nanoprobes for bioimaging. *Theranostics* **2016**, *6*, 2488–2524. [CrossRef]
46. Lécuyer, T.; Bia, N.; Burckel, P.; Loubat, C.; Graillot, A.; Seguin, J.; Corvis, Y.; Liu, J.; Valéro, L.; Scherman, D. Persistent luminescence nanoparticles functionalized by polymers bearing phosphonic acid anchors: Synthesis, characterization, and in vivo behaviour. *Nanoscale* **2022**, *14*, 1386–1394. [CrossRef]
47. Lécuyer, T.; Seguin, J.; Balfourier, A.; Delagrange, M.; Burckel, P.; Lai-Kuen, R.; Mignon, V.; Ducos, B.; Tharaud, M.; Saubaméa, B. Fate and biological impact of persistent luminescence nanoparticles after injection in mice: A one-year follow-up. *Nanoscale* **2022**, *14*, 15760–15771. [CrossRef]
48. Luchechko, A.; Zhydachevskyy, Y.; Ubizskii, S.; Kravets, O.; Popov, A.I.; Rogulis, U.; Elsts, E.; Bulur, E.; Suchocki, A. Afterglow, TL and OSL properties of Mn^{2+}-doped $ZnGa_2O_4$ phosphor. *Sci. Rep.* **2019**, *9*, 9544. [CrossRef]
49. Richard, C.; Viana, B. Persistent X-ray-activated phosphors: Mechanisms and applications. *Light Sci. Appl.* **2022**, *11*, 123. [CrossRef] [PubMed]
50. Li, Y.; Chen, C.; Jin, M.; Xiang, J.; Tang, J.; Li, Z.; Chen, W.; Zheng, J.; Guo, C. External-field-dependent tunable emissions of Er^{3+}-In^{3+} Co-doped $Cs_2AgBiCl_6$ for applications in anti-counterfeiting. *Mater. Today Phys.* **2022**, *27*, 100830. [CrossRef]

51. Wu, Y.; Zhao, X.; Zhang, Z.; Xiang, J.; Suo, H.; Guo, C. Dual-mode dichromatic SrBi$_4$Ti$_4$O$_{15}$: Er^{3+} emitting phosphor for anti-counterfeiting application. *Ceram. Int.* **2021**, *47*, 15067–15072. [CrossRef]
52. Giordano, L.; Cai, G.; Seguin, J.; Liu, J.; Richard, C.; Rodrigues, L.C.V.; Viana, B. Persistent Luminescence Induced by Upconversion: An Alternative Approach for Rechargeable Bio-Emitters. *Adv. Opt. Mater.* **2023**, 2201468. [CrossRef]

Disclaimer/Publisher's Note: The statements, opinions and data contained in all publications are solely those of the individual author(s) and contributor(s) and not of MDPI and/or the editor(s). MDPI and/or the editor(s) disclaim responsibility for any injury to people or property resulting from any ideas, methods, instructions or products referred to in the content.

Article

The Optimization of Radiation Synthesis Modes for YAG:Ce Ceramics

Victor Lisitsyn [1,*], Dossymkhan Mussakhanov [2], Aida Tulegenova [3], Ekaterina Kaneva [4], Liudmila Lisitsyna [5], Mikhail Golkovski [6] and Amangeldy Zhunusbekov [7]

1. Department of Materials Science, Engineering School, National Research Tomsk Polytechnic University, 30, Lenin Avenue, Tomsk 634050, Russia
2. Department of Radio Engineering, Electronics and Telecommunications, Eurasian National University L.N. Gumilyov, 2, Satpaev Str., Astana 010008, Kazakhstan; dos_f@mail.ru
3. Department of Solid State and Nonlinear Physics, Al-Farabi Kazakh National University, 71, Al-Farabi Ave., Almaty 050040, Kazakhstan; tulegenova.aida@gmail.com
4. X-ray Analysis Laboratory, Vinogradov Institute of Geochemistry SB RAS, 1A, Favorsky Str., Irkutsk 664033, Russia; kev604@mail.ru
5. Department of Physics, Chemistry and Theoretical Mechanics, Tomsk State University of Architecture and Building, 2, Solyanaya Sq., Tomsk 634003, Russia; lisitsyna@mail.ru
6. Budker Institute of Nuclear Physics, SB RAS, 11, Lavrentiev Ave., Novosibirsk 630090, Russia; golkovski@mail.ru
7. Department of Technical Physics, Eurasian National University L.N. Gumilyov, 2, Satpaev Str., Astana 010008, Kazakhstan; zhunusbekov_am@enu.kz
* Correspondence: lisitsyn@tpu.ru; Tel.: +79-138242469

Citation: Lisitsyn, V.; Mussakhanov, D.; Tulegenova, A.; Kaneva, E.; Lisitsyna, L.; Golkovski, M.; Zhunusbekov, A. The Optimization of Radiation Synthesis Modes for YAG:Ce Ceramics. *Materials* 2023, 16, 3158. https://doi.org/10.3390/ma16083158

Academic Editor: Dirk Poelman

Received: 12 March 2023
Revised: 12 April 2023
Accepted: 14 April 2023
Published: 17 April 2023

Copyright: © 2023 by the authors. Licensee MDPI, Basel, Switzerland. This article is an open access article distributed under the terms and conditions of the Creative Commons Attribution (CC BY) license (https://creativecommons.org/licenses/by/4.0/).

Abstract: Synthesis in the radiation field is a promising direction for the development of materials transformation processes, especially those differing in melting temperature. It has been established that the synthesis of yttrium–aluminum ceramics from yttrium oxides and aluminum metals in the region of a powerful high-energy electron flux is realized in 1 s, without any manifestations that facilitate synthesis, with high productivity. It is assumed that the high rate and efficiency of synthesis are due to processes that are realized with the formation of radicals, short-lived defects formed during the decay of electronic excitations. This article presents descriptions of the energy-transferring processes of an electron stream with energies of 1.4, 2.0, and 2.5 MeV to the initial radiation (mixture) for the production of YAG:Ce ceramics. YAG:Ce ($Y_3Al_5O_{12}$:Ce) ceramics samples in the field of electron flux of different energies and power densities were synthesized. The results of a study of the dependence of the morphology, crystal structure, and luminescence properties of the resulting ceramics on the synthesis modes, electron energy, and electron flux power are presented.

Keywords: synthesis; YAG:Ce ceramics; structure; high-power electron flux; energy loss; luminescence

1. Introduction

Materials based on metal oxides are widely used as dosimetric, scintillation, phosphors, and optical active media due to their high functional and operational properties [1–4]. It is possible to create many optical materials with complex compositions for various applications from metal oxides. It is possible to correct their properties by introducing modifiers, a combination of activators. Materials based on metal oxides of various structures are used: crystals, powders, and ceramics.

Most initial materials for material synthesis based on metal oxides are refractory; their melting points are usually in the range of 1000–4000 °C. Therefore, the synthesis of oxide materials is a difficult task. It is even more difficult to fulfill the conditions for the materials synthesis of complex compositions. Synthesis is realized through the exchange of elements of the initial substances, the melting points of which can vary greatly. Therefore, direct melting of initial materials to initiate element exchange reactions in the liquid state is possible only in very rare cases. For synthesis, it is necessary to use complex, multi-stage methods to use additional substances that contribute to the process. Various methods are used for the synthesis of YAG:Ce phosphors [5–7]. The most widespread and used in industry is the solid-phase reaction method of thermal synthesis and its modifications [8,9]. To implement the exchange of elements between the particles of the initial oxide powders Al_2O_3 (melting temperature t_m = 2044 °C), Y_2O_3 (t_m = 2410 °C), and Ce_2O_3 (t_m = 2177 °C) in a mixture of stoichiometric composition, synthesis is carried out in a liquid melt, for example, BaF_2 (t_m = 1368 °C). With prolonged exposure at temperatures of 1400–1500 °C, particles stick together, and partial exchange of elements of different phases occurs. Then, at a higher temperature, BaF_2 is removed by repeated annealing, and the formation of $Y_3Al_5O_{12}$:Ce is completed. The synthesis procedure is complex, lengthy, and laborious. It is possible to obtain a good quality phosphor with good repeatability of luminescence properties, with strict observance of the synthesis regulations.

Other methods for the synthesis of luminescent materials based on YAG have been developed and are being improved. For example, the sol–gel method [10–12] and coprecipitation [13,14] produce ceramics and phosphors by using a combination of chemical reactions between substances containing the elements necessary for the formation of YAG. All these methods are complex; elements of the auxiliary materials used for the synthesis are preserved in the final products. They are interested in the synthesis methods of phosphors, ceramics in the flame of burners [15], and mixtures of combustible materials with metal oxide powders in approximately equal amounts [16]. Synthesis in the flame is realized in a short time. However, the synthesis process is difficult to control. Obviously, the resulting product may contain residues of combustible substances. It is assumed that it is possible to synthesize transparent ceramic materials by the SPS method (spark plasma sintering) [17].

For the first time, we proposed and implemented the synthesis method of YAG ceramics with activators from a mixture of Y_2O_3 and Al_2O_3 powders in the field of a powerful flux of high-energy electrons [18,19]. The method seems promising: the synthesis is realized in 1s, without the use of any substances facilitating the synthesis, with high productivity without the use of additional energy sources. It has been established that the set of processes ensuring high efficiency of radiation synthesis is completely different from those occurring when using other methods, in which the structure formation is stimulated by heat. It is assumed that the efficiency of synthesis in the field of powerful radiation streams is determined by the high density of ionization of matter and reactions in the electron–ion plasma.

There is an obvious need to establish the basic regularities of ceramics formation in the field of radiation of high density in order to optimize radiation exposure modes and to determine the requirements for the initial raw material, the processing methods of the resulting ceramics, expansion of the nomenclature of similar ceramics, etc. The possibility of express synthesis of materials contributes to the solution of these problems.

The present work is aimed at studying one of the most important tasks: studying the influence of radiation exposure modes on the initial mixture on the result and obtaining YAG ceramics with high functional properties.

It is necessary to study the dependence of the luminescence properties of this ceramic on various factors that may affect the result of synthesis. This article presents the results of such a study: the energy-transferring processes of the electron flux to the initial substance (mixture) to obtain YAG:Ce ceramics; the dependence of the synthesis result (morphology) on the synthesis modes, electron energy, and electron power flux; and the influence of synthesis modes on the crystal structure of the emerging ceramic's luminescence properties.

2. Materials and Methods

2.1. Energy Losses of the Electron Beam in the Material

The distribution of absorbed energy in matter under the action of a spatially limited beam of high-energy electrons is inhomogeneous. The maximum absorbed energy is concentrated along the beam axis. The inhomogeneity of the distribution is even more pronounced when using beams with a Gaussian power distribution over the cross-section.

Figure 1 defines the distribution profiles of electron losses calculated using the CASINO V2.5 program during their passage through a mixture with a bulk density of 1.2 g/cm^3 from Y_2O_3 (57%) and Al_2O_3 (43%) powders used for the synthesis of $Y_3Al_5O_{12}$ ceramics. The calculations were carried out by the Monte Carlo method for beams with a Gaussian flux income over the cross-section and velocity of incident electrons, with a density of 10,000, at energies of 1.4, 2.0, and 2.5 MeV, as used in experiments.

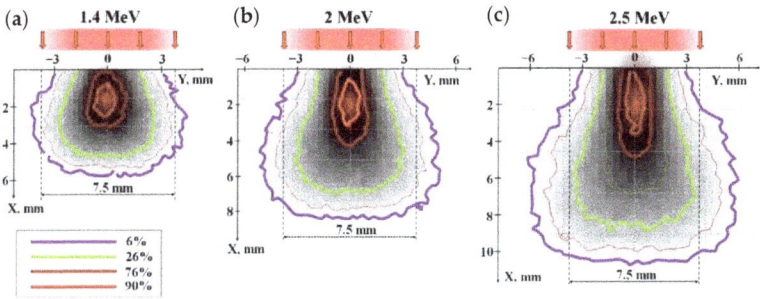

Figure 1. Energy loss distribution of electrons with E = 1.4 (**a**), 2.0 (**b**), 2.5 (**c**) MeV in a mixture with a bulk density of 1.2 g/cm^3 for the synthesis of $Y_3Al_5O_{12}$ ceramics. Colored lines of equal loss are given in units relative to the loss in the center.

The electron beam entering the material is scattered by atoms and ions of the substance and transfers its energy to ionization and generation of secondary electrons. As a result of these processes, as the electrons pass through the substance, the spatial structure of the beam energy transfer changes. Part of the energy is transferred to matter outside the beam section. There is a concentration of energy losses along the beam axis. About 50% of the loss of the entire beam energy falls on the charge region along the beam axis with a cross-section of 0.3–0.4 of the beam area on the surface. This leads to a characteristic loss distribution along the beam axis for an electron beam.

Figure 2a shows the calculated profiles of the distribution of electron energy losses dE/dx in the mixture over depth for an equal number of incident electrons with energies of 1.4, 2.0, and 2.5 MeV. Here, the energy losses of the electrons in an electron beam during their passage through a substance are understood as the magnitude of losses at a depth X in the entire region perpendicular to the beam axis. The maximum of the absorbed energy is located at a certain depth from the surface, which depends on the energy of the electrons. The positions of the energy loss maxima dE/dx of the beams fall at 2.8, 3.7, and 4.6 mm for the indicated electron energies. The value of energy losses in the maxima is 30–40% higher than at the target surface.

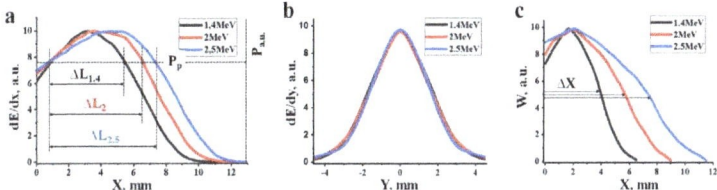

Figure 2. Energy loss distribution profiles dE/dx (**a**) and dE/dy (**b**) of electrons with energies of 1.4, 2.0, and 2.5 MeV in the mixture and absorbed energy density W (**c**).

Figure 2b shows the calculated distribution profiles of electron energy losses dE/dy in the direction perpendicular to the beam axis. It can be seen from the results of calculating the energy loss distribution of electrons of different energies presented in Figure 1 that the profiles dE/dy change with depth. These changes are different for electron beams of different energies. The figure shows the profiles dE/dy for the depths corresponding to the maximum absorbed energy density Wr of the electron beam. The absorbed energy density W is understood as the amount of energy loss per unit volume of a substance. The dE/dy profiles match. This means that under the condition adopted for the calculation, an equal number of incident electrons, the values of Wr are the same.

The energy loss densities along the axis of the passing beam are always higher than outside the axis and have the form of a curve with a maximum. Figure 2 c shows the calculated profiles of the dependence of W on the depth of electron passage. Dependences have the form of curves with maxima at 1.8, 2.1, and 2.3 mm for electrons with energies of 1.4, 2.0, and 2.5 MeV. The region length Δx with an equal density of absorbed energy along the axis increases on average by a factor of 2 with an increase in the electron energy from 1.4 to 2.5 MeV. Note that the energy loss maxima dE/dx are located at depths of 2.8, 3.7, and 4.6 mm at energies E equal to 1.4, 2.0, 2.5 MeV, and the region length with the same absorbed energy along the axis with increasing electron energy from 1.4 to 2.5 MeV increases by only 25% on average. This is due to the fact that by the end of the run, an expansion of the energy loss region is observed. Thus, the maxima of energy loss in depth (Figure 2a) (dE/dx) and energy loss densities W (Figure 2c) do not coincide.

Obviously, the areas of maximum energy loss densities should determine subsequent processes. It is in these areas that the ionization density is maximum, and the temperature to which the material is heated is maximum. In these regions, during the time of beam exposure, when the energy loss exceeds the threshold value, the crystal structure of yttrium and aluminum oxides is transformed into yttrium aluminum garnet. First of all, synthesis should be carried out at depths corresponding to the maximum energy loss densities along the beam axis, but not energy losses.

With a change in the electron beam power P, the distribution profile does not change, but the absolute values of the energy losses dE/dx and dE/dy change proportionally. Synthesis is realized in the material when the energy losses dE/dx and dE/dy in a specific region of the material with XY coordinates exceed a certain threshold value P_p of the beam power. The threshold P_p at which the synthesis can be realized depends only on the composition of the initial mixture, that is, on the composition of the synthesized material. Synthesis may not occur at the target surface and at great depths, but it may occur in the depth range ΔL, at which the energy loss exceeds the value required for synthesis. The range of depths ΔL at which synthesis can be realized increases with increasing E. Synthesis is realized in the depth range in which the energy loss density W exceeds the value required for synthesis. With an increase in the electron beam power P, the length increases, and the diameter of the region of maximum energy loss W along the beam axis increases. Therefore, synthesis can be realized in a larger volume, in which the energy loss densities exceed the synthesis realization threshold P_p.

Thus, the volumetric energy loss density changes significantly in the longitudinal and transverse to the direction of propagation of the electron flux in the substance. With an increase in the power density of the input electron flux, the energy losses increase proportionally. In the region of maximum energy losses, the process of material synthesis is most likely to be realized. With an increase in the power of the electron beam, the volume of the material in which synthesis can be realized increases, the upper limit of this volume can reach the surface of the mixture, and the lower limit can reach depths equal to the length of the extrapolated electron path Xe. The extrapolated depth of the free path of electrons in the mixture for the synthesis of YAG:Ce ceramics (see Figure 2a) is 9, 10, and 11 mm for electrons of 1.4, 2.0, and 2.5 MeV, respectively.

2.2. Synthesis of YAG Ceramics

A cycle of studies of the efficiency dependence of radiation synthesis of YAG:Ce ceramics on the electron energy and beam power was completed. The concentration of

Ce introduced for activation was 0.5%. Such an amount of the activator does not affect the main regularities of energy losses but allows using luminescence methods to make sure that Ce is included in the crystalline structure of ceramics. The prepared mixture with the composition of Y_2O_3 (56%), Al_2O_3 (43%), and Ce_2O_3 (0.5% by weight of the mixture) had a bulk density of 1.2 g/cm^3. Synthesis of YAG:Ce ceramics was carried out in copper crucibles with a depth of 14 mm, exceeding the total electron path at E = 2.5 MeV and dimensions of 50 × 100 mm.

Synthesis was carried out by direct action on the mixture in the crucible of an electron beam with energies of 1.4, 2.0, and 2.5 MeV extracted into the atmosphere from the ELV-6 accelerator of the INP named after. Budker SB RAS. The electron beam with a Gaussian flux distribution had a diameter of 1 cm on the mixture surface. Two modes of action of the electron beam on the mixture were used: "without scanning", when the crucible was pulled relative to the beam; and "with scanning", when the crucible with dimensions of 50 × 100 mm was pulled relative to the beam scanning in the transverse direction at a frequency of 50 Hz. The crucible was stretched along its entire length under the electron beam for 10 s. To obtain equality of the absorbed energy, the beam power in the "with scanning" mode was 5 times greater than in the "without scanning" mode.

Since the distribution of the absorbed energy of an electron beam in a substance is heterogeneous, it is necessary to designate the criteria for choosing the irradiation conditions under which the comparison of the results of synthesis by beams with different energies will be correct. The densities of maximum beam energy losses in the irradiated region of the substance should be close. Our previous studies have shown that when using an electron beam with E = 1.4 MeV, the synthesis of YAG:Ce ceramics in the "without scanning" mode is successfully implemented when transferring a charge with a bulk density of 1.2 g/cm^3 to a substance with an energy of 4 kJ/s cm^3 in the central region, to which 50% of the absorbed energy is transferred. Such an absorbed energy density is provided under the used irradiation conditions by a beam with a power of 5 kW/cm^2. When irradiated with electrons with higher energies, 50% of the absorbed energy at the center of the beam passage occurs in a larger volume. Based on the study of the dependence of the absorbed energy distribution on the electron energy (Figures 1 and 2), we have shown that the electron beam power should be 1.4 times higher for electrons with E = 2.0 MeV and 1.8 times higher for electrons with E = 2.5 MeV. The correction of modes during synthesis was carried out experimentally.

Photographs of ceramic samples in crucibles synthesized under the influence of electron fluxes with E = 1.4 MeV, E = 2.0 MeV, and E = 2.5 MeV at different power densities P are shown in Figure 3. The synthesis was carried out in the "without scanning" mode, which makes it possible to visually compare the results of the analysis of energy losses and synthesis.

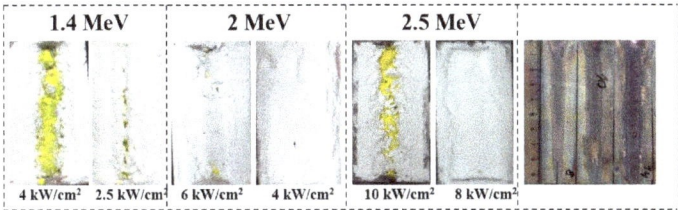

Figure 3. Photographs of ceramic samples synthesized under the exposure to electron fluxes with E = 1.4 MeV (P = 4 and 2.5 kW/cm^2), E = 2.0 MeV (P = 6 and 4 kW/cm^2), and E = 2.5 MeV (P = 10 and 8 kW/cm^2), and traces of the impact of electron flows with E = 1.4 MeV (P = 8, 10, 14 kW/cm^2) on the copper plate.

It can be seen from the above images that during the exposure time of 10 s, electron beams form ceramic samples in the mixture in the form of rods of yellow color characteristic of YAG:Ce. At large P, the rod samples are on the irradiated surface or close to the surface. With a decrease in P, the formed samples can be hidden under a layer of mixture. The depth of the formed sample in the mixture is greater the higher the value of E. This regularity corresponds to the conclusion made above about the dependence of the position of the region of maximum energy loss of the electron beam on E and P.

The same figure shows photographs of traces of the impact of an electron beam with E = 1.4 MeV in the "without scanning" mode on a thick copper plate. In the experiment, the upper surface of the plate was placed at the same distance from the accelerator outlet, where the outer surface of the charge was located during synthesis. The images clearly show that the width of the trace of the impact of the flux in the middle of the image reaches 7–10 mm, and the flow power in the center is much higher. At a flux density of $P = 8$ kW/cm^2, only a trace of oxidation is visible in the image; at $P = 12$ kW/cm^2, melting of the outer surface is observed. We emphasize that the melting temperature of copper (1085 °C) is used for the synthesis of oxides: Al_2O_3 (2044 °C) and Y_2O_3 (2410 °C). The synthesis of YAG:Ce ceramics is realized under the same conditions of exposure to an electron beam at $P < 4$ kW/cm^2, which is explained by the difference in the processes of dissipation of the absorbed energy of hard radiation in metals and dielectrics.

We also note the following effect. The photographs shown in Figure 3 show that the synthesis of YAG:Ce ceramics is realized almost uniformly along the entire length of the crucible, which is moved relative to the electron beam. The trace of the impact of the electron beam on the copper plate has a variable width. As the plate moves (or with time after the beginning of the beam impact), the trace expands. The expansion of the track is due to the fact that, over time, the temperature of the entire volume of the copper plate increases due to its high thermal conductivity (401 W/m* C). The thermal conductivity of the mixture for synthesis is 0.15–0.16 W/m* C, three orders of magnitude lower than in copper [19,20]. The time of passage of the irradiated section of the charge is 1 s. During this time, heat from the irradiated area does not have time to be transferred to the environment [17].

Figure 4 shows photographs of the samples taken out of the crucibles. The first three were completely covered by the mixture; the last two were open. All samples have the form of rods of different diameters. Sample 1 was synthesized at E = 1.4 MeV, P = 2.5 kW/cm^2; sample 2 at E = 2.0 MeV, P = 4 kW/cm^2; and sample 3 at E = 2.5 MeV, P = 8 kW/cm^2. Sample 4 was only slightly covered by the mixture from above (E = 2.0 MeV, P = 6 kW/cm^2), whereas sample 5 was almost completely open (E = 2.5 MeV, P = 10 kW/cm^2).

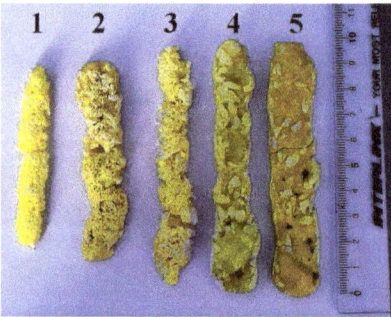

Figure 4. Photographs of YAG:Ce ceramic samples synthesized under the influence of electron fluxes of different E and P: 1—E = 1.4 MeV, P = 2.5 kW/cm^2; 2—E = 2.0 MeV, P = 4 kW/cm^2; 3—E = 2.5 MeV, P = 8 kW/cm^2; 4—E = 2.0 MeV, P = 6 kW/cm^2; 5—E = 2.5 MeV, P = 10 kW/cm^2.

The samples formed inside the mixture, at a relatively low power density P, have a shorter length and a porous surface. Ceramic samples that reached the surface of the mixture during the formation of ceramics have a solid surface but are porous inside. Note that the light spots in the photographs of the samples are the mixture traces, which are difficult to remove without damaging the sample.

As P decreases, the diameter of the forming sample decreases, and the solid rod turns into a dotted one. The smallest ceramic samples in the form of rare dotted particles with sizes of about 3 mm in diameter and up to 10 mm in length were obtained by exposure to electron beams with E = 1.4 MeV and P = 1.5 kW/cm^2. The samples are friable and crumble under slight pressure. However, they have a characteristic yellow color for YAG:Ce ceramics.

Figure 5 shows a photograph of a YAG:Ce sample synthesized by the "scanned" method at E = 2.5 MeV, P = 37 kW/cm^2. The sample has the form of a plate with dimensions of 90 × 45 mm. The plate surface is uneven. The sample thickness is 6 mm on average. The weight of the plate is 83 g. Inside the plate is porous, but the porosity is much lower than that of the samples obtained by the "without scanning" method at the same electron energy and absorbed energy. The pores are large and located parallel to the outer surfaces of the plate. The thickness of dense layers is 2–2.5 mm. The porosity of samples obtained at lower E and P is much higher.

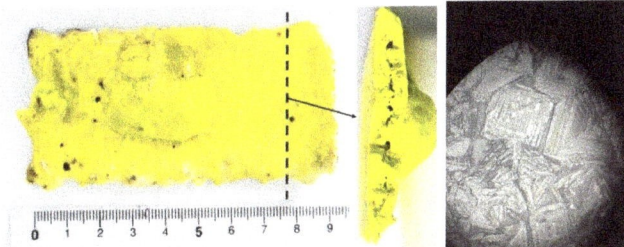

Figure 5. Photographs of YAG:Ce ceramic sample synthesized at E = 2.5 MeV, P = 37 kW/cm^2. On the right is a photograph of the outer surface of the ceramic sample taken by optical microscope.

A visual representation of the synthesized ceramic's surface structure is given in a photograph taken by an optical microscope, the "XJP 146 Trinocular Microscope" (Ningbo Wason Optical Instrument Co., Ltd., Zhejiang, China). The outer surface image taken by the optical microscope is shown in Figure 5 on the right. The surface looks like a set of regular shape crystallites bound together by a binding phase. The surface layer is hard but brittle. When an indenter is pressed, the surface breaks. The crystallites are up to 0.1 mm in size.

Figure 6 shows a photograph of a part of a YAG:Ce ceramic sample synthesized by the "scanned" method at E = 1.4 MeV, P = 25 kW/cm^2. The sample has the form of a plate with uneven edges, with a total area of 80 × 45 mm. The plate has a variable area thickness from 0 to 5 mm. The cross-section of the ceramic is more porous than that obtained at E = 2.5 MeV. The thickness of the dense layers near the surfaces is 1–1.5 mm.

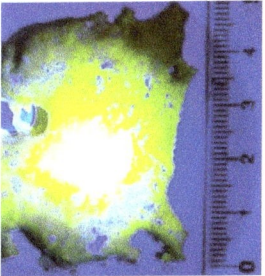

Figure 6. Photo of YAG:Ce ceramic sample synthesized at E = 1.4 MeV, P = 25 kW/cm^2, illuminated by chip radiation λ = 450 nm.

The photo was taken when the sample was illuminated by a concentrated lens with chip radiation with λ = 450 nm. This radiation excites the luminescence of YAG:Ce ceramics. The white color of the central part of the sample is due to luminescence and reflected radiation from the chip. The near-yellow color is due to the reflection of the white light of the central region by the ceramics. The blue color is due to the reflection of the chip radiation from the white paper on which the sample was placed.

3. Results
3.1. Structure of Synthesized Ceramics

The structure of the synthesized ceramics was studied by X-ray diffraction using a D8 ADVANCE Bruker diffractometer equipped with a scintillation detector in a step-by-step shooting mode in the range of diffraction angles 2θ from 10 to 80 degrees using a CuKα radiation source. The experiments were carried out at room temperature in the Bragg–Brentano geometry with a flat sample. The experimental conditions were as follows: 40 kV, 40 mA, exposure time—1 s, and step size—0.02° 2θ. The received data were processed using the DIFFRACplus software package. Samples were identified using the PDF-2 Powder Diffraction Database (ICDD, 2007) and indexed using EVA software (Bruker, 2007). In the TOPAS 4 program (Bruker, 2008), using the Rietveld method, the parameters of the YAG unit cell and the relative content of the main and accompanying phases were refined. The phase detection limit and the error for semi-quantitative analysis are 1–3% and 1–5%, respectively. The diffraction patterns of the samples are shown in Figure 7.

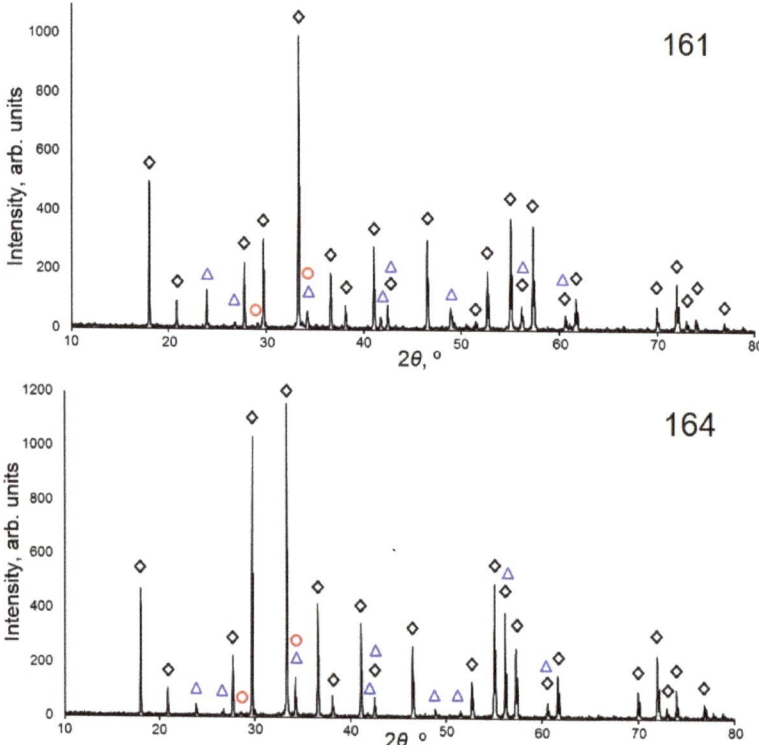

Figure 7. Cont.

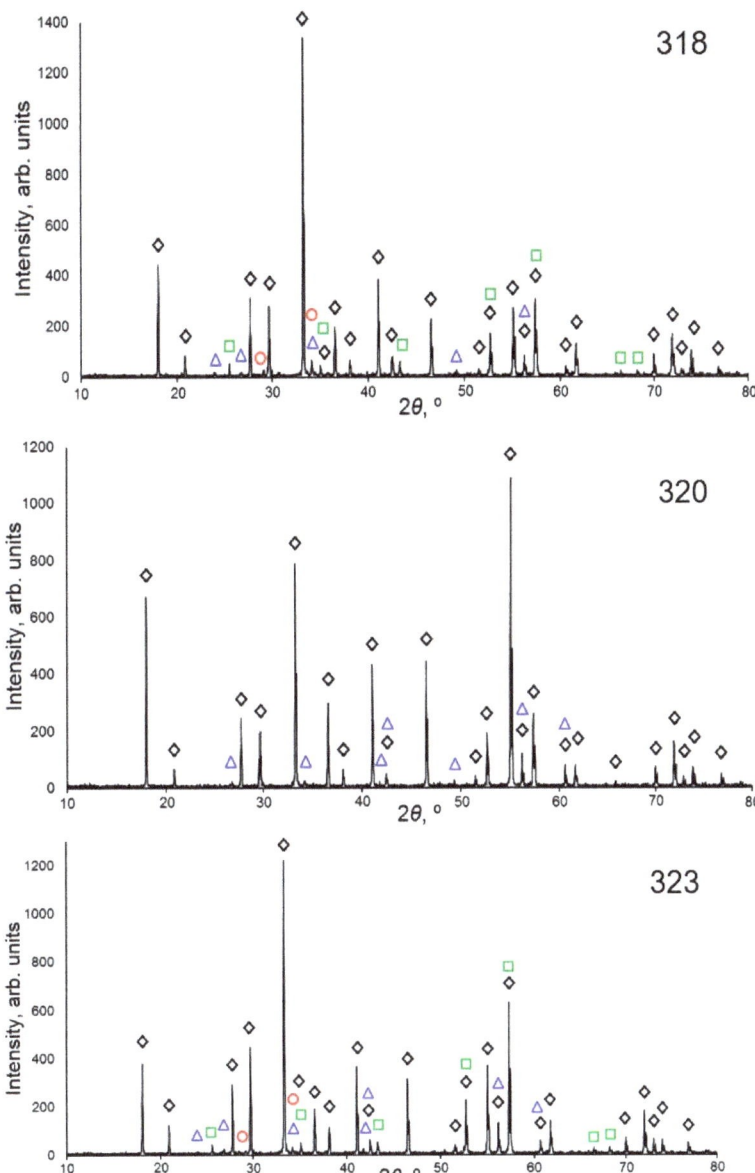

Figure 7. X-ray diffraction patterns of $Y_3Al_5O_{12}$ ceramic samples. Designations: black rhombus—$Y_3Al_5O_{12}$ reflections, blue triangle—$YAlO_3$, red circle—Y_2O_3, green square—Al_2O_3, black star—$Y_4A_{12}O_9$. The serial number in the figures is the sample number in the accounting system used by the authors.

For qualitative phase analysis and identification of diffraction patterns, the following data from the PDF-2 file (ICDD, 2007) were used: PDF 00-033-0040 "Aluminum Yttrium Oxide ($Al_5Y_3O_{12}$)", PDF 01-070-1677 "Yttrium Aluminum Oxide ($YAlO_3$)", PDF 00-046-1212 "Aluminum Oxide (Al_2O_3)", PDF 00-041-1105 "Yttrium Oxide (Y_2O_3)", and PDF 01-083-0933 "Aluminum Yttrium Oxide ($Al_2Y_4O_9$)".

The results of studying the phase composition of the samples are shown in Table 1.

Table 1. Results of studying the phase composition of the samples.

	Main Phase	Related Phase	Rwp(%)
161	$Y_3Al_5O_{12}$ (~91%) $Ia-3d$; a = 12.005(2) Å; V = 1730.3(3) Å^3	$YAlO_3$ (~7%) Y_2O_3 (~2%)	5.0
164	$Y_3Al_5O_{12}$ (~92%) $Ia-3d$; a = 12.009(4) Å; V = 1732.1(3) Å^3	$YAlO_3$ (~6%) Y_2O_3 (~2%)	5.6
318	$Y_3Al_5O_{12}$ (~91%) $Ia-3d$; a = 11.999(2) Å; V = 1727.4(3) Å^3	$YAlO_3$ (~4%) Al_2O_3 (~3%) Y_2O_3 (~2%)	4.3
320	$Y_3Al_5O_{12}$ (~97%) $Ia-3d$; a = 12.002(2) Å; V = 1728.9(3) Å^3	$YAlO_3$ (~3%)	5.0
323	$Y_3Al_5O_{12}$ (~90%) $Ia-3d$; a = 12.005(2) Å; V = 1730.1(3) Å^3	$YAlO_3$ (~5%) Al_2O_3 (~3%) Y_2O_3 (~2%)	5.3

Samples 161 and 164 are almost identical in their phase composition: they contain YAG and accompanying phases ($YAlO_3$ and Y_2O_3) in approximately the same ratio. The powder diffraction patterns of samples 318 and 323, in addition to those listed above, also have Al_2O_3 as an accompanying phase and, at the same time, are very close to each other in composition. The purest sample is YAG 320, which contains about 3% $YAlO_3$ impurities (~3%). In sample 161, the amount of the $YAlO_3$ phase is 7%. The obtained unit cell parameters for $Y_3Al_5O_{12}$ in all samples are very close (Δa = 0.006 Å).

Thus, the synthesized ceramics have the $Y_3Al_5O_{12}$ phase as the main one. The proportion of this phase is the same in all samples obtained using electron energies of 1.4, 2.0, and 2.5 MeV and flux powers (in the "without scanning" mode) from 2.5 to 14 kW/cm^2.

3.2. Luminescence of Synthesized Ceramics

A series of studies of the luminescence properties of YAG:Ce ceramics synthesized under different irradiation modes has been carried out. The synthesis was carried out at electron energies of 1.4, 2.0, and 2.5 MeV and electron beam powers in the range from 2.5 to 14 kW/cm^2 (in the "without scanning" mode) and from 12 to 40 kW/cm^2 (in the "with scanning" mode). We measured the excitation and luminescence spectra under stationary conditions at room temperature using an Agilent Cary Eclipse spectrofluorometer for all YAG:Ce ceramic samples synthesized under different irradiation conditions. Examples of spectra measured in samples synthesized with different E and P are shown in Figures 8 and 9. Sample synthesis modes are shown in Table 2.

Table 2. Synthesis modes. The serial number in the table is the sample number in the accounting system used by the authors.

No.	E, MeV	P, kW	
318	2.5 MeV	8 kW	Without Scan
320	2.5 MeV	10 kW	Without Scan
321	2.5 MeV	37 kW	With Scan
325	2.0 MeV	30 kW	With Scan
371	1.4 MeV	25 kW	With Scan

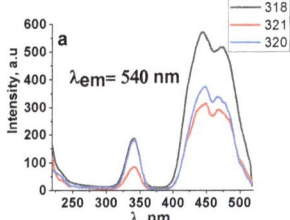 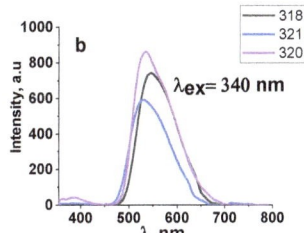 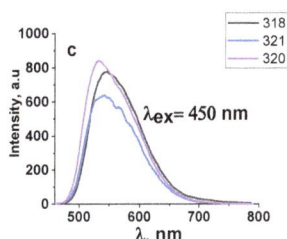

Figure 8. Excitation (**a**) and luminescence spectra (**b**,**c**) of ceramic samples synthesized under the exposure to an electron beam with E = 2.5 MeV and P = 8 and 10 kW/cm² in the "without scanning" mode.

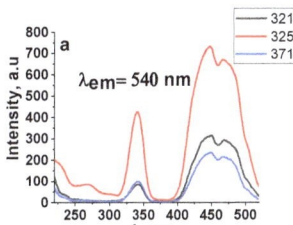 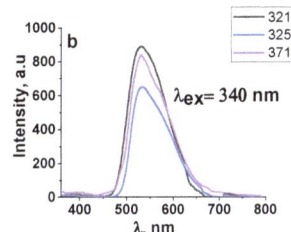 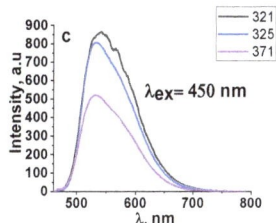

Figure 9. Excitation (**a**) and luminescence (**b**,**c**) spectra of ceramic samples synthesized when exposed to an electron beam with E = 1.4, 2.0, 2.5 MeV and P = 25, 33, 37 kW/cm² in "with scanning" mode.

Figure 8 shows the excitation and luminescence spectra of ceramic samples synthesized under exposure to an electron beam with E = 2.5 MeV and P = 8 and 10 kW/cm² in the "without scanning" mode. The spectra are similar. Additionally, the excitation and luminescence spectra of ceramic samples synthesized under exposure to an electron beam with E = 2.5 MeV and P = 37 kW/cm² in the "with scanning" mode are shown. At E = 2.5 MeV and P = 8 kW/cm² in the "without scanning" mode and P = 37 kW/cm² in the "with scanning" mode, the maximum values of the absorbed energy density Wr are equal. The difference is that in the "without scanning" mode, each irradiated area of the substance is exposed to the Gaussian electron beam passing through this area. In the "with scanning" mode, each area of the substance is exposed to a series of pulses of a scanning beam with a frequency of 50 Hz and a duration of 2 ms with a Gaussian envelope. The excitation and luminescence spectra of all measured samples are similar. Consequently, under the specified synthesis conditions, a change in the power and method of introducing the beam energy into the substance does not affect the entry of the activator into the crystal structure of ceramic crystallites.

Figure 9 shows the excitation and luminescence spectra of ceramic samples synthesized under exposure to an electron beam with E = 1.4, 2.0, and 2.5 MeV and P = 25, 33, and 37 kW/cm² in the "with scanning" mode. At the indicated powers P of the electron beam, the maximum values of the absorbed energy density W_r are equal.

As can be seen from the presented measurement results, the luminescence excitation spectra of ceramic samples synthesized when exposed to an electron beam of different energies are similar. In particular, the luminescence spectra upon excitation in the bands with radiation at 340 and 450 nm are similar. Note that the spectra in Figures 8 and 9 are completely similar to those published in many works devoted to the study of the spectral characteristics of YAG:Ce phosphors and ceramics and are explained by the existence of absorption and emission levels in cerium ions [21–23]. It is assumed that two Ce^{3+} excitation bands at 460 and 340 nm are due to $^4F_{5/2} \rightarrow {}^5D_0, {}^5D_1$ transitions, and the broad luminescence band at 520 and 580 nm is due to $^5D_0 \rightarrow {}^4F_{5/2}, {}^4F_{7/2}$ transitions.

Thus, the radiation synthesis modes, the power of the high-energy electron beam, and their energy do not affect the spectral luminescence characteristics of the resulting ceramics.

4. Discussion

Electron beams are widely used for spraying, sputtering materials [24,25], melting refractory materials [26,27], property modification [28,29], creating coatings [30], and making nanopowders [31].

The promising use of electron beams is due to the high efficiency of conversion of the input energy into the energy of the electron beam, the efficiency of energy transfer to the workpiece and surface, the simplicity of beam control, and the purity of processing procedures. Most frequently, electron fluxes with energies in the range of 10^2–10^5 eV are used for this purpose. For sterilization and introscopy, electron fluxes with energies up to 10 MeV are used. Of particular interest is the use of electrons with energies between 1.0 and 3.0 MeV. Electron accelerators of such energies are simple, although large-sized devices capable of generating electrons with flux power up to 100 kW, which allows the instantaneous transfer of high energy density to matter. Such accelerators have found application in the creation of multi-layer metal articles of great thickness.

We have shown the possibility of obtaining, using such electron beams, dielectric materials with new structural phases from initial substances with the same elemental composition. In the field of a high-energy electron flux, YAG ceramics were obtained from Y and Al oxides with the desired stoichiometric composition. It turned out that the synthesis of high-temperature (refractory) YAG ceramics from Al_2O_3 (t_m = 2044 °C) and Y_2O_3 (t_m = 2410 °C). It has been established that this effect is explained by the dominant role of ionization processes in dielectric materials and their relaxation after creation. Therefore, it is extremely important to understand the elementary processes of electron flux energy transfer to matter.

A quantitative analysis of the energy transfer of an electron beam with a spatially limited cross-section with energies of 1.4, 2.0, and 2.5 MeV has been performed. It is shown that the beam energy is transferred to the substance inhomogeneously. In a mixture of Y_2O_3 (57%) and Al_2O_3 (43%) powders with a bulk density of 1.2 g/cm^3 used for the synthesis of $Y_3Al_5O_{12}$ ceramics, the energy maximum dE/dx is transferred to the substance along the beam propagation axis at a depth of 2.8, 3.7, and 4.6 mm at extrapolated electron path lengths of 9, 10, and 11 mm at energies E equal to 1.4, 2.0, and 2.5 MeV. At these depths, the diameter of the energy loss cross-sectional area exceeds 8, 10, and 12 mm, respectively.

The maxima of the absorbed energy density W at E = 1.4, 2.0, and 2.5 MeV are located along the beam axis and are at depths of 1.8, 2.1, and 2.3 mm. In these areas, the ionization density, the temperature to which the material is heated, is maximum. Obviously, the regions of maximum absorbed energy densities W determine the subsequent processes of structural transformations. In these areas, during the exposure to the beam, energy loss densities sufficient to transform the crystal structure of yttrium and aluminum oxides into yttrium aluminum garnet can be achieved.

The absorbed energy density W is proportional to the electron beam power density P. Structural transformations are realized when P exceeds the threshold P_p. As P increases above P_p, the range ΔL increases (Figure 2a), which can reach ΔL_e, the extrapolated electron range. The distribution of the absorbed energy density W and the position of the maximum absorbed energy density W_r depend on the electron energy, as shown in Figure 2c.

For experimental studies of the synthesis dependence on E and P, a series of samples with the same composition of Y_2O_3 (57%) and Al_2O_3 (43%) + Ce_2O_3 (0.5% of the total mass) was obtained. The P ranges (without scanning) were chosen as 1.5–6, 4–8, and 8–10 kW/cm^2 for E = 1.4, 2.0, and 2.5 MeV, respectively, which fit the W ratios expected from the modeling results. Synthesis under all irradiation modes was realized at a rate above 1 cm/s, completely—in the entire crucible—for 10 s. In all modes, samples of YAG:Ce ceramics of a characteristic yellow color were obtained. At low P, the samples in the form of rods obtained in the mode (without scanning) were covered from above with a layer of charge. As P increased, specimens with an open surface and greater thickness were obtained. As P decreases, the thickness of the forming sample decreases, and the solid rod turns into a dotted one. The smallest samples of ceramics in the form of rare dotted particles with sizes of about 3 mm in diameter and up to 10 mm in length were obtained by exposure to electron beams with E = 1.4 MeV and P = 1.5 kW/cm^2 (without scanning mode). The samples are friable and crumble under slight pressure. These dimensions, apparently, can be considered the smallest, below which radiative synthesis does not occur. The largest dimensions of the YAG:Ce ceramic sample were obtained at E = 2.5 MeV and P = 37 kW/cm^2 ("with scanning" mode). The sample has the form of a plate with dimensions of 90 × 45 mm; the plate weight is 83 g.

The results of the XRD study showed that the dominant phase in the obtained samples was $Y_3Al_5O_{12}$, and the accompanying phase was the $YAlO_3$ phase in an amount from 3 to 7%. The proportion of the main $Y_3Al_5O_{12}$ phase exceeds 90% in all samples obtained using electron energies of 1.4, 2.0, and 2.5 MeV and flux powers (in the "without scanning" mode) from 2.5 to 14 kW/cm^2.

The evidence for the formation of YAG:Ce ceramics is the results of studying the luminescence spectra of the synthesized samples. All synthesized samples of YAG:Ce ceramics have characteristic luminescence bands at 540 nm and excitation bands at 340 and 450 nm in their spectra. Consequently, the activator ions, in a short synthesis time of less than 1 s, have time to integrate into the nodes of the emerging lattice.

5. Conclusions

This paper presents the results of optimal modes search of radiation synthesis of dielectric materials by the example of YAG:Ce ceramics synthesis. Based on the conducted research, the flux rates in the ranges of 3–5, 5–8, and 7–12 kW/cm^2 ("without scanning" mode) using electron energies of 1.4, 2.0, and 2.5 MeV are optimal for the radiation synthesis of YAG:Ce ceramics. The use of E > 3–4 MeV electrons for synthesis may allow for obtaining samples of greater thickness. However, synthesis with electrons of these energies requires a significant increase in power density to compensate for the decrease in the average ionization density due to the increase in the volume of the energy loss region. When using electrons with E < 0.5 MeV, the region of high ionization density becomes so narrow that part of the absorbed energy will go beyond the optimal synthesis region. It should be emphasized that the conclusions and patterns presented are characteristic of dielectric materials only. In metals, the absorbed energy of radiation flux is immediately transferred to the lattice and leads to material heating. Due to the high thermal conductivity of metals, which is two orders of magnitude higher than in dielectrics and three orders of magnitude higher than in dielectric powders, the absorbed energy quickly leaves the region with the maximum absorbed energy density. In metals, only processes stimulated by radiation heating of the material are initiated.

Author Contributions: Conceptualization, V.L. and L.L.; methodology, M.G.; software, A.Z.; validation, L.L., D.M. and A.Z.; formal analysis, V.L.; investigation, V.L.; resources, M.G.; data curation, E.K.; writing—original draft preparation, V.L.; writing—review and editing, A.T.; visualization, L.L.; supervision, V.L.; project administration, V.L.; funding acquisition, A.Z. All authors have read and agreed to the published version of the manuscript.

Funding: This research was funded by the Science Committee of the Ministry of Science and Higher Education of the Republic of Kazakhstan (Grant No. AP14871114) and the Russian Science Foundation of the Russian Federation. (Grant No. 23-73-00108).

Institutional Review Board Statement: Not applicable.

Informed Consent Statement: Not applicable.

Data Availability Statement: The data presented in this study are available on request from the corresponding author.

Acknowledgments: This research was supported by the Russian Science Foundation of the Russian Federation. (Grant No. 23-73-00108) and was performed within the framework of the Agreement on Scientific and Technical Cooperation No. 1736 from 28.02.2022 between the INP of SB RAS and TPU.

Conflicts of Interest: The authors declare no conflict of interest.

References

1. Xiao, Z.; Yu, S.; Li, Y.; Ruan, S.; Kong, L.B.; Huang, Q.; Huang, Z.; Zhou, K.; Su, H.; Yao, Z.; et al. Materials Development and Potential Applications of Transparent Ceramics: A Review. *Mater. Sci. Eng. R Rep.* **2020**, *139*, 100518. [CrossRef]
2. Xia, Z.; Meijerink, A. Ce3+-Doped Garnet Phosphors: Composition Modification, Luminescence Properties and Applications. *Chem. Soc. Rev.* **2017**, *46*, 275–299. [CrossRef] [PubMed]
3. Smet, P.; Parmentier, A.; Poelman, D. Selecting Conversion Phosphors for White Light-Emtting Diodes. *J. Electrochem. Soc.* **2011**, *158*, R37–R54. [CrossRef]
4. Ye, S.; Xiao, F.; Pan, Y.X.; Ma, Y.Y.; Zhang, Q.Y. Phosphors in Phosphor-Converted White Light-Emitting Diodes: Recent Advances in Materials, Techniques and Properties. *Mater. Sci. Eng. R Rep.* **2010**, *71*, 1–34. [CrossRef]
5. Pan, Y.; Wu, M.; Su, Q. Comparative Investigation on Synthesis and Photoluminescence of YAG:Ce Phosphor. *Mater. Sci. Eng. B* **2004**, *106*, 251. [CrossRef]
6. Sharma, S.K.; James, J.; Gupta, S.K.; Hussain, S. UV-A,B,C Emitting Persistent Luminescent Materials. *Materials* **2022**, *16*, 236. [CrossRef]
7. Wang, X.; Li, J.; Shen, Q.; Shi, P. Flux-Grown Y3Al5O12:Ce3+ Phosphors with Improved Crystallinity and Dispersibility. *Ceram. Int.* **2014**, *40*, 15313–15317. [CrossRef]
8. Shi, H.; Zhu, C.; Huang, J.; Chen, J.; Chen, D.; Wang, W.; Wang, F.; Cao, Y.; Yuan, X. Luminescence Properties of YAG:Ce, Gd Phosphors Synthesized under Vacuum Condition and Their White LED Performances. *Opt. Mater. Express OME* **2014**, *4*, 649–655. [CrossRef]
9. Yang, Y.-G.; Wang, X.-P.; Liu, B.; Zhang, Y.-Y.; Lv, X.-S.; Li, J.; Wei, L.; Yu, H.-J.; Hu, Y.; Zhang, H.-D. Molten Salt Synthesis and Luminescence of Dy3+-Doped Y3Al5O12 Phosphors. *Luminescence* **2020**, *35*, 580–585. [CrossRef]
10. Murai, S.; Fujita, K.; Iwata, K.; Tanaka, K. Scattering-Based Hole Burning in Y3Al5O12:Ce3+ Monoliths with Hierarchical Porous Structures Prepared via the Sol–Gel Route. *J. Phys. Chem. C* **2011**, *115*, 17676–17681. [CrossRef]
11. Dippong, T.; Andrea Levei, E.; Cadar, O.; Grigore Deac, I.; Lazar, M.; Borodi, G.; Petean, I. Effect of Amorphous SiO2 Matrix on Structural and Magnetic Properties of Cu0.6Co0.4Fe2O4/SiO2 Nanocomposites. *J. Alloys Compd.* **2020**, *849*, 156695. [CrossRef]
12. Abdullin, K.A.; Kemel'bekova, A.E.; Lisitsyn, V.M.; Mukhamedshina, D.M.; Nemkaeva, R.R.; Tulegenova, A.T. Aerosol Synthesis of Highly Dispersed Y3Al5O12:Ce3+ Phosphor with Intense Photoluminescence. *Phys. Solid State* **2019**, *61*, 1840–1845. [CrossRef]
13. Dai, P.; Ji, C.; Shen, L.; Qian, Q.; Guo, G.; Zhang, X.; Bao, N. Photoluminescence Properties of YAG:Ce3+,Pr3+ Nano-Sized Phosphors Synthesized by a Modified Co-Precipitation Method. *J. Rare Earths* **2017**, *35*, 341–346. [CrossRef]
14. Zhang, L.; Li, X.; Hu, D.; Dobrotvorska, M.; Yavetskiy, R.; Dai, Z.; Xie, T.; Yuan, Q.; Chen, H.; Liu, Q.; et al. Fine-Grained Tb3Al5O12 Transparent Ceramics Prepared by Co-Precipitation Synthesis and Two-Step Sintering. *Magnetochemistry* **2023**, *9*, 47. [CrossRef]
15. Serrano-Bayona, R.; Chu, C.; Liu, P.; Roberts, W.L. Flame Synthesis of Carbon and Metal-Oxide Nanoparticles: Flame Types, Effects of Combustion Parameters on Properties and Measurement Methods. *Materials* **2023**, *16*, 1192. [CrossRef]
16. Huczko, A.; Kurcz, M.; Baranowski, P.; Bystrzejewski, M.; Bhattarai, A.; Dyjak, S.; Bhatta, R.; Pokhrel, B.; Kafle, B.P. Fast Combustion Synthesis and Characterization of YAG:Ce3+ Garnet Nanopowders. *Phys. Status Solidi (B)* **2013**, *250*, 2702–2708. [CrossRef]
17. Le Godec, Y.; Le Floch, S. Recent Developments of High-Pressure Spark Plasma Sintering: An Overview of Current Applications, Challenges and Future Directions. *Materials* **2023**, *16*, 997. [CrossRef]
18. Lisitsyn, V.; Tulegenova, A.; Kaneva, E.; Mussakhanov, D.; Gritsenko, B. Express Synthesis of YAG:Ce Ceramics in the High-Energy Electrons Flow Field. *Materials* **2023**, *16*, 1057. [CrossRef]
19. Lisitsyn, V.M.; Golkovsky, M.G.; Musakhanov, D.A.; Tulegenova, A.T.; Abdullin, K.A.; Aitzhanov, M.B. YAG Based Phosphors, Synthesized in a Field of Radiation. *J. Phys. Conf. Ser.* **2018**, *1115*, 052007. [CrossRef]
20. Volchenko, T.S.; Yalovets, A.P. Calculation of the Effective Thermal Conductivity of Powders Formed by Spherical Particles in a Gaseous Atmosphere. *Tech. Phys.* **2016**, *61*, 324–336. [CrossRef]
21. Dorenbos, P. 5d-Level Energies of Ce3+ and the Crystalline Environment. IV. Aluminates and "Simple" Oxides. *J. Lumin.* **2002**, *99*, 283–299. [CrossRef]
22. Ueda, J.; Tanabe, S. Review of Luminescent Properties of Ce3+-Doped Garnet Phosphors: New Insight into the Effect of Crystal and Electronic Structure. *Opt. Mater. X* **2019**, *1*, 100018. [CrossRef]
23. Lisitsyn, V.M.; Korepanov, V.I.; Yakovlev, V.Y. Evolution of Primary Radiation Defects in Ionic Crystals. *Russ. Phys. J.* **1996**, *39*, 1009–1028. [CrossRef]
24. Shin, J.K.; Jo, S.-H.; Kim, T.-H.; Oh, Y.-H.; Yu, S.; Son, Y.-S.; Kim, T.-H. Removal of NOx Using Electron Beam Process with NaOH Spraying. *Nucl. Eng. Technol.* **2022**, *54*, 486–492. [CrossRef]

25. Yushkov, Y.G.; Tyunkov, A.V.; Oks, E.M.; Zolotukhin, D.B. Electron Beam Evaporation of Boron at Forevacuum Pressures for Plasma-Assisted Deposition of Boron-Containing Coatings. *J. Appl. Phys.* **2016**, *120*, 233302. Available online: https://aip.scitation.org/doi/10.1063/1.4972268 (accessed on 11 April 2023). [CrossRef]
26. Klimov, A.S.; Bakeev, I.Y.; Dvilis, E.S.; Oks, E.M.; Zenin, A.A. Electron Beam Sintering of Ceramics for Additive Manufacturing. *Vacuum* **2019**, *169*, 108933. [CrossRef]
27. Gao, Y. Surface Modification of TA2 Pure Titanium by Low Energy High Current Pulsed Electron Beam Treatments. *Appl. Surf. Sci.* **2011**, *257*, 7455–7460. [CrossRef]
28. Hegelmann, E.; Jung, A.; Hengst, P.; Zenker, R.; Buchwalder, A. Investigations Regarding Electron Beam Surface Remelting of Plasma Nitrided Spray-Formed Hypereutectic Al–Si Alloy. *Adv. Eng. Mater.* **2018**, *20*, 1800244. [CrossRef]
29. Valkov, S.; Ormanova, M.; Petrov, P. Electron-Beam Surface Treatment of Metals and Alloys: Techniques and Trends. *Metals* **2020**, *10*, 1219. [CrossRef]
30. Zolotukhin, D.B.; Kazakov, A.V.; Oks, E.M.; Tyunkov, A.V.; Yushkov, Y.G. Electron Beam Synthesis of Silicon-Carbon Coatings in the Forevacuum Pressure Range. *Ceram. Int.* **2022**, *48*, 13890–13894. [CrossRef]
31. Kim, J.-U.; Cha, S.-H.; Shin, K.; Jho, J.Y.; Lee, J.-C. Synthesis of Gold Nanoparticles from Gold(I)–Alkanethiolate Complexes with Supramolecular Structures through Electron Beam Irradiation in TEM. *J. Am. Chem. Soc.* **2005**, *127*, 9962–9963. [CrossRef]

Disclaimer/Publisher's Note: The statements, opinions and data contained in all publications are solely those of the individual author(s) and contributor(s) and not of MDPI and/or the editor(s). MDPI and/or the editor(s) disclaim responsibility for any injury to people or property resulting from any ideas, methods, instructions or products referred to in the content.

Article

Express Synthesis of YAG:Ce Ceramics in the High-Energy Electrons Flow Field

Victor Lisitsyn [1,*], Aida Tulegenova [2], Ekaterina Kaneva [3], Dossymkhan Mussakhanov [4] and Boris Gritsenko [1]

[1] Department of Materials Science, Engineering School, National Research Tomsk Polytechnic University, 30 Lenin Ave., Tomsk 634050, Russia
[2] Department of Solid State and Nonlinear Physics, Al-Farabi Kazakh National University, Al-Farabi Ave. 71, Almaty 050040, Kazakhstan
[3] X-ray Analysis Laboratory, Vinogradov Institute of Geochemistry SB RAS, 1A Favorsky Str., Irkutsk 664033, Russia
[4] Department of Radio Engineering, Electronics and Telecommunications, Eurasian National University L.N. Gumilyov, Str. Satpaeva 2, Astana 010008, Kazakhstan
* Correspondence: lisitsyn@tpu.ru; Tel.: +79-13-8242-469

Abstract: YAG:Ce ceramics by the direct action of an electron beam with 1.4 MeV energy were synthesized on a mixture of a stoichiometric composition of Y, Al, and Ce oxides without adding any substances to facilitate the process. The synthesis is realized in a time less than 1 s. The main structural phase of the obtained ceramics is YAG and YAP can be additional. The luminescence characteristics of the synthesized samples, the excitation, luminescence, decay time, and pulsed cathodoluminescence spectra, are similar to those known for YAG:Ce phosphors. The conversion efficiency of the excitation energy into the luminescence of the samples reaches 60–70% of those used for the manufacture of LED phosphors. The set of processes that determine the rate and efficiency of radiation synthesis differs from those occurring during thermal methods by the existence of a high degree of the initial compositions' ionization under the influence of a radiation flux and a high probability of the decay of electronic excitations into short-lived radiolysis products.

Keywords: radiation synthesis; ceramics; luminescence; oxides; yttrium–aluminum garnet

1. Introduction

Metal oxides phosphors are the most promising for use in LEDs. They are able to withstand the impact of a high excitation density by the chip radiation for a long time. The synthesis of such materials is difficult due to the feedstock high melting point. The synthesis complexity of multicomponent oxide phosphors is determined by the large difference in the melting points of the starting materials. For example, to obtain a phosphor based on yttrium–aluminum garnet (YAG), the oxides Y_2O_3 and Al_2O_3 with melting points of 2410 and 2044 °C are required. Therefore, the existing methods implement a synthesis at temperatures lower than those necessary to melt the component with the lowest temperature. The most widespread is the solid-phase method for the synthesis of YAG phosphors, with various modifications [1–4]. The initial procedure in this method is the sintering of ceramics from Y_2O_3 and Al_2O_3 powders with activators, for example, Ce_2O_3 using Ba, Na, K, and H_3BO_3 fluorides as binders at 1300–1600 °C. Then, long-term repeated annealing is carried out at temperatures of 1500–1750 °C to form the YAG phase and evaporate all the residues of the binder materials.

The developed synthesis methods, sol-gel [5,6], hydrothermal [7], and others, do not relieve the need for a long-term high-temperature treatment to form a phosphor powder with the desired structure and morphology and a purification from the substances used. The combustion method [8,9] provides the possibility of a rapid synthesis, but the need for a subsequent purification from residues of combustible materials requires annealing at high temperatures.

Thus, all the existing synthesis methods of YAG phosphors have two main disadvantages: a long synthesis time of 20–60 h and the use of additional substances in the synthesis process, the removal of which is very difficult. The disadvantages include the need for a thorough mixing of the initial powders prior to the synthesis.

The impact of hard radiation beams during the synthesis process can contribute to the necessary reactions between the medium elements and increase the efficiency of the formation of a new structure [10,11]. In [12,13], the possibility of synthesizing ceramics based on refractory metal fluorides and oxides was shown for the first time by directly exposing the mixture to flows of high-energy electrons with a high-power density.

This paper is intended to summarize the obtained results of the radiation synthesis of ceramic YAG samples of different compositions in a flux of 1.4 MeV electrons with a power density of 15–25 kW/cm^2, to discuss the possible nature of the processes that provide a high rate and efficiency of the formation of new structural phases.

2. Material Synthesis

The mixture used for the synthesis of cerium-activated yttrium–aluminum garnet (YAG:Ce) ceramics had a stoichiometric composition, consisting of 57 wt% of Y_2O_3 and 43 wt% of Al_2O_3. Cerium oxide Ce_2O_3 was added in an amount of 0.2–2 wt% of the total weight of the mixture. All the starting materials had a purity degree which was not less than PMA and were thoroughly mixed. The mixture for the synthesis of cerium-activated yttrium–aluminum–gallium garnet (YAGG:Ce) ceramics consisted of oxides Y_2O_3, Al_2O_3, and Ga_2O_3 with 1 wt% Ce_2O_3 with a different Al/Ga ratio. Ga entered the lattice by the replacement of Al ions during the synthesis. The ratio of the Al/Ga ions varied from 0 to 1. Additionally, the synthesized ceramic YAGG:Ce was obtained with the replacement of yttrium by gadolinium Gd up to 15%.

The synthesis was carried out by a direct irradiation of the prepared mixture with an electron beam of 1.4 MeV energy and a power density P = 15–25 kW/cm^2 of the ELV-6 accelerator of the INP named Budker SB RAS. Massive copper crucibles had a recess with dimensions of 100 × 50 × 6 mm in the upper part, which was completely covered with an even layer of the charge. Electrons with an energy of 1.4 MeV were completely absorbed by the 6 mm charge layer. The electron beam with a Gaussian flux distribution had a diameter of 1 cm on the charge surface. The flow was scanned in the perpendicular direction of the rectangular surface of the crucible at a frequency of 50 Hz. The crucible moved relative to the scanning beam at a speed of 1 cm·s^{-1}. Thus, each portion of the charge surface in the crucible was subjected to multiple flow effects in 1 s during its movement. The impact of the radiation flux on each area in this mode of exposure can be represented as the impact of a series of rising then falling pulses with a duration of 2 ms. The total time of the action of the electron flow on the entire mixture in a crucible with an area of 50 cm^2 was 10 s. At P = 20 kW/cm^2, with the specified beam sizes, scanning modes, and charge layer thickness, the absorbed energy by the charge was 20 kJ/cm^2 or 15 kJ/g. A typical view of the samples series of YAG:Ce ceramics in a crucible obtained as a result of the synthesis is shown in Figure 1a. The samples have a hard shell, are up to 2 mm thick, and have a porous structure inside; the total weight of the samples in the crucible is 20–30 g. The number of samples in the crucible can vary from one to ten, with dimensions from 40 × 90 × 7 mm^3 to balls with a diameter of 3 mm, depending on the preliminary compaction, the characteristics of the starting materials, and the irradiation modes. The weight loss of the mixture during synthesis, though it depends on the prehistory of the initial substances, can reach 10%, mainly due to the spraying of fine particles of powders when they are charged with electrons.

Figure 1. Photographs of YAG:Ce samples synthesized in the mode with scanning ((**a**), P = 20 kW/cm^2), without scanning ((**b**,**c**) P = 5 and 4 kW/cm^2)), in crucibles (**a**–**c**), taken out from crucibles (**b$_1$**,**c$_1$**), and traces of impact of beams with P = 27(7), 20(5), 3, 15(4) kW/cm^2 on steel plate (**d**). On figure (**d$_3$**) shows the structure of the beam distribution on the surface of the sample.

To study the processes of the radiation synthesis of ceramics, the second method was used to influence the flow of electrons on the charge, which is called the "without scanning" method. In this method, the crucible is pulled under the electron beam without scanning for 10 s. Through each section of the irradiated surface of the mixture passes an increasing–decreasing Gaussian electron beam. To preserve the amount of absorbed energy used in the scanning mode, the power density was reduced by a factor of 4. In this mode, the energy absorbed during irradiation by each section of the charge was equal to the integral of all pulsed actions on the same section in the scanning mode. As a result of the charge irradiation in the "without scanning" mode, a porous ceramic sample in the form of a rod is formed in the crucible along its length. As P decreases, the of the rod diameter decreases, and the sample is formed under the outer surface of the charge. The axis of the rod is at a depth of 2–3 mm from the surface.

Figure 1 shows, for example, photographs of ceramic and steel samples obtained in the two modes.

The same figure shows a photograph of the traces of exposure to a steel plate of electron beams in the "with scanning" mode, with P = 27, 20, 15 kW/cm^2 corresponding to irradiation with P = 7, 5, 4 kW/cm^2 in the "without scanning" mode and at P = 3 kW/cm^2. When exposed to a flow with P = 7 kW/cm^2, the metal melts, the thickness of the melt is 3–4 mm. At P = 5 kW/cm^2, a thin melt film of about 0.1 mm thick is formed on the metal surface. The impact of a flow with P = 4 kW/cm^2 leaves only a trace of the melt film; at 3 kW/cm^2, a trace is formed on the metal due to oxidation and an insignificant only in the central part of the trace of the melt film, since the flux distribution in the beam has a Gaussian shape. Metal melting occurs only at P > 6.5 kW/cm^2 (P > 25 kW/cm^2 in the "with scanning" mode); a layer of melt is formed with a thickness of more than 3 mm.

The difference in the thickness of the YAG:Ce ceramic and metal layers transformed by the radiation flux is probably due to the difference in the thermal conductivity of the substances. Under the influence of 1.4 MeV electrons on the mixture for the synthesis of YAG:Ce ceramics, the maximum energy loss occurs at a depth of 2.2 mm from the surface. The heat released from the zone of the maximum absorption of the energy of the electron flow is removed outside it due to thermal conductivity. The characteristic length of the thermal front displacement for the selected time is determined from the relationship:

$$l = \left(\frac{\lambda}{pC} \cdot t \right)^{\frac{1}{2}} \quad (1)$$

where l is the temperature front displacement length, t is the temperature propagation time, λ is the thermal conductivity coefficient; C is the heat capacity; and p is the density of the material. The thermal conductivity coefficients in W/mK: steel—75, copper—390, and YAG—1.4. The thermal conductivity of a mixture of Al_2O_3 powder and Y_2O_3 with a bulk density of about 1.5 g/cm^3 is 0.15–0.16 W/mK [14]. For 1 s, the front displacement length l in the YAG ceramics is 0.72 mm; in steel, it is 1.5 mm; in a charge for the YAGs synthesis with a bulk density of 1.15 g/cm^3, it is 0.28 mm. During the time of the exposure to the radiation flux in steel, the heat has time to reach the surface but it does not have time in the charge for the YAGs synthesis. Therefore, the totality of the processes in a charge of Al_2O_3 and Y_2O_3 powder should be considered as taking place in a closed region bounded by the walls of a "cold" charge, or as in a closed "reactor".

2.1. Distribution of Energy Losses of an Electron Flux in a Substance under Irradiation

Using the Casino v2.51 program, the numerical simulation of electron energy losses during the passage through the YAG by Monte Carlo methods was performed under the following conditions: electron energy of 1.4 MeV; beam diameter of 7.5 mm; and a YAG density of 4.56 g/cm^3. The results of the calculations for the passage of 10,000 electrons in the YAG are shown in Figure 2. In the experiments performed, the synthesis was carried out by the action of an electron beam on a mixture of stoichiometric powders. Therefore, the electron path depth is given for a bulk density of a mixture of 1.2 g/cm^3 of oxides Y_2O_3, Al_2O_3.

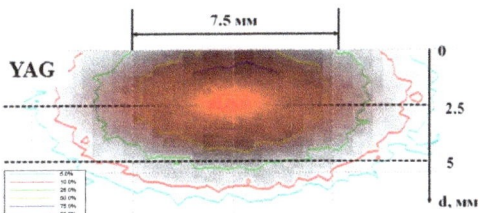

Figure 2. Distribution of the absorbed dose in the mixture during the passage of electrons with energy of 1.4 MeV.

As follows from the results presented in Figure 2, when an electron beam limited in a cross section passes through the substance, the redistribution of energy losses takes place. Part of the energy is transferred to matter outside the beam-confining hole. Part of the energy is redistributed towards the center of the beam during its passage. Due to this, the energy loss density along the beam axis exceeds the peripheral one. The energy loss of the electron flow grows up to a certain depth, then decreases. The result of these effects is the following redistribution of the energy losses in matter. The largest share of energy losses falls on the region remote from the surface and is concentrated along the beam axis. The curves in the figure show the areas of energy loss of an equal magnitude in the relative units. For clarity, the area of the matter with maximum losses is marked with a solid fill. For a mixture of Y_2O_3, Al_2O_3 with a bulk density of 1.2 g/cm^3, prepared for the synthesis of YAG, and of about 0.5 of the total energy is absorbed in a region 4.0 mm in diameter in a cross-section perpendicular to the direction of the electron incidence and at a path depth of 1.4 to 2.9 mm from the surfaces. The energy loss density in the central part is at least 5 times higher than the volume average. This corresponds to the results of the ceramic synthesis described above in the "no-scan" mode.

The following should be noted. The electrons distribution in the beam in our experiments has a Gaussian shape; along the entry axis, the beam density is much higher than at the periphery. Therefore, the described effect of the concentration of absorbed energy along the axis of the beam passage in matter in a real situation should be even more pronounced.

2.2. The Synthesized Ceramics Structure

The ceramic structure was studied using a D8 Advance Bruker diffractometer with a CuKα radiation source. The experiments were performed at room temperature in the Bragg–Brentano geometry with a flat sample in the following modes: 40 kV, 40 mA, an exposure time of 2 s, and a step size of 0.01° 2θ. The received data were processed using the Diffrac^{plus} software package. The samples were identified using the PDF-2 database

(ICDD, 2007) and indexed using the EVA software (Bruker, 2007, Germany) and Topas 4 (Bruker, 2008, Germany). The phase detection limit is 1–3%.

The diffraction patterns of the typical series samples are shown in Figure 3; a summary is in Table 1. From the presented research results, it follows that the structural type of yttrium–aluminum garnet (YAG) is dominant for all the studied samples. Samples 1 and 2 are monophasic. Samples 3 and 4 contain yttrium–aluminum perovskite (YAP) as a minor phase with a content of about 7 and 11%, respectively. Sample 4 with the same composition as 1 was cooled and irradiated again after the synthesis in order to determine the possibility of a reconstruction.

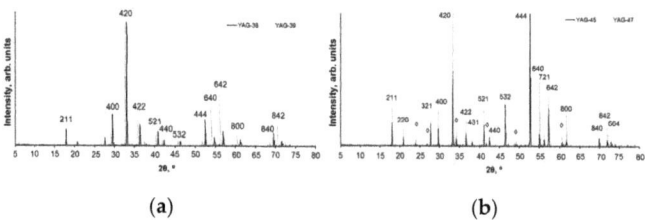

(a) (b)

Figure 3. X-ray diffraction patterns of YAG samples: (**a**) 1 (solid line) and 2 (dashed line): (**b**) 3 (solid line) and 4 (dashed line). Reflections belonging to accompanying phases are marked with a ◊.

Table 1. Results of studying the phase composition of the samples.

	Compound	Main Phase	Accompanying Phase	R_{wp} (%)
1	$Y_3Al_4GaO_{12}$: 1%Ce_2O_3	$Y_3Al_4GaO_{12}$:1%Ce_2O_3 Ia–$3d$ $a = 12.086(2)$ Å $V = 1765.4(1)$ Å^3	No other phases	8.1
2	$Y_3AlGa_4O_{12}$: 1%Ce_2O_3	$Y_3AlGa_4O_{12}$:1%Ce_2O_3 Ia–$3d$ $a = 12.224(3)$ Å $V = 1826.6(1)$ Å^3	No other phases	7.3
3	$Y_3Al_5O_{12}$: 1% Ce_2O_3	$Y_3Al_5O_{12}$:1%Ce_2O_3 (~93%) Ia–$3d$ $a = 12.008(2)$ Å $V = 1731.7(1)$ Å^3	$YAlO_3$ (~7%) $Pnma$ $a = 5.323(5)$ Å $b = 7.349(4)$ Å $c = 5.183(4)$ Å $V = 202.8(1)$ Å^3	6.8
4	$Y_3Al_5O_{12}$: 1% Ce_2O_3 Re-Irradiated	$Y_3Al_5O_{12}$:1%Ce_2O_3 (~89%) Ia–$3d$ $a = 12.010(2)$ Å $V = 1732.4(1)$ Å^3	$YAlO_3$ (~11%) $Pnma$ $a = 5.322(3)$ Å $b = 7.375(3)$ Å $c = 5.185(2)$ Å $V = 203.5(1)$ Å^3	6.7

For a qualitative phase analysis and indexing of the diffraction patterns, the following data from the PDF-2 file (ICDD, 2007) were used: PDF 01-089-6659 "Yttrium Gallium Aluminum Oxide ($Y_3Ga_2Al_3O_{12}$)", PDF 00-033-0040 "Aluminum Yttrium Oxide ($Al_5Y_3O_{12}$)", PDF 01-070-1677 "Yttrium Aluminum Oxide ($YAlO_3$)", and PDF 00-046-1212 "Aluminum Oxide (Al_2O_3)". The unit cell parameters of YAG and YAP are shown in Table 1.

YAG crystallizes in the cubic system, has an elementary I-cell, and a space group of Ia–$3d$. Ce^{3+} ions partially replace Y^{3+}. In samples 3 and 4, the Al^{3+} ions occupy both the tetrahedral and octahedral structural positions, while in samples 1 and 2, the octahedral position is occupied by Al^{3+} and Ga^{3+} cations (with a ratio of ~50/50), and the tetrahedral position is occupied mainly by Al^{3+} ions in sample 1 and Ga^{3+} in sample 2. A similar situation was observed in [15].

Due to the difference in the ionic radii of Al and Ga [16], the volume of the YAG unit cell of the composition $Y_3AlGa_4O_{12}$ (sample 2) significantly exceeds the volume calculated for the unit cell of $Y_3AlGa_4O_{12}$ (1). This is also reflected in Figure 1: the diffraction peaks of sample 2 are shifted towards small angles 2θ. Samples 3 and 4 are almost identical in their phase composition as they contain YAG and YAP in approximately the same ratio; the unit cell parameters for $Y_3Al_5O_{12}$ and $YAlO_3$ are very close. It can be concluded that the phase composition of the resulting sample depends on the composition and morphology of the charge.

2.3. Luminescence of Ceramics Synthesized in a Radiation Field

A series of luminescence properties studies of the synthesized YAG:Ce ceramics was performed. Photoexcitation and photoluminescence (PL) spectra under stationary conditions using a Cary Eclipse spectrofluorometer and time-resolved cathodoluminescence (CL) spectra using a pulsed electron accelerator with an energy of 250 keV for excitation [17–20]. In a generalized form, the spectral-kinetic properties of the obtained YAG:Ce ceramics are shown in Figure 4.

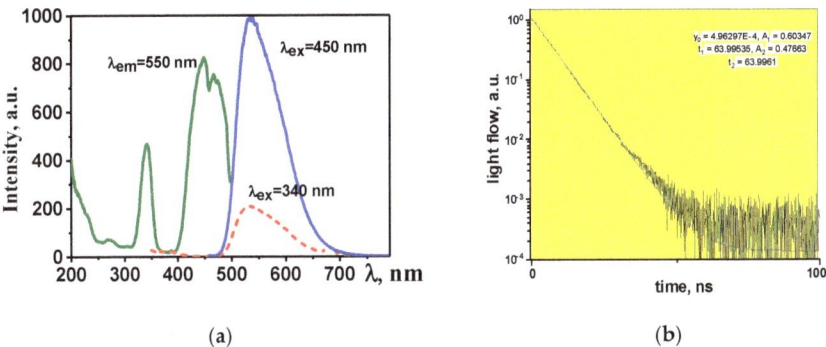

Figure 4. Excitation, luminescence (**a**), and luminescence decay kinetics spectra (**b**) of YAG:Ce ceramics.

Luminescence is effectively excited by UV radiation in the region of 350 and 450 nm. The maximum of the broad luminescence band falls at 550–560 nm upon excitation with λ = 450 nm. The positions of the excitation and luminescence band maxima may differ depending on the prehistory of the starting materials used for the synthesis and the synthesis modes. For example, depending on the synthesis modes, YAM and YAP crystal phases are detected in addition to the basic YAG in the material [21,22]. The introduction of Gd and Ga ions as a modifier into the lattice leads to a shift of the luminescence bands to the long and short wavelength regions of the spectrum [23,24]. The probable cause of the shift in the position of the bands is a change in the lattice parameter and, accordingly, a change in the mutual arrangement of the levels in the activator ion. The CL in the samples of synthesized YAG:Ce ceramics has a typical luminescence spectrum of YAG:Ce phosphors, with a dominant band in the region of 550 nm, and a characteristic luminescence decay time of 60–65 ns.

The luminescence decay kinetic curves are well described by the function:

$$I = A_1 * e^{\left(-\frac{t}{\tau_1}\right)} + A_2 * e^{\left(-\frac{t}{\tau_2}\right)} \qquad (2)$$

As can be seen from the presented results, the qualitative characteristics of the luminescence (spectra, dynamics of their relaxation) are similar to those known for the YAG:Ce phosphors.

Essential for the synthesized luminescent ceramics is a quantitative characteristic, the efficiency of the conversion of the excitation energy Φ_{ex} into luminescence Φ_{em}: $\eta = \Phi_{em}/\Phi_{ex}$ The quantitative measurements of the optical radiation are complex: the measurement result depends on the luminescence and excitation spectra and light distribution in space.

For operational quantitative measurements, it is possible to use methods for comparing luminescence brightness' [25]. The luminescence brightness of diffusely scattering media, which are powders, is proportional to the radiation flux. Industrial luminophores from well-known companies can serve as a reference for comparison: their quantitative characteristics of radiation do not change for at least several years. Therefore, a comparison of the brightness of the prepared series of luminophores (ceramics) under the same excitation conditions gives an objective relative estimate of the efficiency of converting the excitation energy into the luminescence of the samples under study.

In the measurements of the relative brightness of the samples $L_{rel} = L_i/L_s = \Phi_{ei}/\Phi_{es}$, L_i, L_s, Φ_{ei}, and Φ_{es} are the brightness and luminescence beams of the test and reference samples during excitation. The measurements were carried out with a Chrom Metr CS-200 luminance meter when excited by a blue LED with $\lambda = 449$ nm. The phosphor "Grand Lux Optoelectronics Co. Ltd." Shenzhen, China: YAG 01 was used as a reference phosphor. As emphasized above, the synthesized ceramic samples consist of dense closed shells with a porous cavity inside. For the measurements, the samples were split and the outer and inner surfaces were measured. The table shows the numbers of the series of samples synthesized by the authors and the average values of the brightness of a series of measurements are given.

The results of the measurements of the brightness of the outer (L_{iex}) and inner (L_{iin}) surfaces, the X and Y color coordinates, and information on the composition of ceramic samples are presented in Table 2.

Table 2. Brightness and luminescence color coordinates of ceramic samples.

№	The Composition of the Mixture	$L_{iex\ cd/m^2}$	$L_{iin\ cd/m^2}$	L_{iinau}, %	X	Y
YAG 01		10,800		100	0.4604	0.5276
285	Al_2O_3(43%), Y_2O_3(57%), Ce_2O_3 (1%)	6490	7502	69	0.4219	0.5556
286	Al_2O_3 (43%) Y_2O_3(57%), Ce_2O_3 (1%)	6517	6690	61	0.4273	0.5542
288	Al_2O_3 (43%) Y_2O_3(57%), Ce_2O_3 (0.5%)	4550	5730	53	0.4111	0.5610
289	Al_2O_3 (43%) Y_2O_3(57%), Ce_2O_3 (0.2%)	3506	6250	57	0.4100	0.5579
302	Al_2O_3 (43%) Y_2O_3(57%), Ce_2O_3 (0.2%)	4060	6340	56	0.4198	0.5537

As follows from the presented results, the efficiency of converting the blue radiation of the chip into luminescence reaches 70% of that of industrial phosphors. There is a large scatter in the values of the conversion efficiency of the samples from different series and samples of the same series. Obviously, by optimizing the choice of starting materials, their quality, and irradiation regimes, it is possible to increase the conversion efficiency. Note that the color coordinates of the radiation of the samples of all the series are close but differ from those measured for YAG 01.

3. Discussion

It has been established that the synthesis of luminescent YAG:Ce ceramics from refractory powders of yttrium, aluminum, and cerium oxides is possible by means of radiation exposure. All samples have the YAG phase as the dominant one; in some samples, the YAP phase is the additional one. The absence of other phases indicates that, in the process of radiation synthesis, there is an effective mixing of the elements in a charge consisting of Y_2O_3, Al_2O_3, and Ga_2O_3. Synthesis is realized only due to the energy of the high-energy electron flow, only from the charge materials, without the addition of any other materials that facilitate the synthesis, in a time less than 1 s. The conducted studies have shown that the efficiency of the synthesis does not depend on the method of beam exposure. The synthesis result is determined only by the irradiation dose: it is the same in the stationary (without scanning) and pulsed (with scanning) irradiation modes. Consequently, the synthesis is realized completely within the exposure time of a single pulse, that is, within a time less than 2 ms.

The flow of electrons with P = 15–25 kW/cm^2 leads to the formation of YAG:Ce ceramics from Y_2O_3, Al_2O_3, and Ga_2O_3 powders with T_{mp} = 2410, 2044, and 1725 °C. The melting of steel having T_{mp} = 1450–1520 °C occurs when exposed to a flow with P > 25 kW/cm^2. This fact suggests that the synthesis of the YAG:Ce ceramics is realized through processes that differ significantly from those induced thermally. This is confirmed by the results of the synthesis of ceramics based on metal fluorides. The radiation synthesis of fluoride ceramics from powders with T_{mp} = 1300–1400 °C is realized when the charge is exposed to electron flows with P = 12–23 kW/cm^2.

The set of processes stimulated by the action of a radiation flux in dielectric and metallic materials differ in the relaxation of excited states [26,27].

The whole set of processes of radiation energy dissipation in dielectric materials can be represented schematically (Figure 5) and described as follows: 99% of the energy of the high-energy radiation flow is spent on the ionization of the material and electrons pass from the valence (VB) to the conduction band (CB). For 1 act of creating an electron-hole pair (EHP), energy is consumed equal to 2–3 band gaps, e.g., the EHP creation time is no more than 10^{-15} s. Then, it happens that:

- For the relaxation to the lowest states of EHP with the transfer of 0.5–0.7 energy to the lattice for heating, the relaxation time is less than $\tau = 10^{-12}$ s;
- For the decay of electronic excitations, radiative or non-radiative into pairs of short-lived defects (SD), for example, Frenkel pairs, radicals, and ions (I), the time range of these processes is from 10^{-12} s to 10^{-9} s; part of the energy is transferred to the lattice; the decay of electronic excitations is into SD pairs; and their transformation into stable ones is facilitated by the high temperature of the substance;
- For the recombination or transformation of primary pairs into stable ones, the formation of stable complexes and new phases (NP) happens during the time τ from 10^{-9} to 10^{-3} s, and part of the energy is transferred during this forming phase when it is heating;
- The cooling of the material (transfer of energy to the environment by radiation, heat conduction, and convection) occurs in a time longer than 1s;
- In metals, the electronic excitations created under the action of high-energy radiation disappear non-radiatively and without decay into defects in a time less than 10^{-12} s. The energy released in this case is immediately transferred to the lattice and the material is heated.

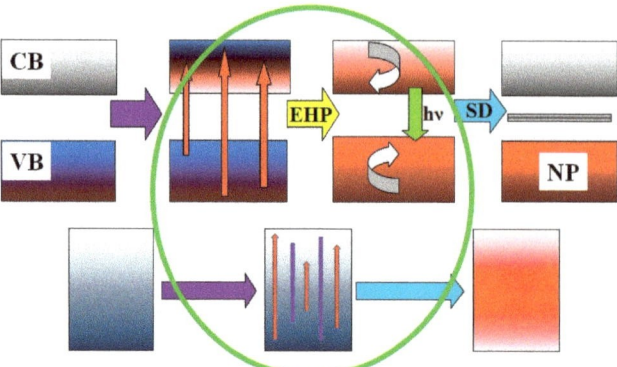

Figure 5. Schematic representation of excitation energy relaxation in dielectrics and metals.

Thus, the main difference between the excitation energy dissipation processes in dielectric and metallic materials is the existence of short-lived radiolysis products in dielectric materials. In metals, there are no processes associated with the decay of electronic excitations into pairs of defects and the formation of radicals, which are mobile intermediate components capable of participating in the transformation of the structure during their existence.

Under the used regimes of radiation exposure, a high ionization density is created in the substance. For example, at P = 20 kW/cm^2, energy W = 7.8 × 10^{22} J/cm^3 is transferred to the charge substance for the YAGs synthesis during an exposure time of less than 1 s. This energy is sufficient to create $N_{el.excitation}$ = 10^{22} cm^{-3} of electronic excitations. During the impact of the flow, a number of electronic excitations is created that exceeds the number of molecules (4.7 × 10^{21} cm^{-3}) and elementary cells (6 × 10^{20} cm^{-3}); that is, it is enough to decompose the lattice of the entire volume of the substance, to create high concentrations of short-lived radiolysis products. The band widths of the materials based on the metal oxides and fluorides used by us for the synthesis are in the range from 6 to 12 eV. Therefore, the concentration of the electronic excitations in them will differ by no more than two times.

The high rate of the synthesis of materials from refractory metal oxides in the field of powerful flows of high-energy electrons suggests the existence of a high efficiency of the mutual exchange of elements between charge particles for the formation of a new phase. Obviously, the exchange of elements between crystalline particles is impossible in 1 s; it is unlikely in the liquid phase after the instantaneous melting of all the particles. The exchange of elements between the particles is possible in 1 s in an electron-ion plasma. It can be assumed that at high excitation densities in dielectric materials, radicals are formed that are highly reactive and capable of providing the formation of new phases corresponding to a given stoichiometric composition.

4. Conclusions

The presented results demonstrate the possibility of a new method for the synthesis of luminescent YAG:Ce ceramics by a direct irradiation with electrons with an energy of 1.4 MeV and a flux density of 15–23 kW/cm^2 of a prepared mixture of stoichiometric composition of yttrium and aluminum oxides. The synthesis is realized in a time less than 1 s without the addition of substances facilitating the synthesis. During the synthesis, there is an effective mixing of the initial powder Y_2O_3, Al_2O_3, Ga_2O_3, and Ce_2O_3, an effective exchange of elements between the particles of the mixture powders. Radiation fusion is implemented for the first time. The resulting ceramics have qualitative radiative characteristics (luminescence and excitation spectra) quite similar to those known for YAG:Ce phosphors. The conversion efficiency of UV radiation with λ = 450 nm into luminescence in the visible region of the spectrum is 60% of that achieved in industrial phosphors. The short duration of the synthesis makes it possible to perform studies of many variants of dependencies on various factors to optimize the process.

A large amount of research remains to be done aimed at establishing the nature of the processes in materials in the field of powerful radiation beams that ensure a high synthesis efficiency, establishing the dependence of the radiative properties of ceramics on the prehistory of the starting materials (purity and fineness), optimizing the composition of activators, synthesis modes (electron energy and power beam), and others. Establishing the possibility of using the radiation synthesis of materials based on YAG with a complex composition of activators, modifiers, and other refractory dielectric materials seems promising.

Author Contributions: Methodology, A.T. and B.G.; Formal analysis, D.M.; Investigation, V.L. and E.K. All authors have read and agreed to the published version of the manuscript.

Funding: This research was supported by TPU development program.

Institutional Review Board Statement: Not applicable.

Informed Consent Statement: Not applicable.

Data Availability Statement: The data presented in this study are available on request from the corresponding author.

Conflicts of Interest: The authors declare no conflict of interest.

References

1. Pan, Y.; Wu, M.; Su, Q. Comparative investigation on synthesis and photoluminescence of YAG:Ce phosphor. *Mater. Sci. Eng. B* **2004**, *106*, 251–256. [CrossRef]
2. Smet, P.F.; Anthony, B. Parmentier; and Dirk Poelman. Selecting Conversion Phosphors for White Light-Emitting Diodes. *J. Electrochem. Soc.* **2011**, *158*, R37. Available online: https://iopscience.iop.org/article/10.1149/1.3568524 (accessed on 4 April 2011).
3. Ye, S.; Xiao, F.; Pan, Y.X.; Ma, Y.Y.; Zhang, Q.Y. Phosphors in phosphor-converted white light-emitting diodes: Recent advances in materials, techniques and properties. *Mater. Sci. Eng. R Rep.* **2010**, *71*, 1–34. [CrossRef]
4. Silveira, L.G.D.; Cotica, L.F.; Santos, I.A.; Belancon, M.P.; Rohling, J.H.; Baesso, M.L. Processing and luminescence properties of Ce:$Y_3Al_5O_{12}$ and Eu:$Y_3Al_5O_{12}$ ceramics for white-light applications. *Mater. Lett.* **2012**, *89*, 86–89. [CrossRef]
5. Jiao, H.; Ma, Q.; He, L.; Liu, Z.; Wu, Q. Low temperature synthesis of YAG:Ce phosphors by LiF assisted sol-gel combustion method. *Powder Technol.* **2010**, *198*, 229–232. [CrossRef]
6. Kareiva, A. Aqueous Sol-Gel Synthesis Methods for the Preparation of Garnet Crystal Structure Compounds. *Mater. Sci.* **2011**, *17*, 428–437. [CrossRef]
7. Xu, M.; Zhang, Z.; Zhao, J.; Zhang, J.; Liu, Z. Low temperature synthesis of monodispersed YAG:Eu crystallites by hydrothermal method. *J. Alloys Compd.* **2015**, *647*, 1075–1080. [CrossRef]
8. Huczko, A.; Kurcz, M.; Baranowski, P.; Bystrzejewski, M.; Bhattarai, A.; Dyjak, S.; Bhatta, R.; Pokhrel, B.; Kafle, B.P. Fast combustion synthesis and characterization of YAG:Ce^{3+} garnet nanopowders. *Phys. Status Solidi B* **2013**, *250*, 2702–2708. [CrossRef]
9. Ohyama, J.; Zhu, C.; Saito, G.; Haga, M.; Nomura, T.; Sakaguchi, N.; Akiyama, T. Combustion synthesis of YAG:Ce phosphors via the thermite reaction of aluminum. *J. Rare Earths* **2018**, *36*, 248–256. [CrossRef]
10. Boldyrev, V.V.; Bystrykh, L.I. The chemical action of ionising radiations on inorganic crystals. *Russian Chem. Reviews.* **1963**, *32*, 426–435. Available online: https://iopscience.iop.org/article/10.1070/RC1963v032n08ABEH001352 (accessed on 4 April 2011).
11. Boldyrev, V.V.; Zakharov, Y.A.; Konyshev, V.P.; Pinaevskaya, E.N.; Boldyreva, A. On Kinetic Factors which Determine Specific Mechano-chemical Processes in Inorganic Systems. *Kinet. Katal.* **1972**, *13*, 1411–1421.
12. Lisitsyn, V.M.; Golkovsky, M.G.; Musakhanov, D.A.; Tulegenova, A.T.; Abdullin, K.A.; Aitzhanov, M.B. YAG based phosphors, synthesized in a field of radiation. *IOP Conf. Ser. J. Phys. Conf. Series.* **2018**, *1115*, 052007. Available online: https://iopscience.iop.org/article/10.1088/1742-6596/1115/5/052007 (accessed on 4 April 2011).
13. Lisitsyn, V.M.; Lisitsyna, L.A.; Dauletbekova, A.K.; Golkovskii, M.G.; Karipbayev, Z.; Musakhanov, D.A.; Akilbekov, A.; Zdorovets, M.; Kozlovskiy, A.; Polisadova, E. Luminescence of the tungsten–activated MgF_2 ceramics synthesized under the electron beam. *Nucl. Instrum. Methods Phys. Res. Sect. B Beam Interact. Mater. At.* **2018**, *435*, 263–267. [CrossRef]
14. Volchenko, T.S.; Yalovets, A.P. Calculation of the effective thermal conductivity of powders formed by spherical particles in a gaseous atmosphere. *Techol. Phys.* **2016**, *61*, 324–336. [CrossRef]
15. Xia, M.; Gu, S.; Zhou, C.; Liu, L.; Zhong, Y.; Zhang, Y.; Zhou, Z. Enhanced photoluminescence and energy transfer performance of $Y_3Al_4GaO_{12}$:Mn^{4+}, Dy^{3+} phosphors for plant growth LED lights. *RSC Adv.* **2019**, *9*, 9244. [CrossRef] [PubMed]
16. Shannon, R.D. Revised effective ionic radii and systematic studies of interatomic distances in halides and chalcogenides. *Acta Cryst.* **1976**, *A32*, 751–767. [CrossRef]

17. Lisitsyn, V.M.; Lisitsyna, L.A.; Golkovskii, M.G.; Musakhanov, D.A.; Ermolaev, A.V. Formation of Luminescing High-Temperature Ceramics Upon Exposure To Powerful High-Energy Electron Flux. *Russ. Phys. J.* **2021**, *63*, 1615–1621. [CrossRef]
18. Ermolayev, A.V.; Tulegenova, A.T.; Lisitsyn, L.A.; Korzhneva, T.G.; Lisitsyn, V.M. The Influence of the Initial Charge Compaction on the Radiation Synthesis of YAG:Ce Ceramics. *Russ. Phys. J.* **2022**, *64*, 1692–1696. [CrossRef]
19. Karipbayev, Z.T.; Lisitsyn, V.M.; Mussakhanov, D.A.; Alpyssova, G.K.; Popov, A.I.; Polisadova, E.F.; Elsts, E.; Akilbekov, A.T.; Kukenova, A.B.; Kemere, M.; et al. Time-resolved luminescence of YAG:Ce and YAGG:Ce ceramics prepared by electron beam assisted synthesis. *Nucl. Instrum. Methods Phys. Res. B* **2020**, *479*, 222–228. [CrossRef]
20. Lisitsyn, V.M.; Lisitsyna, L.A.; Ermolaev, A.V.; Musakhanov, D.A.; Golkovskii, M.G. Optical Ceramics Synthesis in the Field of a Powerful Radiation Flux. *Russ. Phys. J.* **2021**, *64*, 1067–1073. [CrossRef]
21. Fedorov, P.; Maslov, V.A.; Usachev, V.A.; Kononenko, N.E. Synthesis of laser ceramics based on nanodispersed powders of yttrium aluminum garnet $Y_3Al_5O_{12}$. *Vestn. MGTU Im. N.E. Bauman. Ser. Instrum. Mak.* **2012**, *8*, 28–44. Available online: http://engjournal.ru/articles/315/html/files/assets/basic-html/toc.html (accessed on 17 September 2012).
22. Qun, Z.; Guangjie, X.; Fei, W. Preparation and Properties of Ce:YAG Transparent Ceramics. *Chin. J. Lumin.* **2016**, *37*, 650–654. [CrossRef]
23. Xiu-Chen, S.; Sheng-Ming, Z.; Yan-Ru, T.; Xue-Zhuan, Y.; De-Ming, H.; Jie, C. Luminescence Characteristics of Ce:YAG Ceramic Phosphors with Gd^{3+} Doping for White Light-emitting Diodes. *J. Inorg. Mater.* **2018**, *33*, 1119–1124. [CrossRef]
24. Zorenko, Y.; Zorenko, T.; Malinowski, P.; Sidletskiy, O.; Neicheva, S. Luminescent properties of $Y_3Al_{5-x}Ga_xO_{12}$:Ce crystals. *J. Lumin.* **2014**, *156*, 102–107. [CrossRef]
25. Alpyssova, G.; Lisitsyn, V.; Golkovski, M.; Mussakhanov, D.; Karipbayev, Z.; Grechkina, T.; Karabekova, D.; Kozlovskiy, A. Luminescence efficiency of cerium-doped yttrium aluminum garnet ceramics formed by radiation assisted synthesis. *East. -Eur. J. Enterp. Technol.* **2021**, *6*, 39–48. [CrossRef]
26. Lushchik, C.B.; Lushchik, A.C. *Decay of Electron Excitations with the Formation of Defects in Solids*; Nauka: Moscow, Russia, 1989; 264p.
27. Lisitsyn, V.M.; Korepanov, V.I.; Yakovlev, V.Y. Evolution of primary radiation defectiveness in ionic crystals. *Russ. Phys. J.* **1996**, *39*, 5–29. [CrossRef]

Disclaimer/Publisher's Note: The statements, opinions and data contained in all publications are solely those of the individual author(s) and contributor(s) and not of MDPI and/or the editor(s). MDPI and/or the editor(s) disclaim responsibility for any injury to people or property resulting from any ideas, methods, instructions or products referred to in the content.

Article

The Effect of Glass Structure on the Luminescence Spectra of Sm^{3+}-Doped Aluminosilicate Glasses

Andreas Herrmann [1,*], Mohamed Zekri [2], Ramzi Maalej [2] and Christian Rüssel [3]

[1] Department of Inorganic-Nonmetallic Materials, Institute of Materials Science and Engineering, Ilmenau University of Technology, Gustav-Kirchhoff-Str. 5, 98693 Ilmenau, Germany
[2] LaMaCoP, Faculty of Sciences of Sfax, Sfax University, Sfax 3018, Tunisia
[3] Otto Schott Institute of Materials Research, Jena University, Fraunhoferstr. 6, 07743 Jena, Germany
* Correspondence: andreas.herrmann@tu-ilmenau.de

Abstract: Peralkaline Sm^{3+}-doped aluminosilicate glasses with different network modifier ions (Mg^{2+}, Ca^{2+}, Sr^{2+}, Ba^{2+}, Zn^{2+}) were investigated to clarify the effect of glass composition and glass structure on the optical properties of the doped Sm^{3+} ions. For this purpose, the Sm^{3+} luminescence emission spectra were correlated with the molecular structure of the glasses derived by molecular dynamics (MD) simulations. The different network modifier ions have a clear and systematic effect on the peak area ratio of the Sm^{3+} emission peaks which correlates with the average rare earth site symmetry in the glasses. The highest site symmetry is found for the calcium aluminosilicate glass. Glasses with network modifier ions of lower and higher ionic radii show a notably lower average site symmetry. The symmetry could be correlated to the rare earth coordination number with oxygen atoms derived by MD simulations. A coordination number of 6 seems to offer the highest average site symmetry. Higher rare earth coordination probabilities with non-bridging oxygen result in an increased splitting of the emission peaks and a notable broadening of the peaks. The zinc containing glass seems to play a special role. The Zn^{2+} ions notably modify the glass structure and especially the rare earth coordination in comparison to the other network modifier ions in the other investigated glasses. The knowledge on how glass structure affects the optical properties of doped rare earth ions can be used to tailor the rare earth absorption and emission spectra for specific applications.

Keywords: aluminosilicate glass; glass structure; hypersensitivity; rare earth site symmetry; samarium; molecular dynamics simulations; luminescence

Citation: Herrmann, A.; Zekri, M.; Maalej, R.; Rüssel, C. The Effect of Glass Structure on the Luminescence Spectra of Sm^{3+}-Doped Aluminosilicate Glasses. *Materials* **2023**, *16*, 564. https://doi.org/10.3390/ma16020564

Academic Editor: Dirk Poelman

Received: 22 November 2022
Revised: 15 December 2022
Accepted: 23 December 2022
Published: 6 January 2023

Copyright: © 2023 by the authors. Licensee MDPI, Basel, Switzerland. This article is an open access article distributed under the terms and conditions of the Creative Commons Attribution (CC BY) license (https:// creativecommons.org/licenses/by/ 4.0/).

1. Introduction

Rare earth-doped glasses are the backbone of today's telecommunication. Erbium (Er^{3+})-doped silica-based glass fibers are used for amplification of optical signals in long distance fiberoptic cables which currently span the whole globe and make high-speed Internet, world-wide broadcasting, and communication possible. However, rare earth-doped glasses are also used as optical amplification medium for lasers, or can be used as optical sensors or wavelength converters for lighting and solar applications. Crucial for the use of rare earth-doped materials are their optical properties, which strongly depend on the host material. Crystalline materials are, in this respect, well investigated and understood; however, the effect of glass composition and molecular structure on the properties of the doped rare earth ions is not yet fully decoded. The main reasons are the amorphous, irregular molecular structure of glasses and the low rare earth doping concentrations that usually do not exceed about 1 mol%.

In our previous publications [1,2], among others, we tried to clarify the effect of glass composition and glass structure of aluminosilicate glasses on the molecular level on the optical properties of doped rare earth ions. Sm^{3+}- and Er^{3+}-doped glasses and their optical properties were respectively investigated in these publications. Aluminosilicate

(AS) glasses offer some advantages in comparison to other glass types, e.g., in many cases better mechanical properties and lower coefficients of thermal expansion, and also notably broadened rare earth absorption and emission spectra [3,4]. All these properties are beneficial for their potential application as laser-active material, especially for generation of high-power laser pulses; however, their mechanical properties also strongly depend on their actual composition.

In principle, only a few methods are suitable to study the incorporation of rare earth elements into glasses. Raman and infrared spectroscopy are not sensitive enough for taking the small rare earth concentration usually present in luminescent glasses into account. Despite the paramagnetism of rare earth ions, Electron Paramagnetic Resonance (EPR) spectroscopy does not provide useful information (with the exception of Eu^{2+}) and Nuclear Magnetic Resonance (NMR) spectroscopy is strongly affected by the strong paramagnetic shift induced by the rare earth elements. Mößbauer spectroscopy can only be applied for europium (^{151}Eu) due to the lack of other suitable radionuclides. A remaining favorable method seems to be extended X-ray absorption fine structure (EXAFS) which, however, requires synchrotron radiation and hence is not widely available. The only favorable method to gain insight into the incorporation of rare earth ions to the glass structure at low doping concentrations is the method of molecular dynamic (MD) simulations, which gives the actual three-dimensional atomic structure of the glasses, including the rare earth dopants' coordination with neighboring atoms. By using this method, we were so far able to correlate the increased splitting of absorption and emission peaks to increased coordination numbers of the doped rare earth ions with so called non-bridging oxygen (NBO) atoms $^-O–Si\equiv$, i.e., oxygen atoms that are not bond to two network former ions (e.g., $\equiv Si–O–(Al, Si)\equiv$) and therefore represent a "break" in the glass network [2,5,6]. On the other hand, we proposed a correlation of the overall average rare earth coordination number with the rare earth site symmetry and the intensity of the so called "hypersensitive" optical rare earth transitions in the investigated glasses [2]. These absorption or emission transitions show a much higher sensitivity in their oscillator strengths, i.e., their intensity, than other non-hypersensitive transitions to changes of the host material or even single ligands [7,8]. In fact, the correlation between rare earth site symmetry and intensity of the hypersensitive transitions is already proven in crystalline materials [7]; however, up to now, it is unclear, how exactly the rare earth site symmetry is affected in glasses. In the following, we report on luminescence spectra of Sm^{3+}-doped alkaline earth aluminosilicate glasses and attributed MD simulations. Previous results from luminescence spectroscopy [1] and MD simulations [2] are revisited and included in the discussion. In comparison to alkali AS glasses, alkaline earth-containing AS glasses show lower melting temperatures and are therefore easier to produce and possess in most cases better mechanical properties [3].

For a short or a more comprehensive introduction to aluminosilicate glass structure, the reader is referred to [1] and [9], respectively.

2. Materials and Methods

Glass preparation: For each glass composition, 100 to 200 g of glass was prepared by using oxides and carbonates of the molar compositions given in Table 1. The used raw materials were MgO (Merck KG, Germany), $CaCO_3$ (Merck KG, Germany), $SrCO_3$ (Reachim, Ukraine), $BaCO_3$ (Reachim, Ukraine), ZnO (Merck KG, Germany), Al_2O_3 (Pengda Munich GmbH, Germany), SiO_2 (Sipur A1, Schott Jenaer Glas GmbH, Germany), and Sm_2O_3 (Auer-Remy GmbH, Germany). All glasses are peralkaline aluminosilicate glasses with a molar ratio of 35/10/55 of network modifier oxide, Al_2O_3, and SiO_2, respectively. Consequently, the samples are denoted as XAS3510, where X stands for the used network modifying ion. The doping concentration of Sm^{3+} was kept constant at 1×10^{20} ions/cm^3. This corresponds to about 0.2 mol% Sm_2O_3, depending on the density of the glass. A constant rare earth volume concentration is important to avoid different luminescence concentration quenching intensities for different glass samples. The batches were thoroughly mixed, transferred to a platinum crucible, and then placed in an electric furnace for melting between 1550 and

1630 °C, depending on glass composition. To assure a good glass quality, the melt was shaken from time to time. Usually, the melting time was between 2 and 3 h. Afterwards, the glass melts were removed from the furnace, poured on a brass block, and transferred to a cooling furnace preheated to around 20 K above the glass transition temperature T_g. Subsequently, the cooling furnace was switched off, which allowed the samples to cool down gradually over night. In this way, homogeneous and bubble-free glasses were obtained. From the glass blocks, samples of 1 cm thickness were cut and polished for the optical measurements.

Table 1. Chemical compositions, densities, transition temperatures T_g, refractive indices n_e, theoretical optical basicities Λ, peak areas A_{645}, A_{600} and peak area ratios A_{645}/A_{600} of the transitions at 645 nm ("hypersensitive") and the transitions at 600 nm of the studied samples.

Sample Name	MgAS3510	ZnAS3510	CaAS3510	SrAS3510	BaAS3510
Network modifier oxide (mol%)	(MgO) 35	(ZnO) 35	(CaO) 35	(SrO) 35	(BaO) 35
Al_2O_3 (mol%)	10	10	10	10	10
SiO_2 (mol%)	55	55	55	55	55
density (g/cm^3)	2.62	3.24	2.80	3.28	3.75
transition temperature T_g (°C)	811	705	812	795	771
refractive index n_e	1.562	1.614	1.594	1.599	1.621
theoretical optical basicity Λ	0.561	0.592	0.605	0.625	0.635
peak area A_{600} (cm^{-1})	404.140	417.650	404.985	414.462	424.554
peak area A_{645} (cm^{-1})	236.895	234.086	220.666	228.934	236.162
peak area ratio A_{645}/A_{600}	0.586	0.560	0.545	0.552	0.556

Glass characterization: The glass densities were measured using an AccPyc 1330 pycnometer (Micromeritics GmbH, Unterschleißheim, Germany). The error of these measurements was lower than ± 0.03 g/cm^3. The transition temperatures were derived by using a differential scanning calorimeter (DSC PT 1600, Linseis GmbH, Selb, Germany) at a heating rate of 10 K/min. For the determination of the exact transition temperature, T_g, the onset point of the DSC curve was used. The error of this method is estimated to about ± 5 K.

The refractive indices were measured using a Pulfrich-Refractometer PR2 (VEB Carl Zeiss Jena, Jena, Germany). The measurement error depends strongly on the optical quality of the glasses and is about ± 0.002. The luminescence emission and excitation spectra were recorded using a luminescence spectrometer RF-5301PC (Shimadzu, Kyoto, Japan) operated in reflection mode.

The optical basicities Λ of the glasses were calculated from their molar compositions according to Duffy [10] using the averaged basicity values of Duffy and Lebouteiller/Courtine published in [11]. The used values are $\Lambda_{SiO_2} = 0.48$, $\Lambda_{Al_2O_3} = 0.603$, $\Lambda_{MgO} = 0.78$, $\Lambda_{ZnO} = 0.935$, $\Lambda_{CaO} = 1.00$, $\Lambda_{SrO} = 1.10$, and $\Lambda_{BaO} = 1.15$.

Molecular dynamic simulations: The molecular dynamic (MD) simulations were conducted as stated in detail in our previous publications on the topic [5,12]. We used a specific method of MD simulations, the so-called "inherent structure (IS) sampling". This method uses relatively small sets of atoms (around 400 in our case) but simulates a large number of them (3000 in our case). The molecular structures of these 3000 IS are then statistically analyzed by weighting them by their potential energy per atom. That means an inherent structure of relatively high potential energy is regarded with less weight than a structure with comparably low potential energy, since a low energy structure is assumed to better represent the real atomic structure of the glass. With this method, we obtained the coordination of 6000 rare earth ions for each of the different glass compositions. At a doping concentration of about 0.5 at% (two rare earth atoms per 400 atoms in each IS), a MD simulation of about 1,200,000 atoms with the conventional method would be needed, which would probably take years to be calculated with recent computational power. In [12], we have also shown that the inherent structure sampling results in similar molecular structures as the conventional MD simulation method (here compared to a simulation size of 6460 atoms).

For the MD simulations, the Large-Scale Atomic/Molecular Massively Parallel Simulator (LAMMPS, Sandia National Laboratories, Albuquerque, NM, USA) [13] along with the empirically parameterized interatomic potential functions of Pedone et al. [14] were employed. This parametrization already proved to accurately predict the atomic structure of various silica-based glasses [15–18]. Computation of short-range interactions used a cutoff of 15 Å, and for summation of long-range Coulomb interactions, the particle-particle-particle-mesh method [19] was applied. Integration of the equations of motion used the velocity Verlet algorithm [20] with a time step of 1 fs. At first, initial structures were generated by randomly placing ions according to the molar glass composition ($Gd_2NM_{50}Al_{29}Si_{79}O_{253}$, Σ = 413 atoms, NM = network modifier atom) into a cubic unit cell with a volume corresponding to the experimentally observed mass density of the glass composition [1]. Next, the initial structures were geometrically optimized and equilibrated at T = 3000 K for 6.5 ns using the canonical (NVT, constant particle number N, volume V and temperature T) ensemble along with the Nosé-Hoover thermostat [21,22] and employing periodic boundary conditions. During the last 6 ns, geometry optimizations under constant (zero) pressure conditions were applied to structures taken from the MD trajectory every 2 ps yielding the 3000 inherent structures for each glass composition. In the simulations Gd^{3+} was chosen as model rare-earth ion because there are no potential functions published so far for Sm^{3+}, and because Gd^{3+} has a similar ionic radius as Sm^{3+} (R_{Gd} = 0.94 Å, R_{Sm} = 0.96 Å) [23].

3. Results and Discussion

3.1. Peak Area Ratio and Sm^{3+} Site Symmetry

The left diagram in Figure 1 shows the emission spectra of Sm^{3+}-doped peralkaline aluminosilicate glasses with different network modifier (NM) ions, Mg^{2+}, Ca^{2+}, Sr^{2+}, and Ba^{2+}. The doping concentration is constant at 1×10^{20} Sm^{3+} ions/cm^3. The sample data have already been published in [1] and the spectrum of sample BaAS3510 has already been shown there; however, here the spectra of different samples are added for comparison. Sm^{3+} shows a strong orange-red luminescence emission if excited in the UV-blue spectral region. The typical Sm^{3+} emission spectrum consists of four distinct emission peaks at around 565, 600, 645, and 710 nm due the transitions from the energy level $^4G_{5/2}$ to the levels $^6H_{5/2}$, $^6H_{7/2}$, $^6H_{9/2}$, and $^6H_{11/2}$, respectively [24,25]. The spectra in Figure 1 are normalized to their highest emission peaks at around 600 nm. As shown in Figure 1 (left), the relative intensity of the peak at 645 nm changes with glass composition and shows a minimum for the CaAS3510 composition. For more clarity, the spectral region around this peak is shown in more detail in the inset of Figure 1 (left). The transition $^4G_{5/2} \rightarrow {}^6H_{9/2}$ at about 645 nm has $\Delta J = 2$ and therefore is a pure electric dipole transition [25]. Although typically not considered as a hypersensitive transition, it is heavily affected by the electrical field surrounding the Sm^{3+} ion. The transition $^4G_{5/2} \rightarrow {}^6H_{7/2}$ at about 600 nm has $\Delta J = 1$ and a notable magnetic dipole contribution [25]. Therefore, this transition is less affected by changes in the local electrical field. As already shown in [1] and elsewhere (e.g., [26–29]), the ratio of these two Sm^{3+} emission peaks is highly sensitive to structural changes in the host composition. For this reason, the peak area ratio A_{645}/A_{600} of the peaks at 645 nm and 600 nm was calculated by integrating the peaks in energy scale. These data are added to Table 1. The peak area ratio A_{645}/A_{600} corresponds to the asymmetry at the Sm^{3+} site. That means the Sm^{3+} ions in the glass CaAS3510 on average show the lowest asymmetry (=highest symmetry) of the investigated glasses. The glass compositions with NM ions of smaller (Mg^{2+}) and larger ionic radii (Sr^{2+}, Ba^{2+}) show a lower average symmetry at the Sm^{3+} sites. It must be noted that the differences in the peak area ratios of the glasses shown here are comparably small. With different network modifying ions, most notably NM ions of low field strength (e.g., alkali ions), and lower NM/Al_2O_3 ratios, the Sm^{3+} emission spectra can be altered more fundamentally. As an example, the emission spectrum of the

Sm^{3+}-doped peralkaline potassium aluminosilicate glass KAS3010 from [1] is shown in the right diagram of Figure 1. The peak at 645 nm can be even the most intense, thereby changing the overall emission color to a deeper red [1,27,29]. This shows how specific rare earth emission or absorption peaks can be amplified or attenuated by changing the host glass composition.

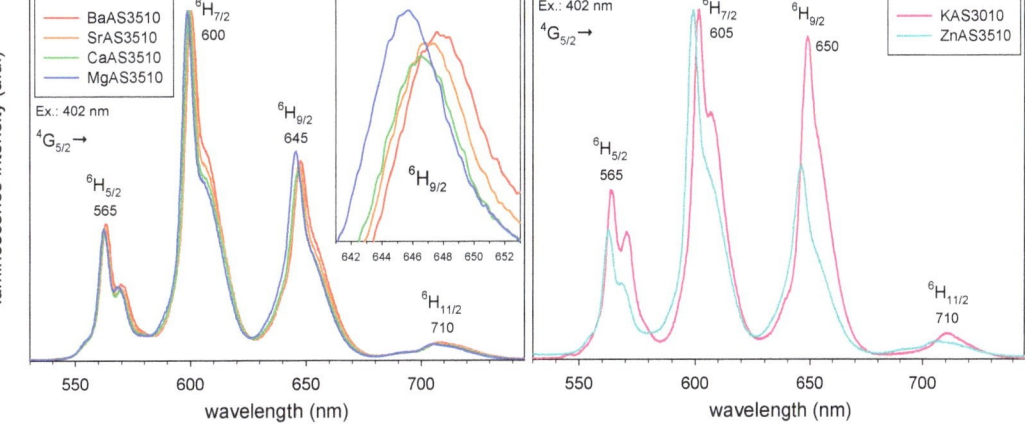

Figure 1. Sm^{3+} emission spectra of different alkaline earth aluminosilicate glasses (**left**), and of a potassium and a zinc aluminosilicate glass (**right**) excited at around 402 nm. The spectra are normalized to their most intense peak at around 600 nm. The peak intensity of the peak at 645 nm shows clear differences in dependence of the host glass (magnified in the inset left).

The left part of Figure 2 shows a glass structure as obtained by the MD simulations, actually one of 3000 inherent structures of the CaAS3510 glass composition. For more clarity, a slice of the three-dimensional structure has been cut out (right picture in Figure 2). Clearly to be seen are the interconnected SiO_4 (yellow) and $[AlO_4]^-$ tetrahedra (gray) which form the three-dimensional glass network. The tetrahedra are connected by bridging oxygen (BO, red). Note that the structure continues in front and behind the shown slice, which is noted by the red BO that are not connected in the picture. The network is interrupted by the network modifying (NM) ions Ca^{2+} (light blue) and Gd^{3+} (pink). Their charge is compensated by non-bridging oxygen (NBO, blue) and $[AlO_4]^-$ groups.

Table 2 shows the coordination numbers (CN) and distances of the doped rare earth ions, Gd^{3+} in this case, with their neighboring atoms in the first (O) and second (Si, Al, NM) coordination sphere derived from the statistical analysis of the MD simulation results. The results of the barium- and magnesium-containing compositions are taken from [2]. The coordination numbers with oxygen systematically decrease with increasing ionic radius of the NM ion, from about 6.06 for MgAS3510 to about 5.60 for the BaAS3510 glass. Additionally, the glass composition ZnAS3510 was investigated, since Zn^{2+} has a similar ionic radius as Mg^{2+} (R_{Zn} = 0.60 Å, R_{Mg} = 0.57 Å, both at CN 4 [23]) and shows a peak area ratio between MgAS3510 and CaAS3510 [1]. Therefore, also a similar glass structure as for MgAS3510 was assumed. However, as the simulations show, the Gd^{3+} coordination number in ZnAS3510 (5.54) is much different from MgAS3510 (6.06) and even slightly smaller than for BaAS3510 (5.60) (Table 2). This shows that the Zn^{2+} ion has a different effect on the glass structure than the alkaline earth ions, despite its similar emission spectrum. For comparison, the emission spectrum of Sm^{3+}-doped ZnAS3510 is added to the right diagram in Figure 1. As already stated, it is very similar to the spectrum of MgAS3510.

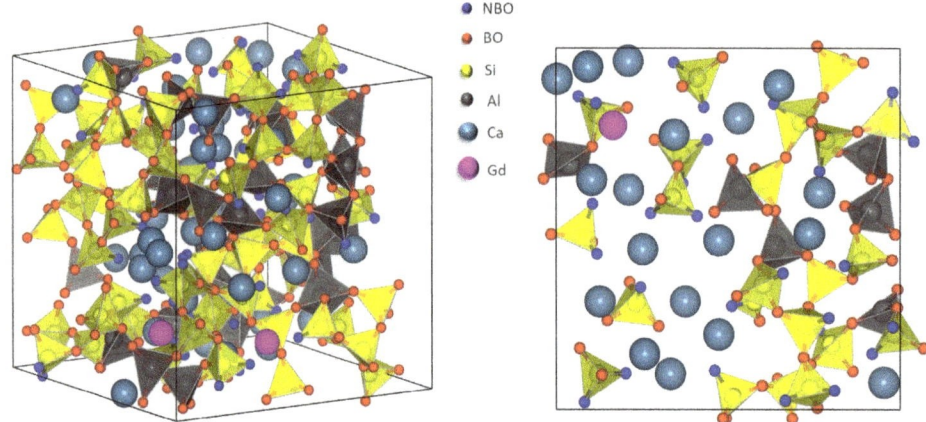

Figure 2. Molecular dynamic simulation results. **Left picture:** An inherent structure of 413 atoms of the glass composition CaAS3510 as obtained by the MD simulations. **Right picture:** A slice of the CaAS3510 composition showing its structure in more detail. SiO_4 tetrahedra (yellow) and $[AlO_4]^-$ tetrahedra (gray) interconnected by bridging oxygen (BO, red) form the three-dimensional glass network. The network modifying ions Ca^{2+} (light blue) and Gd^{3+} (pink) interrupt the network and form depolymerized regions. Their charge is compensated by non-bridging oxygen (NBO, blue) and $[AlO_4]^-$. Note that the structure extends behind and in front of this slice.

Table 2. Gd^{3+} coordination according to molecular dynamic simulations. **Upper part of the table:** the coordination numbers (CN) of Gd^{3+} with different oxygen species in the first coordination sphere in the investigated glasses and the percentages of the different oxygen coordinations. The Gd-O distances are constant at 2.25 Å. NBO: non-bridging oxygen, BO: bridging oxygen, Tri: oxygen triclusters. **Lower part:** the coordination numbers (CN) of Gd^{3+} with different cations Al^{3+}, Si^{4+}, Mg^{2+}, Zn^{2+}, Ca^{2+}, Sr^{2+}, Ba^{2+} in the second coordination sphere, the percentages of the different coordinations and the distance between Gd^{3+} and the respective atoms.

Gd^{3+} Coordination Number (Percentage) Distance	MgAS3510	ZnAS3510	CaAS3510	SrAS3510	BaAS3510
Gd-NBO	4.047 (66.8%)	3.805 (68.7%)	4.209 (70.4%)	4.123 (72.6%)	4.275 (76.4%)
Gd-BO	1.979 (32.7%)	1.712 (30.9%)	1.746 (29.2%)	1.539 (27.1%)	1.310 (23.4%)
Gd-Tri	0.031 (0.5%)	0.020 (0.4%)	0.024 (0.4%)	0.017 (0.3%)	0.012 (0.2%)
Σ (Gd-O)	6.058 2.25 Å	5.538 2.25 Å	5.979 2.25 Å	5.679 2.25 Å	5.598 2.25 Å
Gd-Al	1.906 (16.4%) 3.51 Å	1.883 (17.0%) 3.51 Å	1.749 (14.9%) 3.57 Å	1.779 (15.5%) 3.45 Å	1.715 (14.6%) 3.57 Å
Gd-Si	5.487 (47.2%) 3.57 Å	5.177 (46.7%) 3.57 Å	5.231 (44.7%) 3.57 Å	5.008 (43.6%) 3.57 Å	4.674 (39.9%) 3.57 Å
Gd-NM	(Mg) 4.223 (36.4%) 3.33 Å	(Zn) 4.030 (36.3%) 3.27 Å	(Ca) 4.732 (40.4%) 3.51 Å	(Sr) 4.703 (40.9%) 3.75 Å	(Ba) 5.329 (45.5%) 3.87 Å

Although the MD simulations were conducted with Gd^{3+} as a model rare earth ion, the simulation results can be compared to the spectral measurements of Sm^{3+}-doped samples. It is assumed that both rare earth ions are incorporated to the glass structure in a similar way. However, Gd^{3+} has a slightly smaller ionic radius (R_{Gd} = 0.94 Å) than Sm^{3+} (R_{Sm} = 0.96 Å) (both at a CN of 6) [23]. That means, the actual coordination numbers of Sm^{3+} are probably even slightly higher than the values obtained for Gd^{3+}.

In Figure 3, the peak area ratios A_{645}/A_{600} from the emission spectra are plotted versus the Gd^{3+} coordination numbers obtained from the MD simulations. As already discussed above, the sample CaAS3510 with a coordination number close to 6 offers the highest average symmetry for the doped Sm^{3+} ions. Higher coordination numbers in MgAS3510 and lower coordination numbers in SrAS3510, BaAS3510, and ZnAS3510 offer, on average, a lower site symmetry. Interestingly, the data point of ZnAS3510 is perfectly in line with the other measurements, although not close to MgAS3510 as initially assumed, but rather close to BaAS3510. In our previous publication on this matter, we already found an increasing symmetry for increasing coordination numbers from about 5.4 to 6.0 for Er^{3+}-doped glasses of similar compositions and assumed, that the coordination number 6 could mark a symmetry maximum [2]. Unfortunately, Er^{3+} has a much smaller ionic radius (R_{Er} = 0.89 Å) than Gd^{3+} and Sm^{3+} [23]. Therefore, no higher coordination numbers than 6 could be generated in these glasses and the assumed maximum symmetry at a CN of 6 could not be proven. The larger Sm^{3+} ion, on the other hand, shows a higher coordination number than 6 in MgAS3510 and, as shown in Figure 2, its site symmetry in this glass is much lower than for the other samples. However, it is only one of five data points. From a geometrical point of view, a maximum average symmetry at a coordination of 6 is reasonable. A typical coordination could for instance be a (distorted) octaeder with the rare earth ion in the center.

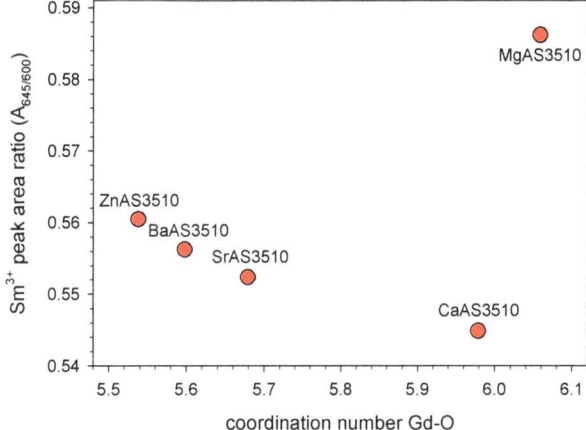

Figure 3. Sm^{3+} emission peak area ratio of the peaks at 645 and 600 nm in dependence of the rare earth (Gd^{3+}) coordination number with oxygen from molecular dynamics simulations for the investigated aluminosilicate glasses. The lowest peak area ratio corresponds to the highest average symmetry at the rare earth sites in the glass structure, which is found at a coordination number of around 6.

The zinc aluminosilicate glass seems to play a special role. Despite the relatively small ionic radius of Zn^{2+}, the rare earth ions in this sample have a comparably low coordination number. The reason for this behavior could be the ZnO's character as a so-called "intermediate" oxide, i.e., an oxide with characteristics between typical network formers, such as SiO_2, and typical network modifiers, such as the alkaline earth oxides. In Table 3, the coordination of the network modifier ions in their first and second coordination sphere derived from the MD simulations are compared. Interestingly, Zn^{2+} shows by far the smallest coordination number with NM ions in the glass despite its higher ionic radius than Mg^{2+}. Zn^{2+} seems to be integrated into the glass forming network of SiO_4 and $[AlO_4]^-$ tetrahedra, preferably forming ZnO_4 tetrahedra itself. This can be assumed from its low CN with oxygen, which is only 4.2, i.e., much lower than for all other NM ions investigated here, e.g., 4.6 for Mg^{2+} (Table 3). Moreover, its bond length of 1.95 Å to neighboring oxygen atoms is the lowest of all divalent cations (Table 3). In [30], Cormier and coworkers also suggest that Zn competes with aluminum in network-forming positions. Therefore, its tendency to coordinate with other network modifying ions in the glass structure is lower. Consequently, the rare earth ions in this glass face a molecular structure which is less modified and rather resembles a metaluminous composition, i.e., a composition where formally all network modifier charges are compensated by $[AlO_4]^-$ groups and where the structure is dominated by a higher polymerization, i.e., a higher percentage of $\equiv$Si–O–(Al, Si)$\equiv$ chains. As reported in [5,12], metaluminous aluminosilicate glasses generally show notably lower rare earth CN than peralkaline ones, which fits to the above explanations. Even the comparably low T_g and melting temperature of ZnAS glasses can probably be explained by this structural model. The $\equiv$Si–O–(Al, Si)$\equiv$ chains are probably weakened by frequent incorporation of ZnO_4 tetrahedra. However, these assumptions must be discussed in more detail in future works.

Table 3. Network modifier (NM) coordination according to molecular dynamic simulations. **Upper part of the table:** the coordination numbers (CN) and distances of the network modifying ions with oxygen in their first coordination sphere. **Lower part:** CN with Al^{3+}, Si^{4+}, and network modifier cations in the second coordination sphere. In the lower part, the percentages of the different cation coordinations are also given.

NM Coordination Number (Percentage) Distance	MgAS3510	ZnAS3510	CaAS3510	SrAS3510	BaAS3510
NM-O	(Mg) 4.635 2.01 Å	(Zn) 4.225 1.95 Å	(Ca) 6.230 2.37 Å	(Sr) 7.404 2.55 Å	(Ba) 9.108 2.73 Å
NM-Al	1.740 (18.4%) 3.21 Å	1.586 (18.7%) 3.15 Å	2.020 (16.4%) 3.21 Å	2.172 (15.6%) 3.39 Å	2.314 (15.2%) 3.51 Å
NM-Si	4.381 (46.5%) 3.27 Å	3.919 (46.1%) 3.21 Å	5.808 (47.1%) 3.57 Å	6.379 (45.8%) 3.63 Å	7.244 (47.7%) 3.57 Å
NM-NM	3.306 (35.1%) 2.97 Å	2.995 (35.2%) 2.79 Å	4.500 (36.5%) 3.39 Å	5.376 (38.6%) 3.75 Å	5.641 (37.1%) 3.99 Å

According to the previous statements on rare earth symmetry, the glass composition CaAS3510 should offer the highest average Sm^{3+} symmetry of all glass compositions in [1]. Interestingly, this is not the case. Some of the peralkaline lanthanum aluminosilicate glasses seem to offer an even slightly higher average Sm^{3+} symmetry. This could either be explained by a Sm^{3+} CN closer to 6 in these glasses or by the chemical similarity of La^{3+} and Sm^{3+}, resulting in a less disturbed surrounding and a lower variation in the second coordination sphere of the doped Sm^{3+} ions. This shows that the coordination with neighboring atoms in the first coordination sphere is not the only parameter that determines symmetry, but probably the most important one. However, more investigations are needed to clarify the role of the atoms in the second coordination sphere of the rare earth ions.

3.2. Peak Position, Optical Basicity, and Peak Broadening

The spectra in Figure 1 do not only give information on Sm^{3+} site symmetry. As clearly seen in the inset of Figure 1 (left), the emission peaks shift to higher wavelengths with increasing ionic radius of the network modifier ion, i.e., the $^4G_{5/2} \rightarrow {^6}H_{9/2}$ peak maximum is at around 645 nm for the MgAS3510 sample but at about 649 nm for the BaAS3510 sample. For ZnAS3510, it is at 646 nm, i.e., between MgAS3510 and CaAS3510 (Figure 1 (right)). The same effect applies to all other peaks, and even for other NM/Al_2O_3 ratios (e.g., [1]). The peak shift is a well-known effect and can be correlated to the optical basicity of the glasses. Higher optical basicity values result in a shift of the emission (and absorption) peaks to longer wavelengths. For this reason, the optical basicity values are added to Table 1.

Another effect seen in Figure 1 is probably more subtle. It can be noted for the emission peaks at 565 and 600 nm. The peaks get broader with increasing ionic radius of the network modifier ion. Since the peak intensities of the 600 nm peaks are normalized, the peak areas A_{600} are showing this effect nicely (Table 1). The A_{600} values increase in the order MgAS3510 < CaAS3510 < SrAS3510 < BaAS3510. The zinc containing glass, again, does not fit, probably due its different structure, as discussed above. Another nice example for this effect is the spectrum of KAS3010 in the right diagram of Figure 1. Its 565 and 600 nm peaks are clearly broader than those of all other spectra in Figure 1. There are two main contributions to this effect, firstly the different variation of rare earth sites in the different glass samples (inhomogeneous broadening) and secondly an increased peak splitting due to an increased electrical field strength at the rare earth positions in the glass network. The second usually also results in a more subdivided peak structure, i.e., additional and more pronounced subpeaks become visible, which is not the case for the glass samples investigated here. However, it can nicely be observed for Er^{3+} [2,31] and Tb^{3+} [6] doped glasses of similar compositions. In our previous publications [2,5], we could clearly correlate the increased peak splitting to an increased coordination of the rare earth ions with non-bridging oxygen (NBO), which have a strongly localized negative charge and therefore expose the neighboring rare earth ion to a stronger electrical field than neighboring bridging oxygen atoms would do. As seen in Table 2, the Gd^{3+} coordination numbers with NBO increase from MgAS3510 (4.05) to BaAS3510 (4.28). The rare earth coordination number with NBO is correlated to the field strength difference between rare earth ion and network modifier ion, since the rare earth ions compete with the NM ions for network positions that can compensate their own charge and have a high local field strength, i.e., the rare earth ions prefer positions with sufficiently high coordination numbers with NBO sites [2,5]. If the NM ions have a high field strength (= small ionic radius), as e.g., Mg^{2+}, the rare earth ions are forced into less preferred positions for charge compensation, probably close to $[AlO_4]^-$ groups with a lower local field strength. Consequently, the Gd^{3+} coordination number with Al in the second coordination sphere is highest in the MgAS3510 and lowest for the BaAS3510 sample (Table 2).

The different variation of rare earth sites in the different glass samples of constant NM/Al_2O_3 ratio is less easy to explain. It is well known that introduction of Al_2O_3 to silicate glasses reduces the rare earth peak splitting, but increases the inhomogeneous broadening of the rare earth spectra (e.g., [32–34]). Al_2O_3 is mostly incorporated in the silicate glass structure as $[AlO_4]^-$ groups (at NM/Al_2O_3 ratios close to or lower than 1, nevertheless also minor quantities of five- and six-fold coordinated Al occur) [5,9,12]. Because of the negative charge of the $[AlO_4]^-$ group, the Al_2O_3 addition decreases the coordination of the rare earth ions with NBO and, due to the addition of an additional coordination partner, i.e., different from SiO_4 tetrahedra, increases the variation of sites. The effect of the NM ions is much smaller and more difficult to determine. For the samples investigated here, the peak splitting is probably the determining factor for the peak broadening, i.e., the subpeaks (shoulders) at the long wavelength side of the Sm^{3+} emission peaks at 565 and 600 nm are split further apart from the main peak and therefore increase the overall peak width. Again, this effect is most obvious for the KAS3010 glass (Figure 1 (right)).

4. Conclusions

By using molecular dynamic (MD) simulations in connection with measurements of the luminescence emission spectra of Sm^{3+}-doped aluminosilicate glasses with different compositions, the effect of different network modifier ions on the glass structure, the local rare earth sites and their emission spectra could be clarified. Coordination of the rare earth ions with an increasing number of non-bridging oxygen atoms increases the peak splitting of the emission peaks and therefore increases the width of the peaks. The overall coordination number of the rare earth ions with oxygen atoms correlates with the rare earth site symmetry in the glass structure. Here, an average coordination number of 6 seems to mark a maximum in the site symmetry. Lower and higher average rare earth coordination numbers result in a lower average rare earth site symmetry. This affects the emission (and absorption) intensities of the so-called hypersensitive peaks.

The MD simulations also show that Zn^{2+} ions are preferably integrated into the glass forming network as ZnO_4 tetrahedra. Therefore, the glass network and rare earth coordination are notably modified in comparison to the alkaline earth containing glasses. In the zinc containing aluminosilicate glass the rare earth ions have unexpectedly low coordination numbers and a relatively low site symmetry.

The knowledge on how glass composition actually affects the optical properties of doped rare earth ions allows to tailor the rare earth spectra for specific applications, i.e., to amplify or attenuate specific (laser) peaks and thereby to increase the laser efficiency, change the color of the rare earth emission, or broaden the emission peaks for an improved compression of rare earth generated laser pulses.

Author Contributions: Conceptualization, A.H.; methodology, A.H. and M.Z.; validation, A.H.; formal analysis, A.H. and M.Z.; investigation, A.H. and M.Z.; resources, C.R. and R.M.; data curation, A.H. and M.Z.; writing—original draft preparation, A.H.; writing—review and editing, C.R., M.Z. and R.M.; visualization, A.H. and M.Z.; supervision, A.H.; project administration, C.R. and R.M.; funding acquisition, C.R., A.H. and R.M. All authors have read and agreed to the published version of the manuscript.

Funding: This work was supported by the German Federal Ministry of Education and Research (BMBF) and the Tunisian Ministry for Higher Education and Scientific Research (MESRS) through the joint German–Tunisian announcement "Regulations Governing the Funding of Scientific and Technological Cooperation (STC) Between Germany and Tunisia" (project number IB-TUNGER15-067).

Institutional Review Board Statement: Not applicable.

Informed Consent Statement: Not applicable.

Data Availability Statement: Not applicable.

Conflicts of Interest: The authors declare no conflict of interest.

References

1. Herrmann, A.; Tewelde, M.; Kuhn, S.; Tiegel, M.; Rüssel, C. The Effect of Glass Composition on the Luminescence Properties of Sm^{3+} doped Alumino Silicate Glasses. *J. Non-Cryst. Solids* **2018**, *502*, 190–197. [CrossRef]
2. Herrmann, A.; Assadi, A.A.; Lachheb, R.; Zekri, M.; Erlebach, A.; Damak, K.; Maâlej, R.; Sierka, M.; Rüssel, C. The effect of glass structure and local rare earth site symmetry on the optical properties of rare earth doped alkaline earth aluminosilicate glasses. *Acta Mater. (under review)*.
3. Tiegel, M.; Hosseinabhadi, R.; Kuhn, S.; Herrmann, A.; Rüssel, C. Young's modulus, Vickers hardness and indentation fracture toughness of alumino silicate glasses. *Ceram. Int.* **2015**, *41*, 7267–7275. [CrossRef]
4. Herrmann, A.; Rüssel, C. New Aluminosilicate Glasses as High-power Laser Materials. *Int. J. Appl. Glass Sci.* **2015**, *6*, 210–219. [CrossRef]
5. Zekri, M.; Erlebach, A.; Herrmann, A.; Damak, K.; Rüssel, C.; Sierka, M.; Maâlej, R. The Structure of Gd^{3+}-Doped Li_2O and K_2O Containing Aluminosilicate Glasses from Molecular Dynamics Simulations. *Materials* **2021**, *14*, 3265. [CrossRef]
6. Assadi, A.A.; Herrmann, A.; Tewelde, M.; Damak, K.; Maalej, R.; Rüssel, C. Tb^{3+} as a probe for the molecular structure of mixed barium magnesium alumino silicate glasses. *J. Lumin.* **2018**, *199*, 384–390. [CrossRef]
7. Blasse, G.; Grabmaier, B.C. *Luminescent Materials*; Springer: Berlin/Heidelberg, Germany, 1994; pp. 41–44. [CrossRef]

8. Henrie, D.E.; Fellows, R.L.; Choppin, G.R. Hypersensitivity in the electronic transitions of lanthanide and actinide transitions. *Coord. Chem. Rev.* **1976**, *18*, 199–224. [CrossRef]
9. Cormier, L. Glasses: Aluminosilicates. *Encycl. Mater. Tech. Ceram. Glasses* **2021**, *2*, 496–518. [CrossRef]
10. Duffy, J.A. A review of optical basicity and its applications to oxidic systems. *Geochim. Cosmochim. Acta* **1993**, *57*, 3961–3970. [CrossRef]
11. Dimitrov, V.; Komatsu, T.J. An interpretation of optical properties of oxides and oxide glasses in terms of the electronic ion polarizability and average single bond strength. *J. Univ. Chem. Technol. Metall.* **2010**, *45*, 219–250.
12. Zekri, M.; Erlebach, A.; Herrmann, A.; Damak, K.; Rüssel, C.; Sierka, M.; Maâlej, R. Structure prediction of rare earth doped BaO and MgO containing aluminosilicate glasses—The model case of Gd_2O_3. *Materials* **2018**, *11*, 1790. [CrossRef]
13. Plimpton, S. Fast parallel algorithms for short-range molecular dynamics. *J. Comput. Phys.* **1995**, *117*, 1–19. [CrossRef]
14. Pedone, A.; Malavasi, G.; Menziani, M.C.; Cormack, A.N.; Segre, U. A new self-consistent empirical interatomic potential model for oxides, silicates, and silica-based glasses. *J. Phys. Chem. B* **2006**, *110*, 11780–11795. [CrossRef] [PubMed]
15. Pedone, A.; Malavasi, G.; Menziani, M.C.; Segre, U.; Cormack, A.N. Role of Magnesium in Soda-Lime Glasses: Insight into Structural, Transport, and Mechanical Properties through Computer Simulations. *J. Phys. Chem. C* **2008**, *112*, 11034–11041. [CrossRef]
16. Linati, L.; Lusvardi, G.; Malavasi, G.; Menabue, L.; Menziani, M.C.; Mustarelli, P.; Pedone, A.; Segre, U. Medium-range order in phospho-silicate bioactive glasses: Insights from MAS-NMR spectra, chemical durability experiments and molecular dynamics simulations. *J. Non-Cryst. Solids* **2008**, *354*, 84–89. [CrossRef]
17. Pedone, A.; Malavasi, G.; Cormack, A.N.; Segre, U.; Menziani, M.C. Insight into Elastic Properties of Binary Alkali Silicate Glasses; Prediction and Interpretation through Atomistic Simulation Techniques. *Chem. Mater.* **2007**, *19*, 3144–3154. [CrossRef]
18. Afify, N.D.; Mountjoy, G. Molecular-dynamics modeling of Eu^{3+}-ion clustering in SiO_2 glass. *Phys. Rev. B* **2009**, *79*, 024202. [CrossRef]
19. Hockney, R.W.; Eastwood, J.W. *Computer Simulation Using Particles*, 1st ed.; CRC Press: Boca Raton, FL, USA, 1988; pp. 22–23, ISBN 978-1-138-41337-5.
20. Swope, W.C.; Andersen, H.C.; Berens, P.H.; Wilson, K.R. A computer simulation method for the calculation of equilibrium constants for the formation of physical clusters of molecules: Application to small water clusters. *J. Chem. Phys.* **1982**, *76*, 637–649. [CrossRef]
21. Nosé, S. A unified formulation of the constant temperature molecular dynamics methods. *J. Chem. Phys.* **1984**, *81*, 511–519.
22. Shinoda, W.; Shiga, M.; Mikami, M. Rapid estimation of elastic constants by molecular dynamics simulation under constant stress. *Phys. Rev. B* **2004**, *69*, 134103. [CrossRef]
23. Haynes, W.M.; Lide, D.R. *CRC Handbook of Chemistry and Physics: A Ready Reference Book of Chemical and Physical Data*, 91st ed.; CRC Press: Boca Raton, FL, USA, 2010; ISBN 978-1439820773.
24. Carnall, W.T.; Fields, P.R.; Rajnak, K. Electronic energy levels of the trivalent lanthanide aquo ions. I. Pr^{3+}, Nd^{3+}, Pm^{3+}, Sm^{3+}, Dy^{3+}, Ho^{3+}, Er^{3+}, and Tm^{3+}. *J. Phys. Chem.* **1968**, *49*, 4424–4442. [CrossRef]
25. May, P.S.; Metcalf, D.H.; Richardson, F.S.; Carter, R.C.; Miller, C.E.; Palmer, R.A. Measurement and analysis of excited-state decay kinetics and chiroptical activity in the $^6H_j \leftarrow {}^4G_{5/2}$ transitions of Sm^{3+} in trigonal $Na_3[Sm(C_4H_4O_5)_3] \cdot 2NaClO_4 \cdot 6H_2O$. *J. Lumin.* **1992**, *51*, 249–268. [CrossRef]
26. Farries, M.C.; Morkel, P.R.; Townsend, J.E. Spectroscopic and lasing characteristics of samarium-doped glass fibre. *IEE Proc. J (Optoelectron.)* **1990**, *137*, 318–322. [CrossRef]
27. Ratnakaram, Y.C.; Thirupathi Naidu, D.; Chakradhar, R.P.S. Spectral studies of Sm^{3+} and Dy^{3+} doped lithium cesium mixed alkali borate glasses. *J. Non-Cryst. Solids* **2006**, *352*, 3914–3922. [CrossRef]
28. Kuhn, S.; Herrmann, A.; Rüssel, C. Judd-Ofelt analysis of Sm^{3+}-doped lanthanum-aluminosilicate glasses. *J. Lumin.* **2015**, *157*, 390–397. [CrossRef]
29. Herrmann, A.; Friedrich, D.; Zscheckel, T.; Rüssel, C. Luminescence properties of Sm^{3+} doped alkali/earth alkali orthoborates of the type $XZBO_3$ with X = Li, Na, Cs and Z = Ca, Sr, Ba. *J. Lumin.* **2019**, *214*, 116550. [CrossRef]
30. Cormier, L.; Delbes, L.; Baptiste, B.; Montouillout, V. Vitrification, crystallization behavior and structure of zinc aluminosilicate glasses. *J. Non-Cryst. Solids* **2021**, *555*, 120609. [CrossRef]
31. Assadi, A.A.; Herrmann, A.; Lachheb, R.; Damak, K.; Rüssel, C.; Maalej, R. Experimental and Theoretical Study of Erbium Doped Aluminosilicate Glasses. *J. Lumin.* **2016**, *176*, 212–219. [CrossRef]
32. Reisfeld, R.; Panczer, G.; Patra, A.; Gaft, M. Time-resolved spectroscopy of Sm^{3+} in silica and silica-Al sol-gel glasses. *Mater. Lett.* **1999**, *38*, 413–417. [CrossRef]

33. Tanabe, S. Optical transitions of rare earth ions for amplifers: How the local structure works in glass. *J. Non-Cryst. Solids* **1999**, *259*, 1–9. [CrossRef]
34. Berneschi, S.; Bettinelli, M.; Brenci, M.; Nunzi Conti, G.; Pelli, S.; Sebastiani, S.; Siligardi, C.; Speghini, A.; Righini, G.C. Aluminum co-doping of soda-lime silicate glasses: Effect on optical and spectroscopic properties. *J. Non-Cryst. Solids* **2005**, *351*, 1747–1753. [CrossRef]

Disclaimer/Publisher's Note: The statements, opinions and data contained in all publications are solely those of the individual author(s) and contributor(s) and not of MDPI and/or the editor(s). MDPI and/or the editor(s) disclaim responsibility for any injury to people or property resulting from any ideas, methods, instructions or products referred to in the content.

Article

Manganese Luminescent Centers of Different Valence in Yttrium Aluminum Borate Crystals

Anastasiia Molchanova [1], Kirill Boldyrev [1], Nikolai Kuzmin [1,2,3], Alexey Veligzhanin [4], Kirill Khaydukov [5], Evgeniy Khaydukov [5], Oleg Kondratev [4], Irina Gudim [6], Elizaveta Miklaeva [7] and Marina Popova [1,*]

[1] Institute of Spectroscopy, Russian Academy of Sciences, Troitsk, 108840 Moscow, Russia
[2] Landau Phystech School of Physics and Research, Moscow Institute of Physics and Technology, 141701 Dolgoprudny, Russia
[3] Faculty of Geology, Lomonosov Moscow State University, 119991 Moscow, Russia
[4] National Research Center "Kurchatov Institute", 123182 Moscow, Russia
[5] Federal Scientific Research Center "Crystallography and Photonics", Russian Academy of Sciences, 119333 Moscow, Russia
[6] Kirensky Institute of Physics, Siberian Branch of the Russian Academy of Sciences, Akademgorodok, 660036 Krasnoyarsk, Russia
[7] Branch "Aprelevka Department of VNIGNI", Federal State Budgetary Institution "All-Russian Research Geological Oil Institute", 143360 Aprelevka, Russia
* Correspondence: popova@isan.troitsk.ru

Citation: Molchanova, A.; Boldyrev, K.; Kuzmin, N.; Veligzhanin, A.; Khaydukov, K.; Khaydukov, E.; Kondratev, O.; Gudim, I.; Mikliaeva, E.; Popova, M. Manganese Luminescent Centers of Different Valence in Yttrium Aluminum Borate Crystals. *Materials* 2023, 16, 537. https://doi.org/10.3390/ma16020537

Academic Editor: Dirk Poelman

Received: 20 November 2022
Revised: 19 December 2022
Accepted: 1 January 2023
Published: 5 January 2023

Copyright: © 2023 by the authors. Licensee MDPI, Basel, Switzerland. This article is an open access article distributed under the terms and conditions of the Creative Commons Attribution (CC BY) license (https://creativecommons.org/licenses/by/4.0/).

Abstract: We present an extensive study of the luminescence characteristics of Mn impurity ions in a $YAl_3(BO_3)_4$:Mn crystal, in combination with X-ray fluorescence analysis and determination of the valence state of Mn by XANES (X-ray absorption near-edge structure) spectroscopy. The valences of manganese $Mn^{2+}(d^5)$ and $Mn^{3+}(d^4)$ were determined by the XANES and high-resolution optical spectroscopy methods shown to be complementary. We observe the R_1 and R_2 luminescence and absorption lines characteristic of the $^2E \leftrightarrow {^4A_2}$ transitions in d^3 ions (such as Mn^{4+} and Cr^{3+}) and show that they arise due to uncontrolled admixture of Cr^{3+} ions. A broad luminescent band in the green part of the spectrum is attributed to transitions in Mn^{2+}. Narrow zero-phonon infrared luminescence lines near 1060 nm (9400 cm^{-1}) and 760 nm (13,160 cm^{-1}) are associated with spin-forbidden transitions in Mn^{3+}: $^1T_2 \to {^3T_1}$ (between excited triplets) and $^1T_2 \to {^5E}$ (to the ground state). Spin-allowed $^5T_2 \to {^5E}$ Mn^{3+} transitions show up as a broad band in the orange region of the spectrum. Using the data of optical spectroscopy and Tanabe–Sugano diagrams we estimated the crystal-field parameter Dq and Racah parameter B for Mn^{3+} in YAB:Mn as $Dq = 1785$ cm^{-1} and $B = 800$ cm^{-1}. Our work can serve as a basis for further study of YAB:Mn for the purposes of luminescent thermometry, as well as other applications.

Keywords: manganese; $YAl_3(BO_3)_4$:Mn crystal; XANES spectroscopy; high-resolution optical spectroscopy; photoluminescence

1. Introduction

Crystals of yttrium-aluminum borate $YAl_3(BO_3)_4$ (YAB) have the structure of the mineral huntite $CaMg_3(CO_3)_4$ with the non-centrosymmetric space group $R32$ of the trigonal system [1]. Figure 1 shows different projections of the YAB unit cell. The crystal structure is formed by layers that are perpendicular to the crystallographic c axis and consist of distorted YO_6 prisms, AlO_6 octahedra, and BO_3 groups of two types ($B1O_3$ and $B2O_3$). Y^{3+} ions in YO_6 prisms are surrounded by six oxygen atoms of one type and occupy sites with the D_3 point symmetry group. The point group of AlO_6 octahedra is C_2. AlO_6 octahedra linked together by their edges form spiral chains running along the c axis. The Y^{3+} ions are situated between three such chains and link the chains together. YO_6 prisms are isolated from each other, having no oxygen atoms in common, which, in the case of a substitution of the Y^{3+} ions by rare-earth or transition metal ions, results in low luminescence quenching [2]. This property, together with high optical nonlinearity and excellent physical characteristics and chemical stability, make YAB extremely interesting for many applications. Doped with various rare-earth and transition metal ions, YAB crystals are well-known phosphors, promising for use as materials for display panels, lasers, scintillators, LEDs, luminescent thermometers, and in medical imaging [3–18]. YAB crystals doped with Nd^{3+} [8,13], Yb^{3+} [11,12,14], Er^{3+}/Yb^{3+} [10], and Yb^{3+}/Tm^{3+} [16] are well-known media for self-frequency doubling, self-frequency summing, and up-conversion lasers. Tunable anti-Stokes ultraviolet–blue light generation was demonstrated using a random laser based on $Nd_{0.10}Y_{0.90}Al_3(BO_3)_4$ [3]. YAB:Eu^{3+}/Tb^{3+} phosphors were proposed for eye-friendly white LEDs [6]. In addition, YAB:Cr is being investigated as a material for LEDs [17]—in particular, as a phosphor for plant growth LEDs—with excellent thermal stability and high luminescent yield [5]. Recently, impressive applications of YAB:Pr^{3+}/Gd^{3+} and YAB:Cr^{3+} in luminescent thermometry were reported [4,15]. In Ref. [15], it was proposed to use several excited levels of the Gd^{3+} ion in YAB doped with Pr^{3+} and Gd^{3+} ions in the UV region of the spectrum to implement a Boltzmann thermometer operating from 30 to 800 K. The UV region allowed detuning from background thermal radiation even at the highest temperatures. In this case, excitation was carried out at a wavelength of 450 nm using an inexpensive commercial LED into the absorption band of the Pr^{3+} ion, followed by the up-conversion energy transfer $Pr^{3+} \rightarrow Gd^{3+}$. In Ref. [4], a combination of optical heating and luminescent thermometry in YAB:Cr^{3+} was realized. Here, the temperature-dependent ratio of emission intensities for the $^4T_2 \rightarrow {}^4A_2$ and $^2E \rightarrow {}^4A_2$ transitions of Cr^{3+} was used to measure the temperature.

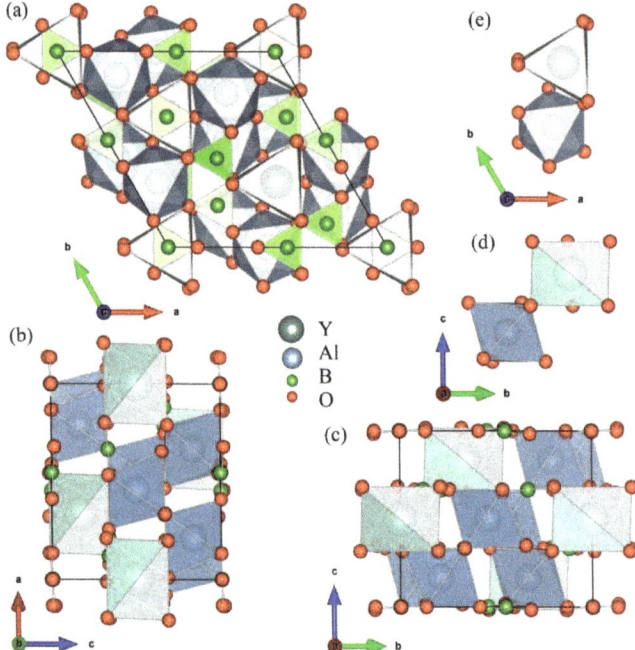

Figure 1. Projections of the $YAl_3(BO_3)_4$ unit cell along the c axis (**a**), the b axis (**b**), and the a axis (**c**). Projections of the YO_6 trigonal prism and AlO_6 distorted octahedron in the $YAl_3(BO_3)_4$ unit cell, along a axis (**d**) and the c axis (**e**).

We note that the Mn^{4+} ion has the same valence electron shell as the Cr^{3+} ion (d^3) and is also used for luminescent thermometry [19]. Compounds with $Mn^{3+}(d^4)$ exhibit broadband, extremely temperature-sensitive luminescence in the near-IR and visible spectral ranges [20,21], due to which compounds with Mn^{3+} are also topical materials for thermoluminescent sensors. Cryogenic luminescence ratiometric thermometry based on the diverse thermal quenching behaviors of Mn^{3+} and Mn^{4+} in manganese-doped garnet-type $Ca_3Ga_2Ge_3O_{12}$ single crystals was explored [22]. Tb^{3+} and Mn^{3+} co-doped $La_2Zr_2O_7$ nanoparticles were recently suggested as a promising material for dual-activator ratiometric optical thermometry [23]. $Mn^{2+}(d^5)$-containing phosphors exhibit bright broadband luminescence with a maximum from the red to green region of the spectrum, depending on the particular matrix [24–26]. In light of all of the above, it is of interest to study the luminescent properties of YAB doped with manganese.

We are aware of only one work on YAB:Mn spectroscopy ([27]). Only the room-temperature spectra were measured in Ref. [27]. Three lines characteristic of $^2E \rightarrow {}^4A_2$ emission of ions with a d^3 electronic configuration were detected in the YAB:Mn room-temperature luminescence spectra [27]. The authors assigned these lines to $Mn^{4+}(d^3)$. The results of electron paramagnetic resonance (EPR) showed that Mn introduced into YAB at low concentrations predominantly occupied the yttrium-ion sites in the crystal structure, its valence in this case being 2+ [28]. Two broad bands peaked at 544 and 637 nm were observed in the room-temperature luminescence spectrum of YAB:Mn and assigned to the transition from the $^4T_1$ state of the Mn^{2+} ion, split by a low-symmetry component of the crystal field, to the ground state $^6A_1$ [27]. Since the Mn^{2+} and Mn^{4+} ions presumably replace the trivalent Y^{3+} and Al^{3+} cations, respectively, the question of charge-compensation arises. The formation of charge-compensating Mn^{2+}–Mn^{4+} dimers was suggested in [27]. In this work, we continue the study of the valence states of manganese in YAB:Mn using XANES spectroscopy and high-resolution broadband temperature-dependent optical spectroscopy, and obtain extensive data on the luminescence of Mn impurity centers of various valences in $YAl_3(BO_3)_4$.

2. Materials and Methods

$YAl_3(BO_3)_4$:Mn crystals were obtained by the flux method of crystal growth in the laboratory of L.N. Bezmaternykh at the Kirensky Institute of Physics of the Siberian Branch of the Russian Academy of Sciences in Krasnoyarsk. They were grown on seeds in platinum crucibles with a volume of 50 mL. The composition of the system during the flux crystal growth was 85 wt.% ($Bi_2Mo_3O_{12} + 2B_2O_3 + 0.5Li_2MoO_4$) + 15 wt.% $YAl_3(BO_3)_4$ with the addition of Mn_2O_3. The temperature regime consisted of heating the solution-melt to 1100 °C and then slowly cooling at a rate of 0.5 °C/h for 48 h. Note that manganese oxide Mn_2O_3 decomposes in air at temperatures above 800 °C to form Mn_3O_4 ($Mn^{2+}Mn^{3+}_2O_4$) [29]. High-purity reagents were used in flux crystal growth. Cr (0.001%) and Pb (0.0005%) impurities in Al_2O_3 as well as Nd_2O_3 and Sm_2O_3 (<0.0001%) in Y_2O_3 have been reported on certificates and are of interest for further discussion.

Powder X-ray diffraction on the grown crystals at room temperature was performed on a Thermo Fisher Scientific ARL X'tra diffractometer (Basel, Switzerland) equipped with a Dectris MYTHEN2 R 1D detector (Cu $K_{\alpha1,2}$ radiation). The operational voltage and current were 40 kV and 40 mA, respectively. Powder diffraction patterns were obtained in continuous mode at a rate of 2°/min in Bragg–Brentano geometry over an angle range of $10° \leq 2\theta \leq 90°$. The unit cell parameters of $YAl_3(BO_3)_4$:Mn were refined by the Le Bail method using the JANA2006 program [30]. All parameters were refined by the least-squares method. The pseudo-Voigt function was used as the peak profile function. The structural data for $YAl_3(BO_3)_4$ (sp. gr. $R32$, a = 9.295(3) Å, c = 7.243(2) Å, $\alpha = \beta = 90°$, $\gamma = 120°$) were used as the initial structural parameters [31].

X-ray fluorescence analysis was carried out on a Bruker M4 Tornado analyzer. Absorption and luminescence spectra in the near-IR and visible ranges (5000–16,000 cm^{-1}) with a spectral resolution up to 0.2 cm^{-1} were recorded on a spectrometer Bruker IFS 125HR (Bruker Optik GmbH, Ettlingen, Germany). Luminescence spectra in the visible and UV ranges (9000–20,500 cm^{-1}) with a spectral resolution up to 3 cm^{-1} were registered using a

OceanInside HDX spectrometer. The sample was cooled down to 5 K using a Cryomech ST403 closed-cycle helium cryostat (Syracuse, NY, USA). X-ray absorption spectra near the manganese K-edge were measured at the "Structural Materials Science" beamline at the Kurchatov Synchrotron Radiation Source [32] by X-ray fluorescence yield. Luminescence excitation spectra were recorded at a liquid nitrogen temperature (77 K) on a Fluorolog®-3 spectrofluorometer at the Institute of Photonic Technologies of the Federal Research Center "Crystallography and Photonics" of the Russian Academy of Sciences.

3. Results and Discussion

3.1. X-ray Diffraction (XRD) Analysis

XRD was used for the fingerprint characterization and investigation of the structural phases in the crystalline state. XRD patterns were analyzed by the Le Bail method in order to extract the parameters of the unit cell. The refined unit cell parameters were a = 9.274(7) Å, c = 7.223(3) Å, $\alpha = \beta = 90°$, and $\gamma = 120°$. The convergence of the Le Bail approximation is shown in Figure 2. It can be seen that the diffraction pattern is well described, as indicated by the low R-factor values and small difference between the calculated and experimental diffraction patterns. The figure shows additional reflections of the Al_2O_3 phase. Their presence is explained by the fact that a corundum mortar was used in the preparation of the powder samples.

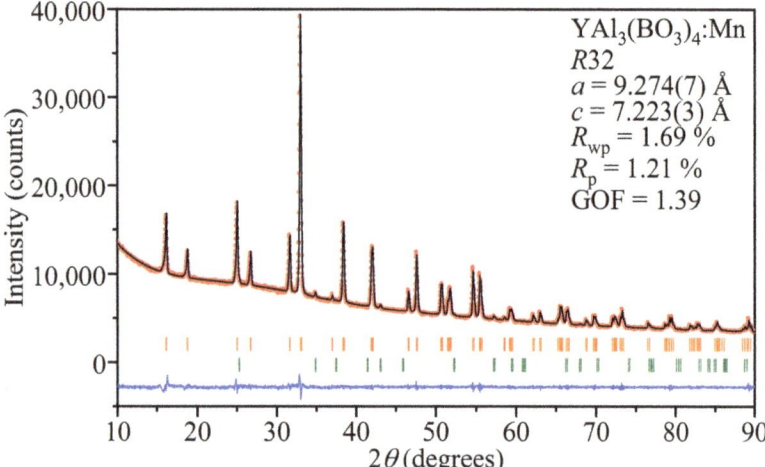

Figure 2. Final convergence of the Le Bail refinement for $YAl_3(BO_3)_4$:Mn. The experimental diffraction pattern is shown by red circles (I_{obs}); the black line (I_{calc}) is the calculated diffraction pattern and the residual intensities (I_{obs}-I_{calc}) are shown as the blue line. The orange bars indicate the $YAl_3(BO_3)_4$ reflections, and the green bars indicate the Al_2O_3 reflections.

3.2. X-ray Fluorescence Analysis

The concentration of manganese ions was determined by X-ray fluorescence analysis to be 0.87 at.%. In addition, the presence of 1.18 at.% Bi was found, which is explained by its presence in the composition of the solvent. Insignificant amounts of potassium, calcium, titanium, and iron impurities were also found (see Table 1).

Table 1. Composition of YAB:Mn determined by X-ray fluorescence analysis. AN—atomic number.

Element	AN	wt.%	Normal. wt.%	Normal. at. %	Error in wt.% (1 Sigma)
Aluminum	13	24.43	21.667	48.17	0.225
Potassium	19	0.08	0.084	0.13	0.006
Calcium	20	0.09	0.090	0.14	0.002
Titanium	22	0.05	0.057	0.07	0.001
Manganese	25	0.75	0.800	0.87	0.009
Iron	26	0.16	0.172	0.18	0.001
Yttrium	39	68.84	73.002	49.25	0.037
Bismuth	83	3.89	4.128	1.18	0.006
		94.3	100	100	

3.3. X-ray Absorption Spectroscopy

To address Mn ion oxidation state and position in the crystal structure, the fine structure of the X-ray absorption spectrum at the K-edge of manganese was measured. The EXAFS (Extended X-ray Absorption Fine Structure) spectrum was processed and analyzed using the software package IFEFFIT, version 1.2.11c [33,34]. The measured XAFS data were first processed by the ATHENA program of this package to merge four independently measured spectra, normalize the spectrum to a unity-height jump, and obtain the oscillating part of the spectrum. The fine structure of the X-ray absorption spectrum obtained in this manner after the K-jump was then used for the structural analysis (Figure 3). The local structure of manganese ions in the crystal was analyzed by fitting the EXAFS spectra at the K-edge of Mn to the model of the local structure based on the crystal structure of YAB [31]. Two distinct models were used for the fitting. The first model includes a manganese atom in the yttrium position. The second model takes into account the partial occupation of aluminum positions by manganese atoms. Since the positions of yttrium and aluminum differ significantly in metal–oxygen distances in the first coordination sphere—2.3 and 2.0 Å, respectively—this was taken into account by introducing an additional Mn-O scattering path of shorter length. To estimate the occupancy of the aluminum position, the coordination numbers for the two nearest oxygen coordination spheres were chosen so that their sum was fixed equal to six (Table 2). The distance for this shorter path was set to 2.052 Å to obtain a stable fit. Other parameters determined by fitting the EXAFS spectra are the distances between the absorbing and neighboring atoms R_j and the Debye–Waller factors σ_j^2 common to atoms of the same type. The errors for the Debye–Waller factors are quite large, since we can only use the spectrum up to $k = 10$ Å^{-1} due to the relatively high noise levels at large k. This leads to a significant correlation of the Debye–Waller factors with the overall amplitude of the EXAFS oscillations and to high uncertainty values. The refinement also included the Fermi energy shift ΔE_0 and the attenuation coefficient of the signal amplitude S_0^2. The fitting ranges in k space and in R space were 2–10 Å^{-1} and 1–4 Å, respectively. The quality of the fit is characterized by the factor R_f, which indicates the percentage mismatch between the data and the model.

Table 2 shows that the two models do not differ in R_f, i.e., manganese in aluminum positions does not contribute much to the EXAFS signal. Thus, one can conclude that the occupation of aluminum sites by manganese atoms is rather small. To estimate this occupation, the coordination numbers for the split coordination sphere of oxygen can be used. For a shorter distance, it was determined to be 0.7, so occupancy can be estimated as no more than 10%. It should be noted that this estimate shows the sensitivity of the EXAFS method for this quantity, since the error bars are also of the same order. The distances determined by the EXAFS fit correspond to the local structure of the yttrium site. The distance to oxygen in the first coordination sphere was determined to be 2.26 ± 0.03 Å, which is slightly smaller than the Y-O distance in the YAB structure (2.313 Å) [31].

The XANES part of the spectrum also provides valuable information. The position of the K-edge can be used to obtain the oxidation state of Mn [35]. Comparing the spectrum with manganese references $Mn(BO_2)_2$, Mn_2O_3, and MnO_2 with the oxidation states Mn^{2+}, Mn^{3+}, and Mn^{4+}, respectively, measured on the same beamline, we can see that the edge position coincides with the Mn^{2+} reference (Figure 4a), which means that most manganese atoms are in the Mn^{2+} state. We cannot decompose the spectrum into a linear combination of references, since they are irrelevant to the local structure of the YAB specimen; the admixture of manganese in higher oxidation states can be roughly estimated as 10%. In addition, we calculated the XANES spectrum using the FDMNES code [36] with two structural models, corresponding to the manganese atom at the Y and Al sites in the YAB crystal structure, respectively (Figure 4b). The experimental data are reproduced only for Mn at the Y position, which confirms the conclusions of the EXAFS data analysis and is consistent with the EPR data [28]. From this point of view, a smaller Y-O distance than in YAB can be explained by a smaller ionic radius of Mn^{2+} (0.83 Å) as compared to Y^{3+} (0.90 Å) [37].

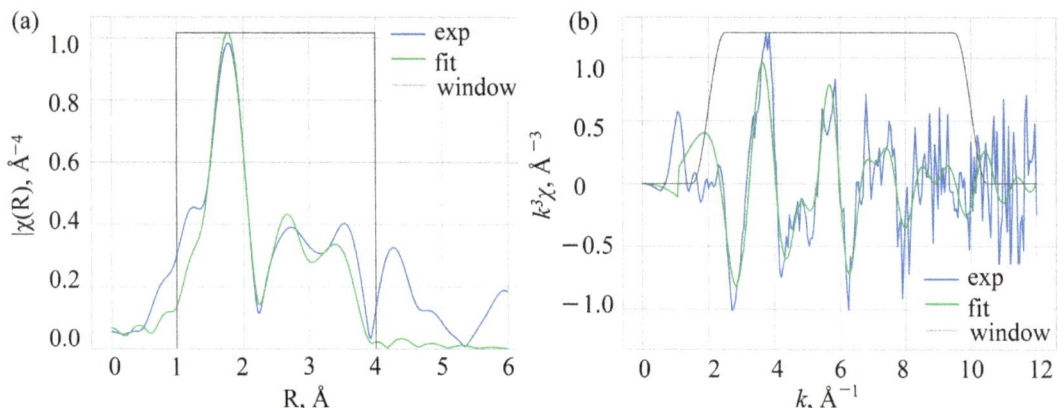

Figure 3. Fourier transform (**a**) and the oscillating part (**b**) of the EXAFS spectrum of Mn in YAB:Mn.

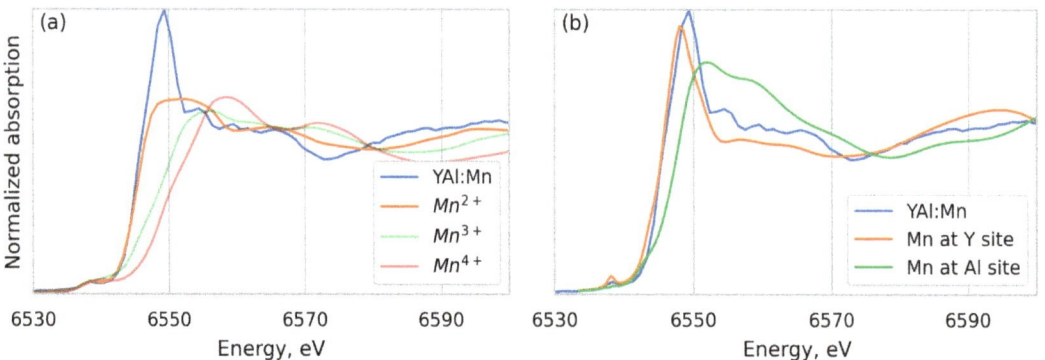

Figure 4. K-edge absorption spectra of Mn in YAB:Mn compared with the "reference" Mn^{2+}, Mn^{3+}, and Mn^{4+} spectra (**a**); calculated XANES spectra for Mn at the Y^{3+} and Al^{3+} sites in YAB crystal structure compared with experimental spectrum (**b**).

Table 2. Parameters of the nearest environment of Mn obtained from EXAFS data.

Model	R_f, %	S_0^2	ΔE_0, eV	Path	N	R, Å	$\sigma^2, 10^{-3}$ Å^2
Mn in Y position	1.7	0.57 ± 0.18	4.4 ± 2.3	Mn-O	6	2.26(3)	6 ± 5
				Mn-B	6	3.0(1)	10 ± 15
				Mn-O	6	3.16(7)	6 ± 5
				Mn-O	6	3.65(8)	6 ± 5
				Mn-Al	6	3.68(4)	3 ± 6
				Mn-O	6	4.24(8)	6 ± 5
Mn in Y position Mn in Al position	1.6	0.56 ± 0.17	2.8 ± 4.1	Mn-O2	0.7 ± 1.3	2.052	4 ± 7
				Mn-O1	5.3 ± 1.3	2.25(3)	4 ± 7
				Mn-B	6	3.0(1)	4 ± 17
				Mn-O	6	3.1(1)	4 ± 7
				Mn-O	6	3.6(1)	4 ± 7
				Mn-Al	6	3.66(5)	1 ± 6
				Mn-O	6	4.23(9)	4 ± 7

3.4. Optical Spectroscopy

Figure 5 shows the photoluminescence (PL) spectrum of YAB:Mn in a broad spectral range. The near-IR luminescence was recorded with the Bruker 125 HR Fourier spectrometer, while for the visible part of the PL spectrum an OceanInside HDX spectrometer was used. Relative intensities of these two parts cannot be compared.

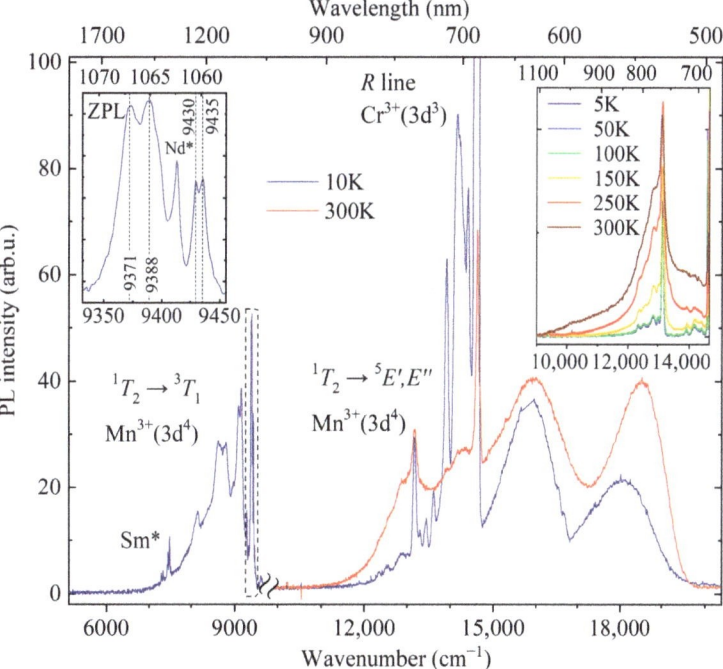

Figure 5. PL spectrum of YAB:Mn in a broad spectral range. The break separates the parts of the spectrum recorded on the Bruker IFS HR125 spectrometer (5100–9900 cm^{-1} at 10 K, λ_{ex} = 450 nm) and on the OceanInside HDX spectrometer (9900–20,500 cm^{-1} at 10 and 300 K, λ_{ex} = 488 nm). The left inset shows the region of zero-phonon lines in an enlarged scale. The right inset presents the region around the line 13,160 cm^{-1} at different temperatures (5–300 K, Bruker IFS HR125 spectrometer). The lines due to uncontrolled Sm^{3+} and Nd^{3+} impurities are marked as Sm* and Nd*, respectively.

A strong relatively narrow peak at about 685 nm is observed in the room-temperature low-resolution PL spectrum. Previously, three narrow peaks with maxima at 682, 684, and 686 nm were reported in the room-temperature luminescence spectrum of YAB:Mn, and two of them were attributed to the R_1 and R_2 lines of Mn^{4+} [27]. The Mn^{4+} ion has the same valence electron shell structure as $Cr^{3+}(d^3)$. Narrow R lines in the spectra of d^3 ions arise due to spin-forbidden transitions from the excited orbital doublet ^2E to the ground orbital singlet $^4A_2$. In a low-symmetry crystal field, the ^2E level, which is doubly degenerate in the cubic crystal field approximation, splits into two components, so that the R_1 and R_2 lines can be observed. We were able to observe peaks at the same wavelengths as in [27], both in the luminescence and absorption room-temperature spectra. However, a more detailed study of the temperature-dependent absorption, PL, and PL excitation spectra led us to the conclusion that those are R lines of uncontrolled Cr^{3+} impurity. Figures 6–8 display these spectra.

Figure 6 shows the absorption and luminescence spectra of YAB:Mn at low temperature (T = 5 K) in the region of the R lines. The spectra have the form of narrow zero-phonon lines (ZPLs) and broad adjacent bands of electron–phonon (vibronic) transitions. Figure 7a demonstrates the evolution of the R absorption lines with temperature. The wavelengths of the R_1 and R_2 lines at room temperature—684 nm and 682 nm, respectively—coincide with those reported for YAB:Cr^{3+} [38,39]. Figure 7b shows very weak lines of a spin-forbidden transition from the ground state $^4A_2$ to the next excited (after the ^2E doublet) level $^2T_1$ in the absorption spectrum of YAB:Mn at 5 K. The excitation spectra of the R lines are presented in Figure 8. All these experimental data allowed us to determine the energies of the ^2E, $^2T_1$, $^4T_2$, and $^4T_1$ levels; they are provided in Table 3. The values in Table 3, within the precision of measurements, coincide with those reported for Cr^{3+} in YAB [38–40]. It is a well-known empirical fact that the strength of the crystal field as well as covalency increases with increased ionic charge [41]. For example, Mn^{4+} in corundum Al_2O_3 demonstrates blue shifts of 364, 413, and 2300 cm^{-1} for the R_1, R_2, and A_1–$^4T_2$ transitions, respectively, as compared to Cr^{3+} in Al_2O_3 (ruby) [41,42]. Both ions substitute for Al^{3+}. We tried to find the R lines of Mn^{4+} in the spectra of YAB:Mn but failed. It is worth noting that Mn^{4+} in Al_2O_3 was introduced together with charge-compensating Mg^{2+}.

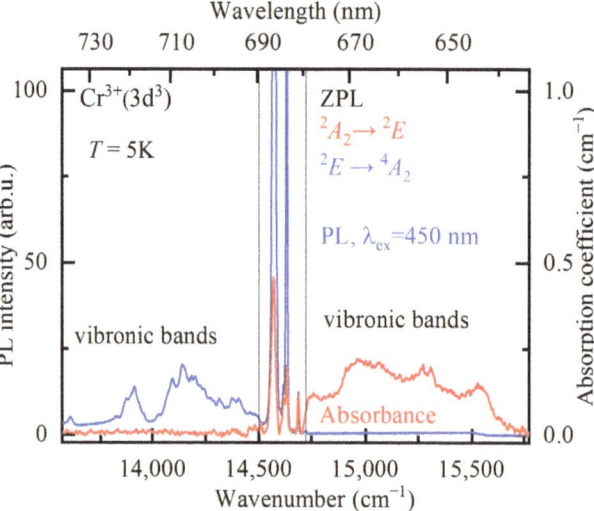

Figure 6. Absorption (red curve) and luminescence (blue curve, excitation wavelength λ_{ex} = 450 nm) spectra of YAB:Mn at the temperature T = 5 K in the region (indicated by gray thin vertical lines) of zero-phonon R lines and the region of associated vibronic bands.

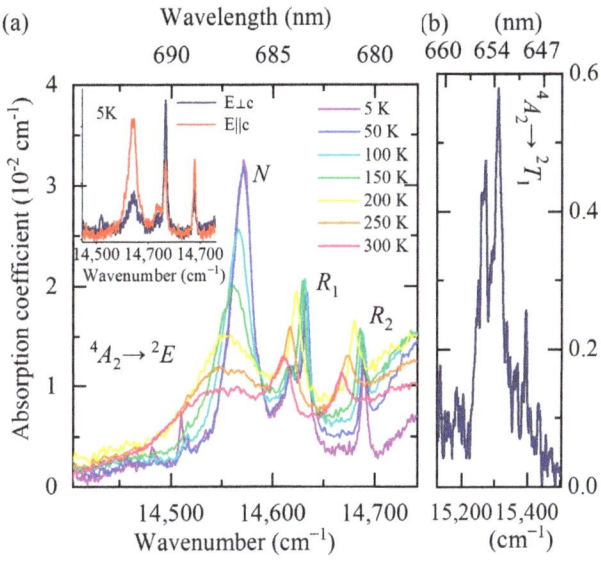

Figure 7. Unpolarized absorption spectra of YAB:Mn at different temperatures in the spectral range of zero-phonon R lines. The inset shows the spectra at $T = 5$ K for two polarization directions of the incident light: $E \parallel c$ (red trace) and $E \perp c$ (blue trace) (**a**). Absorption spectrum at $T = 5$ K in the region of the $^4A_2 \rightarrow {}^2T_1$ transition (**b**).

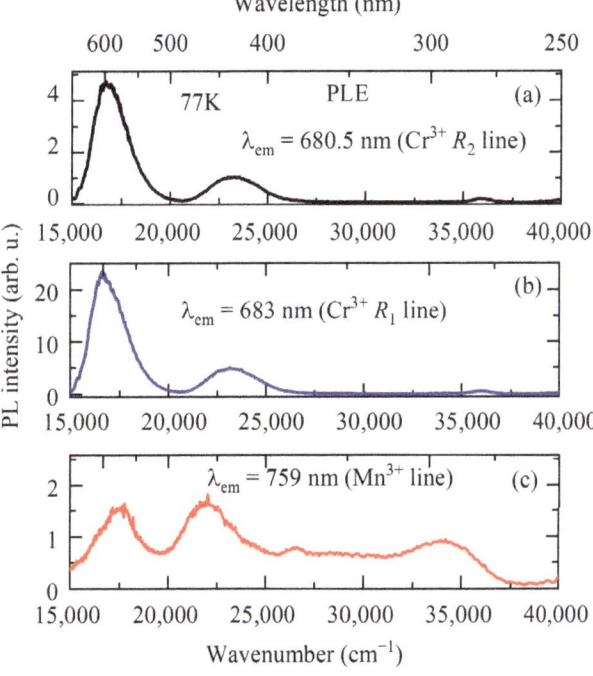

Figure 8. PL excitation spectra for YAB:Mn at $T = 77$ K monitored at 680.5 (**a**), 683 (**b**), and 759 (**c**) nm.

Table 3. Energy values (cm^{-1}) of the ^2E, $^2T_1$ (at 5 K), $^4T_2$, and $^4T_1$ (at 77 K) levels of uncontrolled Cr^{3+} in YAB:Mn determined from the absorption and excitation spectra.

Level	Cr^{3+} in YAB:Mn
^2E	14,633
	14,690
$^2T_1$	15,267
	15,312
	15,396
$^4T_2$	16,730
$^4T_1$	23,320

A broad line at the low-frequency side of the R_1 and R_2 lines of the uncontrolled Cr^{3+} impurity (denoted "N" in Figure 7a) noticeably narrows with decreasing temperature and, at low temperatures ($T < 100$ K), it exceeds in amplitude the R_1 and R_2 lines. At the temperature $T = 5$ K, its frequency is 14,571 cm^{-1}. The N line apparently refers to a transition in exchange-coupled Cr^{3+}-containing pairs. A very similar pattern was observed, for example, in the luminescence spectra of isostructural GdAl$_3$(BO$_3$)$_4$ crystals doped with 1% Cr^{3+} (GAB:Cr^{3+}) [17]. The authors attribute the corresponding transition to the emission from the ^2E state of the Cr^{3+}-Cr^{3+} pairs. In our case, the formation of Cr-Mn pairs could also be possible.

The inset of Figure 7a shows the absorption spectra at the lowest measured temperature ($T = 5$ K) for two directions of incident light polarization, E∥c and E⊥c. The ratio of the amplitudes of the R_1 and R_2 lines is in agreement with the corresponding ratio for YAB:Cr^{3+} [38] (namely, I(R_1)/I(R_2) = 1 for E∥c, I(R_1)/I(R_2) = 2 for E⊥c), which once again confirms the origin of the observed R lines as stemming from the uncontrolled Cr^{3+} impurity. It is also worth noting that we found the same R lines of approximately the same intensity in "pure" YAB crystals grown from the same chemicals in the same laboratory as the YAB:Mn crystals under study. The rest of the spectrum observed for YAB:Mn is absent in YAB, so it is obviously associated with manganese.

EPR measurements revealed Mn^{2+} ions occupying yttrium-ion sites in YAB:Mn [28]. Although Mn^{3+} was introduced into the melt solution in the form of Mn$_2$O$_3$, it must be kept in mind that Mn$_2$O$_3$ decomposes in air at $T > 800\ °$C, losing part of the oxygen— 6(Mn^{3+})$_2$O$_3$ = 4Mn^{2+}(Mn^{3+})$_2$O$_4$ + O$_2$—so that Mn^{2+} ions appear. The charge-compensation can be realized by uncontrolled impurities such as Ti^{4+} (see Table 1). Optical spectra of Mn^{2+} in oxide crystals consist, as a rule, of a single broad band corresponding to the $^4T_1 \rightarrow {}^6A_1$ transition, which for Mn^{2+} in the Y^{3+} position is in the green region of the spectrum [43]. We attribute a broad band peaking at 531 nm ($T = 10$ K, see Figure 5) to the $^4T_1 \rightarrow {}^6A_1$ transition of Mn^{2+} in YAB:Mn.

Mn^{3+} was not found in the EPR studies of YAB:Mn [28]. Note, however, that Mn^{3+} is a non-Kramers ion and can be studied in some cases only by a special high-frequency EPR technique. Such studies on SrTiO$_3$:Mn have shown that Mn^{3+} substitutes for the octahedrally coordinated Ti^{4+} and forms three distinct types of Jahn–Teller centers that differ by charge-compensation mode [44]. The Mn^{3+} ion in octahedral coordination replacing Al^{3+} was found in Al$_2$O$_3$ (corundum) [45] and Y$_3$Al$_5$O$_{12}$ (YAG) [20,21]. Below, we discuss the features observed in our spectra of the YAl$_3$(BO$_3$)$_4$:Mn crystal, which we attribute to the transitions in the octahedrally coordinated Mn^{3+} at the Al^{3+} site.

Low-temperature luminescence of YAB:Mn in the IR range (9500–6500 cm^{-1} or 1055–1500 nm, see Figure 5) consists of relatively narrow (<10 cm^{-1}) ZPLs at 9371, 9388, 9430, and 9435 cm^{-1} and an adjacent vibronic band. In addition, narrow lines of uncontrolled impurities of Nd and Sm ions known from the YAB:Nd [3] and YAB:Sm [46] spectra are observed in the spectrum. A similar spectral pattern with narrow ZPLs with frequencies of about 9400 cm^{-1} and a phonon sideband was observed in a number of Mn^{3+}-doped garnets and was associated by the authors with $^1T_2 \to {}^3T_1$ transitions between excited triplets [20–22]. According to the Tanabe–Sugano diagrams [47], levels $^1T_2$ and $^3T_1$ have the same dependence on the crystal field, so the energy position of the corresponding transition band is practically independent of the strength of the crystal field. The multiple ZPLs observed in this region of the spectrum are most likely due to both the spin-orbit splitting of the $^3T_1$ level and the orbital splitting of excited triplets caused by the low-symmetry component of the crystal field.

One more relatively narrow (~80 cm^{-1}) line associated with manganese is observed in the red part of the low-temperature spectrum at 13,160 cm^{-1} (759 nm) (see Figure 5). It is accompanied by a Stokes vibronic sideband which grows in intensity with rising temperature; simultaneously, an anti-Stokes part appears (see, e.g., [48]). We tentatively assign this line to a transition from the excited orbital triplet $^1T_2$ to the ground Jahn–Teller-split doublet $^5E'$, $^5E''$ in Mn^{3+} [23]. A similar transition (though not as rich in structure) with a peak at 13,700 cm^{-1} was observed in the low-temperature emission spectrum of Y$_3$Al$_5$O$_{12}$ (YAG) doped with Mn^{3+} [20]. The excitation spectrum of the PL line 759 nm is presented in Figure 8c. It shows four bands peaking at 17,450, 22,000, 26,750, and 34,326 cm^{-1}. Bands at 17,450 and 22,000 cm^{-1} can be related to the spin-allowed transition from the ^5E ground state to the excited $^5T_2$ triplet of Mn^{3+}, split by the low-symmetry crystal field, whereas the bands at 26,750 and 34,326 cm^{-1} are apparently associated with the Mn^{3+} transitions to the higher-lying states (^3E, $^3T_1$) [23].

The strongest PL band of Mn^{3+}-doped crystals has the maximum in the region of wavelengths 620–670 nm [20–23] and is associated with the spin-allowed transition $^5T_2 \to {}^5E$. We assign a broad strong emission band peaked at 15,853 cm^{-1} (631 nm) to the $^5T_2 \to {}^5E$ transition of Mn^{3+}. Taking into account positions of the corresponding PLE bands, we find the mean value of 17,725 cm^{-1} as the energy of the $^5T_2$ state.

Based on the experimental values 17,725 cm^{-1} ($^5T_2$) and 13,160 cm^{-1} ($^1T_2$), as well as the Tanabe–Sugano diagram for the d^4 configuration [47], we estimate the crystal-field parameter Dq and Racah parameter B for Mn^{3+} in YAB:Mn as Dq = 1785 cm^{-1} and B = 800 cm^{-1}. The energy difference of ~9400 cm^{-1} between the $^1T_2$ and $^3T_1$ triplets, found from the IR spectra of the $^1T_2 \to {}^3T_1$ transition, agrees with these estimates in the framework of the Tanabe–Sugano diagram, which provides additional verification. The value Dq/B = 2.23 is very close to Dq/B = 2.25 found for Mn^{3+} in garnet-type Ca$_3$Ga$_2$Ge$_3$O$_{12}$ single crystals [22].

4. Conclusions

Using XANES and high-resolution optical spectroscopy, the valence composition of Mn ions in YAB:Mn was determined. According to the EXAFS data, manganese is contained in the crystal mainly in the divalent state Mn^{2+}(d^5), and substitutes for Y^{3+}. This conclusion is in agreement with the EPR results [28]. Luminescence of the Mn^{2+} ions at the $^4T_1 \to {}^6A_1$ transition (near 630 nm) was detected. For charge-compensation reasons, it would be natural to assume that Mn^{4+} is present in a neighborhood of Mn^{2+} [27,49]. It was previously shown for a number of aluminates that Mn^{4+} replaces octahedrally coordinated Al^{3+} [41,49], which is consistent with the proximity of their ionic radii (0.535 Å for Al^{3+} and 0.53 Å for Mn^{4+} [37]). We show that the R lines characteristic of the d^3 configuration (Mn^{4+}, Cr^{3+}), observed both in the absorption spectra ($^4A_1 \to {}^2E$) and in the luminescence spectra ($^2E \to {}^4A_1$) of YAB:Mn, arise not from Mn^{4+} but from the uncontrolled Cr^{3+} impurity. We failed to find the spectra of Mn^{4+}.

During crystal growth, Mn^{3+} was introduced in the form of Mn$_2$O$_3$, so the presence of the Mn^{3+} ions could be anticipated. In the IR range of the luminescence spectra of YAB:Mn at low temperatures, the spin-forbidden transitions $^1T_2 \to {}^3T_1$ and $^1T_2 \to {}^5E'$, $^5E''$ of Mn^{3+}(d^4) were observed. A broad emission band in the orange spectral range (near 630 nm) is associated with the spin-allowed $^5T_2 \to {}^5E$ transition of Mn^{3+}. Using the experimental spectroscopic data and the Tanabe–Sugano diagram for the d^4 configuration, we estimated the crystal-field parameter Dq and Racah parameter B for Mn^{3+} in YAB:Mn.

Further studies are needed to evaluate the application potential of YAB singly doped with manganese or co-doped with chromium. Our work can serve as a basis for these studies.

Author Contributions: Conceptualization, A.M. and M.P.; formal analysis, A.M., N.K., A.V. and E.K.; funding acquisition, A.M. and M.P.; investigation, A.M., K.B., A.V., K.K., E.K., O.K. and E.M.; methodology, K.B. and M.P.; resources, I.G.; validation, A.M., K.B., N.K., E.M., K.K., E.K., A.V. and O.K.; writing—original draft preparation, A.M.; writing—review and editing, M.P.; visualization, A.M. All authors have read and agreed to the published version of the manuscript.

Funding: This work was supported in part by the Russian Science Foundation under Grant No. 21-72-00134. K.B. and M.P. acknowledge financial support from the Ministry of Science and Higher Education of Russia under Grant 0039-2019-0004.

Institutional Review Board Statement: Not applicable.

Informed Consent Statement: Not applicable.

Data Availability Statement: Data can be obtained from the corresponding author upon reasonable request.

Acknowledgments: The authors thank A.A. Aleksandrovsky for important discussions.

Conflicts of Interest: The authors declare no conflict of interest.

References

1. Leonyuk, N.I.; Leonyuk, L.I. Growth and Characterization of $RM_3(BO_3)_4$ Crystals. *Prog. Cryst. Growth Charact. Mater.* **1995**, *31*, 179–278. [CrossRef]
2. Cavalli, E.; Leonyuk, N. Comparative Investigation on the Emission Properties of $RAl_3(BO_3)_4$ (R = Pr, Eu, Tb, Dy, Tm, Yb) Crystals with the Huntite Structure. *Crystals* **2019**, *9*, 44. [CrossRef]
3. Moura, A.L.; Carreño, S.J.M.; Pincheira, P.I.R.; Fabris, Z.V.; Maia, L.J.Q.; Gomes, A.S.L.; de Araújo, C.B. Tunable Ultraviolet and Blue Light Generation from Nd:YAB Random Laser Bolstered by Second-Order Nonlinear Processes. *Sci. Rep.* **2016**, *6*, 27107. [CrossRef] [PubMed]
4. Elzbieciak-Piecka, K.; Marciniak, L. Optical Heating and Luminescence Thermometry Combined in a Cr^{3+}-Doped $YAl_3(BO_3)_4$. *Sci. Rep.* **2022**, *12*, 16364. [CrossRef] [PubMed]
5. Shi, M.; Yao, L.; Xu, J.; Liang, C.; Dong, Y.; Shao, Q. Far-red-emitting $YAl_3(BO_3)_4:Cr^{3+}$ Phosphors with Excellent Thermal Stability and High Luminescent Yield for Plant Growth LEDs. *J. Am. Ceram. Soc.* **2021**, *104*, 3279–3288. [CrossRef]
6. Reddy, G.V.L.; Moorthy, L.R.; Chengaiah, T.; Jamalaiah, B.C. Multi-Color Emission Tunability and Energy Transfer Studies of $YAl_3(BO_3)_4:Eu^{3+}/Tb^{3+}$ Phosphors. *Ceram. Int.* **2014**, *40*, 3399–3410. [CrossRef]
7. Bajaj, N.S.; Koparkar, K.A.; Nagpure, P.A.; Omanwar, S.K. Red and Blue Emitting Borate Phosphor Excited by near Ultraviolet Light. *J. Opt.* **2017**, *46*, 91–94. [CrossRef]
8. Jaque, D. Self-Frequency-Sum Mixing in Nd Doped Nonlinear Crystals for Laser Generation in the Three Fundamental Colours. *J. Alloys Compd.* **2001**, *323–324*, 204–209. [CrossRef]
9. Jamalaiah, B.C.; Jayasimhadri, M.; Reddy, G.V.L. Blue Emitting $YAl_3(BO_3)_4:Tm^{3+}$ Single-Phase Phosphors under UV Excitation. *Phys. Chem. Glas. Eur. J. Glass Sci. Technol. Part B* **2016**, *57*, 68–70. [CrossRef]
10. Tolstik, N.A.; Kisel, V.E.; Kuleshov, N.V.; Maltsev, V.V.; Leonyuk, N.I. $Er,Yb:YAl_3(BO_3)_4$—Efficient 1.5 μm Laser Crystal. *Appl. Phys. B* **2009**, *97*, 357–362. [CrossRef]
11. Dekker, P.; Dawes, J.M.; Piper, J.A.; Liu, Y.; Wang, J. 1.1 W CW Self-Frequency-Doubled Diode-Pumped $Yb:YAl_3(BO_3)_4$ Laser. *Opt. Commun.* **2001**, *195*, 431–436. [CrossRef]
12. Burns, P.A.; Dawes, J.M.; Dekker, P.; Piper, J.A.; Li, J.; Wang, J. Coupled-Cavity, Single-Frequency, Tunable CW Yb:YAB Yellow Microchip Laser. *Opt. Commun.* **2002**, *207*, 315–320. [CrossRef]
13. Bartschke, J.; Knappe, R.; Boller, K.-J.; Wallenstein, R. Investigation of Efficient Self-Frequency-Doubling Nd:YAB Lasers. *IEEE J. Quantum Electron.* **1997**, *33*, 2295–2300. [CrossRef]
14. Jiang, H.; Li, J.; Wang, J.; Hu, X.-B.; Liu, H.; Teng, B.; Zhang, C.-Q.; Dekker, P.; Wang, P. Growth of $Yb:YAl_3(BO_3)_4$ Crystals and Their Optical and Self-Frequency-Doubling Properties. *J. Cryst. Growth* **2001**, *233*, 248–252. [CrossRef]
15. Yu, D.; Li, H.; Zhang, D.; Zhang, Q.; Meijerink, A.; Suta, M. One Ion to Catch Them All: Targeted High-Precision Boltzmann Thermometry over a Wide Temperature Range with Gd^{3+}. *Light Sci. Appl.* **2021**, *10*, 236. [CrossRef]
16. Dominiak-Dzik, G.; Ryba-Romanowski, W.; Lisiecki, R.; Földvári, I.; Beregi, E. $YAl_3(BO_3)_4:Yb\&Tm$ a Nonlinear Crystal: Up- and Down-Conversion Phenomena and Excited State Relaxations. *Opt. Mater.* **2009**, *31*, 989–994. [CrossRef]

17. Malysa, B.; Meijerink, A.; Jüstel, T. Temperature Dependent Luminescence Cr^{3+}-Doped $GdAl_3(BO_3)_4$ and $YAl_3(BO_3)_4$. *J. Lumin.* **2016**, *171*, 246–253. [CrossRef]
18. Du, J.; De Clercq, O.Q.; Korthout, K.; Poelman, D. $LaAlO_3$:Mn^{4+} as Near-Infrared Emitting Persistent Luminescence Phosphor for Medical Imaging: A Charge Compensation Study. *Materials* **2017**, *10*, 1422. [CrossRef]
19. Glais, E.; Đorđević, V.; Papan, J.; Viana, B.; Dramićanin, M.D. $MgTiO_3$:Mn^{4+} a Multi-Reading Temperature Nanoprobe. *RSC Adv.* **2018**, *8*, 18341–18346. [CrossRef]
20. Kück, S.; Hartung, S.; Hurling, S.; Petermann, K.; Huber, G. Emission of Octahedrally Coordinated Mn^{3+} in Garnets. *Spectrochim. Acta. A. Mol. Biomol. Spectrosc.* **1998**, *54*, 1741–1749. [CrossRef]
21. Kück, S.; Hartung, S.; Hurling, S.; Petermann, K.; Huber, G. Optical Transitions in Mn^{3+}-Doped Garnets. *Phys. Rev. B* **1998**, *57*, 2203–2216. [CrossRef]
22. Wang, Y.; Włodarczyk, D.; Brik, M.G.; Barzowska, J.; Shekhovtsov, A.N.; Belikov, K.N.; Paszkowicz, W.; Li, L.; Zhou, X.; Suchocki, A. Effect of Temperature and High Pressure on Luminescence Properties of Mn^{3+} Ions in $Ca_3Ga_2Ge_3O_{12}$ Single Crystals. *J. Phys. Chem. C* **2021**, *125*, 5146–5157. [CrossRef]
23. Jahanbazi, F.; Wang, X.; Mao, Y. Tb^{3+}, Mn^{3+} Co-Doped $La_2Zr_2O_7$ Nanoparticles for Self-Referencing Optical Thermometry. *J. Lumin.* **2021**, *240*, 118412. [CrossRef]
24. Stevels, A.L.N. Red Mn^{2+}-Luminescence in Hexagonal Aluminates. *J. Lumin.* **1979**, *20*, 99–109. [CrossRef]
25. Costa, G.K.B.; Pedro, S.S.; Carvalho, I.C.S.; Sosman, L.P. Preparation, Structure Analysis and Photoluminescence Properties of $MgGa_2O_4$:Mn^{2+}. *Opt. Mater.* **2009**, *31*, 1620–1627. [CrossRef]
26. Majher, J.D.; Gray, M.B.; Strom, T.A.; Woodward, P.M. $Cs_2NaBiCl_6$:Mn^{2+}—A New Orange-Red Halide Double Perovskite Phosphor. *Chem. Mater.* **2019**, *31*, 1738–1744. [CrossRef]
27. Aleksandrovsky, A.S.; Gudim, I.A.; Krylov, A.S.; Temerov, V.L. Luminescence of Yttrium Aluminum Borate Single Crystals Doped with Manganese. *Phys. Solid State* **2007**, *49*, 1695–1699. [CrossRef]
28. Vorotynov, A.M.; Petrakovskiĭ, G.A.; Shiyan, Y.G.; Bezmaternykh, L.N.; Temerov, V.E.; Bovina, A.F.; Aleshkevych, P. Electron Paramagnetic Resonance of Mn^{2+} Ions in Single Crystals of Yttrium Aluminum Borate $YAl_3(BO_3)_4$. *Phys. Solid State* **2007**, *49*, 463–466. [CrossRef]
29. Terayama, K.; Ikeda, M. Study on Thermal Decomposition of MnO_2 and Mn_2O_3 by Thermal Analysis. *Trans. JIM* **1983**, *24*, 754–758. [CrossRef]
30. Petříček, V.; Dušek, M.; Palatinus, L. Crystallographic Computing System JANA2006: General Features. *Z. Für Krist. Cryst. Mater.* **2014**, *229*, 345–352. [CrossRef]
31. Belokoneva, E.L.; Timchenko, T.I. Polytypic relationships in borate structures with the general formula $RAl_3(BO_3)_4$, (R = Y, Nd, Gd). *Kristallografiya* **1983**, *28*, 1118–1123.
32. Chernyshov, A.A.; Veligzhanin, A.A.; Zubavichus, Y.V. Structural Materials Science End-Station at the Kurchatov Synchrotron Radiation Source: Recent Instrumentation Upgrades and Experimental Results. *Nucl. Instrum. Methods Phys. Res. Sect. Accel. Spectrometers Detect. Assoc. Equip.* **2009**, *603*, 95–98. [CrossRef]
33. Newville, M. *IFEFFIT*: Interactive XAFS Analysis and *FEFF* Fitting. *J. Synchrotron Radiat.* **2001**, *8*, 322–324. [CrossRef] [PubMed]
34. Ravel, B.; Newville, M. *ATHENA, ARTEMIS, HEPHAESTUS*: Data Analysis for X-Ray Absorption Spectroscopy Using *IFEFFIT*. *J. Synchrotron Radiat.* **2005**, *12*, 537–541. [CrossRef] [PubMed]
35. Manceau, A.; Marcus, M.A.; Grangeon, S. Determination of Mn Valence States in Mixed-Valent Manganates by XANES Spectroscopy. *Am. Mineral.* **2012**, *97*, 816–827. [CrossRef]
36. Bunău, O.; Joly, Y. Self-Consistent Aspects of x-Ray Absorption Calculations. *J. Phys. Condens. Matter* **2009**, *21*, 345501. [CrossRef]
37. Shannon, R.D. Revised Effective Ionic Radii and Systematic Studies of Interatomic Distances in Halides and Chalcogenides. *Acta Crystallogr. Sect. A* **1976**, *32*, 751–767. [CrossRef]
38. Wang, G.; Gallagher, H.G.; Han, T.P.J.; Henderson, B. Crystal Growth and Optical Characterisation of Cr^{3+}-Doped $YAl_3(BO_3)_4$. *J. Cryst. Growth* **1995**, *153*, 169–174. [CrossRef]
39. Dominiak-Dzik, G.; Ryba-Romanowski, W.; Grinberg, M.; Beregi, E.; Kovacs, L. Excited-State Relaxation Dynamics of Cr^{3+} in $YAl_3(BO_3)_4$. *J. Phys. Condens. Matter* **2002**, *14*, 5229–5237. [CrossRef]
40. Wells, J.-P.R.; Yamaga, M.; Han, T.P.J.; Honda, M. Electron Paramagnetic Resonance and Optical Properties of Cr^{3+} Doped $YAl_3(BO_3)_4$. *J. Phys. Condens. Matter* **2003**, *15*, 539–547. [CrossRef]
41. Geschwind, S.; Kisliuk, P.; Klein, M.P.; Remeika, J.P.; Wood, D.L. Sharp-Line Fluorescence, Electron Paramagnetic Resonance, and Thermoluminescence of Mn^{4+} in α-Al_2O_3. *Phys. Rev.* **1962**, *126*, 1684–1686. [CrossRef]
42. Tanabe, Y.; Sugano, S. The Absorption Spectra of Ruby. *J. Phys. Soc. Jpn.* **1957**, *12*, 556. [CrossRef]
43. Noginov, M.A.; Loutts, G.B.; Warren, M. Spectroscopic Studies of Mn^{3+} and Mn^{2+} Ions in $YAlO_3$. *J. Opt. Soc. Am. B* **1999**, *16*, 475. [CrossRef]
44. Azamat, D.V.; Dejneka, A.; Lancok, J.; Trepakov, V.A.; Jastrabik, L.; Badalyan, A.G. Electron Paramagnetic Resonance Studies of Manganese Centers in $SrTiO_3$: Non-Kramers Mn^{3+} Ions and Spin-Spin Coupled Mn^{4+} Dimers. *J. Appl. Phys.* **2012**, *111*, 104119. [CrossRef]
45. McClure, D.S. Optical Spectra of Transition-Metal Ions in Corundum. *J. Chem. Phys.* **1962**, *36*, 2757–2779. [CrossRef]
46. Kebaïli, I.; Dammak, M. Spectra Energy Levels and Symmetry Assignments of Sm^{3+} Doped in $YAl_3(BO_3)_4$ Single Crystal. *J. Lumin.* **2012**, *132*, 2092–2097. [CrossRef]

47. Tanabe, Y.; Sugano, S. On the Absorption Spectra of Complex Ions II. *J. Phys. Soc. Jpn.* **1954**, *9*, 766–779. [CrossRef]
48. Adachi, S. Review—Temperature Dependence of Transition-Metal and Rare-Earth Ion Luminescence (Mn^{4+}, Cr^{3+}, Mn^{2+}, Eu^{2+}, Eu^{3+}, Tb^{3+}, Etc.) II: Experimental Data Analyses. *ECS J. Solid State Sci. Technol.* **2022**, *11*, 106002. [CrossRef]
49. Noginov, M.A.; Loutts, G.B. Spectroscopic Studies of Mn^{4+} Ions in Yttrium Orthoaluminate. *J. Opt. Soc. Am. B* **1999**, *16*, 3. [CrossRef]

Disclaimer/Publisher's Note: The statements, opinions and data contained in all publications are solely those of the individual author(s) and contributor(s) and not of MDPI and/or the editor(s). MDPI and/or the editor(s) disclaim responsibility for any injury to people or property resulting from any ideas, methods, instructions or products referred to in the content.

Review

UV-A,B,C Emitting Persistent Luminescent Materials

Suchinder K. Sharma [1,*], Jinu James [1], Shailendra Kumar Gupta [1] and Shamima Hussain [2]

1 Amity School of Physical Sciences, Amity University Punjab, IT City, Sector 82A, Mohali 140306, India
2 UGC-DAE Consortium for Scientific Research, Kalpakkam Node, Kokilamedu 603104, India
* Correspondence: suchindersharma@gmail.com or ssharma1@pb.amity.edu; Tel.: +91-172-5203528

Abstract: The nearly dormant field of persistent luminescence has gained fresh impetus after the discovery of strontium aluminate persistent luminescence phosphor in 1996. Several efforts have been put in to prepare efficient, long decay, persistent luminescent materials which can be used for different applications. The most explored among all are the materials which emit in the visible wavelength region, 400–650 nm, of the electromagnetic spectrum. However, since 2014, the wavelength range is extended further above 650 nm for biological applications due to easily distinguishable signal between luminescent probe and the auto-fluorescence. Recently, UV-emitting persistent materials have gained interest among researchers' due to their possible application in information storage, phototherapy and photocatalysis. In the present review, we summarize these recent developments on the UV-emitting persistent luminescent materials to motivate young minds working in the field of luminescent materials.

Keywords: persistent luminescence; UV-emission; optical properties; phototherapy

Citation: Sharma, S.K.; James, J.; Gupta, S.K.; Hussain, S. UV-A,B,C Emitting Persistent Luminescent Materials. *Materials* **2023**, *16*, 236. https://doi.org/10.3390/ma16010236

Academic Editor: Dirk Poelman

Received: 26 November 2022
Revised: 19 December 2022
Accepted: 20 December 2022
Published: 27 December 2022

Copyright: © 2022 by the authors. Licensee MDPI, Basel, Switzerland. This article is an open access article distributed under the terms and conditions of the Creative Commons Attribution (CC BY) license (https://creativecommons.org/licenses/by/4.0/).

1. Introduction

Luminescence or the emission of light is an old phenomenon. The first report of light emission from fireflies and glowworms can be dated back to the period, 1500–1000 B.C., in the Indian holy scriptures "Vedas" and also in Chinese Book of Odes (the Shih Ching). The word for "glowworm" in sanskrit language is "Khadyota" [1]. The reports on the glow of bluish-green color from a stone, by Vincenzo Cascariolo, is another important report in the literature [1–4]. The stone was named "lapis solaris" (also called bolognian stone or solar stone or sun sponge or spingiasolis) [1,2]. Similar reports can also be found for the glow from decaying fish, fungus and bacteria [1,2,5]. Later, in 1612, La Galla wrote first publication on this first man-made material [1,2]. Thus, the term "phosphor" was coined to distinguish it from the elemental phosphorous [2]. Such afterglow which was discovered and reported in different time domains was later known as "phosphorescence" [2]. In 1888, a German physicist named Eilhard Wiedemann coined the term 'luminescence' which included both fluorescence and phosphorescence [2]. The fluorescence and phosphorescence are also termed as short-lived and long-lived luminescence, respectively. The different categories of luminescence were later developed based upon the choice of the different excitation methods [2].

The beginning of modern era can be dated back to the discovery of man-made ZnS:Cu,Co material by Hoogenstraaten and Klasens in the year 1953 [6]. However, the materials could not be commercialized successfully because of its shorter decay time and stability in humid environment. The most intense emission of this material was observed at wavelength 530 nm. In 1971, another important material from the strontium aluminate family, $SrAl_2O_4:Eu^{2+}$, was discovered [7]. Matsuzawa et al., in the year 1996, published first article on the persistent luminescence of $SrAl_2O_4:Eu^{2+},Dy^{3+}$ having maximum emission at wavelength 520 nm [8]. Upon searching for the same paper on Google, one can find that the article has already been cited 2347 times (to date). After this report, the aluminate family became popular and was explored in detail by many researchers [9–12].

The data on the strontium aluminate host family has also been summarized recently by Heggen et al. along with a possible new direction for exploring it further [13]. Among other hosts, $Sr_2MgSi_2O_7:Eu^{2+},Dy^{3+}$ from the silicate family was established in the year 2001 by Lin et al. [14]. These materials have found applications in various domains including biology (bioimaging), chemistry (different synthetic procedures), physics (application of materials for decoration, safety signage, and solid state lighting) and material science (engineered materials). The use of these materials in watches and toys can easily be found around us. The common abbreviation used in the literature for persistent luminescence is PersL, as coined in a recent review article by Xu et al. [15]. Hereafter, we will use the same abbreviation (PersL) for persistent luminescence phenomenon. The term PersL material has been used for materials with phosphorescence from minutes to hours. PersL materials are quite similar to an optical battery where the material is first charged for some duration (few s to min), and the emission of light is observed when the material is kept in the dark. Emergency signage used in case of electricity failures, watch dials, decorative objects, toys, for energy storage and others are some of the uses of these materials [15,16].

As shown in Figure 1, based upon published literature on PersL materials, one can observe three different stages: (a) year 1996 when the first publication by Matsuzawa was published (as discussed in previous section) [8]; (b) year 2007 when another article by Chermont et al. was published where PersL material (silicate host) was used for bio-imaging for the first time [17]; and (c) year 2017 when the first report on the PersL from an organic material was reported [18]. The other articles by Bessiere et al. in 2011 [19] and Maldiney et al. in 2014 [20] are other important publications in the field. Upon looking at the publication year and number of publications/year from the data in Figure 1, one can observe an exponential increase in the number of publications suggesting an increased interest of the scientific community.

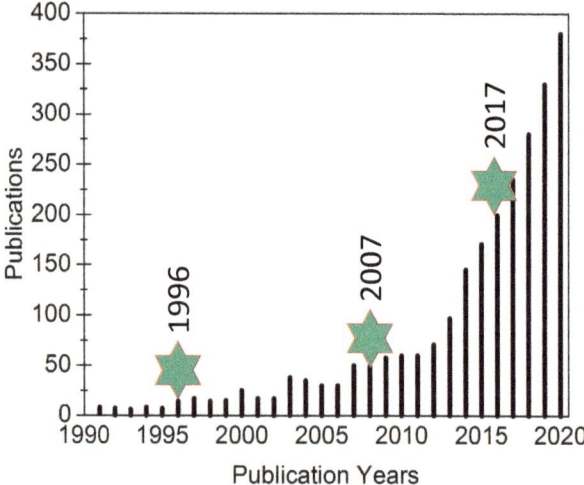

Figure 1. The number of publications on PersL materials as a function of the publication year. Three important years, 1996, 2007 and 2017, are also marked.

2. Crucial Parameters for PersL Materials

The type (nature) of defect, their number and energetics within the host band gap are important for a good PersL material. One can engineer new materials by modulating the host bandgap itself or by changing the defect scenario in the host lattice. The PersL materials are expected to possess following important parameters to get required emission color, efficiency and the long lasting luminance:

1. **Trap depth:** The trap (defect) depth within the host lattice is the first and foremost important parameter for a good PersL material. When the trap depth energies are between 0.5–0.6 eV, the material can be effectively charged (trapping) and discharged (detrapping) at room temperature. In fact, instead of intentional defects (0.5–0.6 eV), materials containing unintended defects with trap depth between 1.0–2.0 eV can be utilized for de-trapping under the influence of thermal or optical energy addressed as thermoluminescence (TL) and optically stimulated luminescence (OSL) respectively. The more details on these aspects can be found in Refs. [21,22].
2. **Minimum light output:** PersL is the light output that is observed when an initial excitation is seized. Hence, the two important parameters that prevail after such seizure of excitation energy are: (a) light output, and (b) its duration. Usually in most of the materials, the light intensity decreases by almost 90 % of the initial value in the initial few minutes limiting their commercial aspects. However, a good PersL material is the one in which the duration below which the photopic intensity decreases to an eye perceivable intensity value of 0.32 mcd/m^2, exists. This minimum threshold value is important for applications too.
3. **Frequency factor (s):** When the charges are detrapped from defects at room temperature, there exists a competition between re-trapping and detrapping processes leading to the delay in phosphorescence. The charges once trapped are released very slowly from traps at room temperature delaying the overall recombination process at the luminescence center. Due to this competition between the trapping–detrapping–retrapping processes, multi-exponential or hyperbolic decay curve is obtained. The frequency factor (s) is an important parameter and its value depends upon the competition between these different processes. The typical value of 's' is between 10^6–10^{14} s^{-1} [21]. However, in literature, a value of 10^{11} s^{-1} is used, which underestimates the overall phenomenon leading to wrong interpretation.

Overall, the key engineering aspect to prepare new/existing PersL materials is the creation of luminescence centers intentionally having a trap depth between energies, 0.5–0.6 eV, which can be effectively charged and/or discharged at the room temperature.

3. PersL Materials: Synthesis

New PersL materials are researched frequently leading to important ideas and innovative applications. In these developments, materials chemistry plays an important role [23]. For example, apart from the preparation method, the influence of crystallinity, particle size distribution and morphology has an impact on the properties and performance of PersL materials. Therefore, the researchers have shown immense interest in nanometre-scale PersL materials. Although there are various methods for the synthesis, the PersL materials with desired spectral and decay time output are challenging. These methods can be classified into two groups:

3.1. Conventional Method

Solid-State synthesis: The synthesis method is one of the widely used technique to prepare PersL materials. For example, AO- and B_2O_3-type compounds, which act as precursors, are used to prepare AB_2O_4-type spinel compounds [19,24]. Such materials have shown promising application in Bio-imaging. In this case, the stochiometric amount of precursors are mixed, either manually or using more sophisticated techniques like ball-milling, to thoroughly mix precursors. Such mixture is then heated at high temperatures for prolonged time. For some materials, repeated grinding and calcining steps are followed for even mixing of the compounds and to obtain the final phase of the product (crystallinity). In the case of AB_2O_4-type spinel compounds, temperature as high as 1400 °C is used to obtain the final product. In some reports, the flux is also preferred to lower the calcining temperature of the samples and to obtain highly crystalline final products.

3.2. Non-Conventional Methods

Non-conventional techniques for the synthesis of PersL materials are defined as those techniques which do not comprise steps such as continuous grinding and/or heating at elevated temperatures. These techniques are also called wet-chemical routes. Some of these routes are:

3.2.1. Sol-Gel Method

Sol-gel synthesis method comprises of atleast six different steps which includes hydrolysis, polymerization, gelation, drying, dehydration and densification. The product formed in sol-gel method is uniform and possesses better control over the crystallite size, dimensions, shape and morphology of the final product. The method involves both the physical and chemical processes. The process can prepare different oxide-based host matrices with ease. The synthesis of PersL materials, $CdSiO_3$ [25], Sr_2SiO_4 [26], $Y_3Al_5O_{12}$ [27], and $LaAlO_3$ [28], are noticeable contributions in the field.

The method involves primarily two important hydrolysis and condensation reactions and the final output depends on the nature of precursor, solvents, reaction temperature and pH of the sol. In literature, from the synthesis point-of-view, $Lu_3Al_{5-x}Ga_xO_{12}:Ce^{3+},Cr^{3+}$ [29], $Gd_3Al_{5-x}Ga_xO_{12}:Ce^{3+},Pr^{3+}$ [30], $CaAl_2O_4:Yb$ [31], $Sr_2MgSi_2O_7:Eu^{3+},Dy^{3+}$ [32], $SrAl_2O_4$, $SrSnO_3:Pr^{3+}$ [33], $Y_2O_3:Dy^{3+}/Nd^{3+}/Sm^{3+}$ [34], $Li_6CaLa_2Nb_2O_{12}:Eu^{3+}$ [35], are important PersL materials synthesized through the sol-gel method. For UV-emitting PersL materials, limited literature based on materials such as Pb^{2+} ions doped $Ca_2La_8(SiO_2)_6O_2$ [36], $Lu_2SiO_5:Pr^{3+}$ [37], $CaMgSi_2O_6$ [38], could be found.

3.2.2. Combustion Method

Combustion method is another interesting technique to prepare nanoparticles of PersL materials. This technique was developed in late 1990s. In this method, the use of a fuel, either of urea, hydrazine or glycine, is preferred. Fuel plays crucial role in propagation of the reaction via layer by layer heat transfer. The final product in combustion synthesis is mostly powders while some other forms like foam and conglomerates are also reported in the literature. As the reaction in combustion synthesis is highly exothermic, the temperature during such heat release is ~2500 K. The method is mostly used to prepare oxides and thus require water soluble precursors (mostly nitrates). The method was modified in later stages to use either heated wire, electric spark or laser beams to provide initial temperature for reaction to initiate. Once initiated, the reaction continues on its own due to the presence of fuel.

The final product in this reaction is obtained within 5 min of the initiation of the reaction. Apart from the use of a fuel for reaction, there are two more important parameters that should be fulfilled: (a) the product to be formed should be refractory in nature, and (b) the solution prepared from precursors should be well dispersed and should possess high chemical energy so that the combustion reaction can initiate. The reaction is initiated at temperature ~500 K in an appropriate atmosphere which is mostly dominant by oxygen gas (usually air) to promote exothermic reaction. Combustion synthesis is characterized by several benefits such as the low reaction duration, low desired initial temperature for reaction initiation, quick synthesis and high final product yield. Moreover, in this method, there is no need for high temperature furnaces as is the case with solid-state synthesis method. Some of the interesting articles in this domain are Refs. [39–41].

3.2.3. Hydrothermal Method

The method is another important non-conventional method to prepare nanoparticles of PersL materials. The method is even capable of preparing nanomaterials which are unstable at high temperatures. The solution of either oxides, hydrides, or metal powders is prepared in the suspension form for an initiation of the reaction. The important parameters, temperature (around 573 K) and pressure (around 100 MPa), are controlled to control the shape and size of the final product (nanoparticles). The reaction starts with a nucleation step followed by the growth under controlled conditions of temperature and pressure. All the type of final products, oxides, metal nanoparticles and non-oxides can be prepared using hydrothermal method. The relevant synthesis process to prepare PersL materials can be obtained from Refs. [42–44].

3.2.4. Co-Precipitation Method

The co-precipitation is a non-conventional method of preparing PersL materials. The main achievement of this synthesis type is that there is a better control over size of the nanoparticles via control over pH, temperature and concentration of the reactants. The nanoparticles prepared are homogeneous and no agglomeration is obtained for the final product. This method has been applied to synthesize variety of PersL materials, for example $Sr_2MgSi_3O_7:Eu^{2+}Dy^{3+}$ [45]. The method can also produce nanotues of $SrAl_2O_4:Eu^{2+}Dy^{3+}$ [46], and other important hosts as mentioned in Refs. [47,48].

3.2.5. Pechini and Citrate Gel Method

Marcilly and coworkers developed this technique in the year 1970 [49]. The pH during this synthesis is controlled between 6 and 7.5 to dissolve nitrates of precursor salts with citric acid solution. Polybasic chelating agent is changed to a resin and later to transparent gel during synthesis process. The gel is then pyrolyzed to obtain nanoparticles of the final product. The method is known to prepare multi-component oxides. More recently, the method is modified and is called modified-Pechini method [50–52]. In this new modified method, the polymerizable complexes are preferred and added to control the growth kinetics.

The above mentioned conventional and non-conventional synthesis procedures show great potential to prepare nanoparticles with reasonable control of the as-prepared PersL materials [53,54].

4. PersL Materials: Wavelength Overview

The most promising application of PersL materials is in bio–imaging as discussed by Maldiney et al. in the year 2014 [20]. PersL materials for such application are red or near-IR emitting around wavelength 700 nm, where there is a very little overlap between emission from the luminescence probe and absorption that of the animal cells, tissue, water and/or melanin. This region is especially important so as to distinguish between the auto-fluorescence from the cells and that of the PersL probe. Moreover, the nanoparticles of such synthetic probes (PersL materials) are generally preferred as they show high carrier mobility in the free-state, enormous specific surface area, as well as exhibit quantum effect. Similarly, visible range emitting PersL materials are preferred for other applications such as signage devices and for other decorative purposes. While most of the discovered PersL materials are emitting within the visible/near-IR wavelength range, very little is published on the UV-emitting PersL materials. The UV-light bands, UV-A (315–400 nm), UV-B (280–315 nm) and UV-C (200–280 nm), play an important role for applications like photocatalysis, anti-counterfeiting and water-disinfecting, etc. [55]. Thus, recently the attention has been evident focusing on the development of novel UV-emitting photoluminescence (PL) materials [56,57]. However, these PL materials are expected to also emit PersL for some specialized applications.

The rays coming from the sun are another important source of UV-radiations especially that of UV-A and UV-B radiations. The amount of UV radiations received on earth from the sun vary from altitude, weather, season of the year, time of the day and latitude. These UV-radiations cause pigmentation or tanning in humans. UV-A and UV-B induces tanning in the basal cell layer and upper layers of the epidermis, respectively [58,59]. The other after effects of high doses of UV-radiations also include DNA damage and photocarcinogenesis [60–62]. On the positive side, photodynamic therapy (PDT) is another novel technique of curing patients suffering from bacterial infection [58]. Depending upon the choice of UV-radiations for exposure (UV-A/B/C), different phototherapy techniques can be classified [58]. Similarly, there are other application domains (as will be discussed in later sections) for which new and promising UV-emitting materials are desired and require immediate attention of the scientific community.

Almost 95% of the emitted UV-radiations from the sun contain UV-A radiations in the wavelength region 315–400 nm. These radiations are capable of affecting the top layer of skin and can cause premature ageing, wrinkles and some skin cancers. There are two important components to consider while thinking about the existing or new UV-emitting PersL materials: (a) the choice of the host lattice, and/or (b) the choice of the dopant ions. The prominent hosts used by materials scientists to prepare PersL materials are garnets, silicates, phosphates and perovskites, as has been summarized recently by Wang and Mao [63]. On the other hand, the choice of luminescence center, defined as the color emitted by the luminescent materials, is mainly focused on Pb^{2+}, Bi^{3+}, Pr^{3+}, Gd^{3+}, Ce^{3+} and Tb^{3+} [63,64].

While Pb^{2+} and Bi^{3+} are post-transition metals, the other prominent dopants in the literature are rare earth (lanthanoid, Ln) ions. The lanthanoids possess electronic configuration, $[Xe]4f^n6s^2$, where the n changes from 1 (for Ce) to 14 (for Lu). Lanthanoid ions can assume the oxidation state of 2+, 3+ or 4+, with respective loss of $6s^2$, $6s^2 + 4f^1$ and $6s^2 + 4f^2$ electrons. Then the outside shielding is performed by the $5s^2$ and $5p^6$ electrons. They also show contraction effect also known as lanthanoid contraction, which causes a decrease in their atomic (and ionic) radii as the atomic number increases from $1 \rightarrow 14$ [65]. Due to this shielding effect, except Ce^{3+}, other Ln^{3+} luminescence is not affected much by the choice of the host matrix. The summary of the choice of lanthanoid ion and corresponding emission in different hosts is shown in Figure 2.

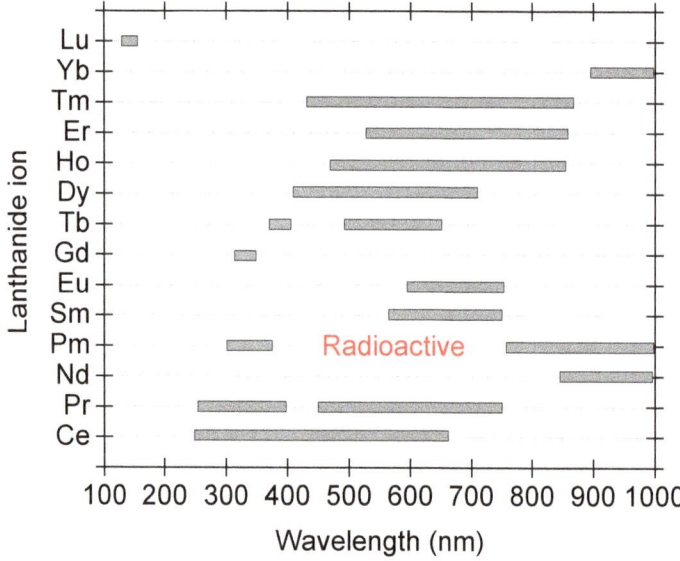

Figure 2. The variation of luminescence emission for all the fourteen lanthanoid ions (Ce to Lu) in different hosts. The marked emission ranges are typical of emission from Ln^{3+} ions only.

Among different 14 lanthanoids, the most promising is Ce^{3+}. Ce^{3+} with one electron in the 4f orbital is shielded by $5s^2$ and $5p^6$ orbitals [66]. The $4f^1$ state is hardly perturbed by the type of compound it is added in to. However, upon excitation to the 5d orbital, a strong interaction exists, which should be taken care of, while interpreting the luminescence spectra. Depending upon the site symmetry, utmost five distinct 4f $\rightarrow$ 5d transitions can be observed upon Ce^{3+} doping. The crystal field leads to a decrease in the lowest of the 5d state by approximately 52,000 cm^{-1} when measured from the top of 4f levels. This shift is also known as 'centroid shift' or 'barycentre'. The combined effect of spin-orbit interactions and crystal field leads to the redshift (denoted by D) of the first 4f → 5d transitions [66–71]. The typical lifetime of Ce^{3+} 5d–4f transitions varies in the range ∼10–60 ns in different hosts. The Ce^{3+} PersL can be observed in different hosts, which includes, oxides, sulfides, silicates, garnets, etc. For only Ce^{3+} doped phosphor hosts, a clear variation in PersL emission maximum from 385 nm to 525 nm can be observed. Among aluminates, $SrAl_2O_4$:Ce^{3+} possesses UV-A PersL emission at 385 nm with afterglow of over 10 h [72,73].

Apart from Ce^{3+} ion, when we look into Figure 2 to find the promising lanthanoid for an emission in the UV-region (200-400 nm), very limited options are available below 400 nm. The ones with such capability are Ce^{3+}, Pr^{3+}, Pm^{3+} (also radioactive), Gd^{3+}, and Tb^{3+}. Lu^{3+} doping is also promising though the emission is in the vacuum ultra-violet region, cf. Figure 2. For all these doping options, the excitation from high energy (for downconversion) is required and thus requires more sophisticated instrumentation (even for the excitation process).

5. UV-Emitting PersL Materials

The UV-emitting PersL materials are promising for diverse applications as discussed in previous sections. The literature published on PersL materials emitting in the three different UV domains, UV-A, UV-B, and UV-C, is compiled in Table 1.

From the data, three different rare earth–based dopant ions, Ce^{3+}, Gd^{3+} and Pr^{3+}, are observed to give PersL emission in the UV-region. Ce^{3+} doping mostly gives PersL emission in the UV-A region of the electromagnetic spectrum, while the Gd^{3+} emits mostly in the UV-B region. For Pr^{3+}, the PersL emission is observed in all the three wavelength regions, UV-A, B and C, of the UV-region. PersL emission is based on the excitation of charges to the higher excited state followed by trapping of these charges into the defect states. The charges in these defect states can be detrapped at the room temperature, and then trapped again in same defects, thereby delaying the overall PersL time. When we look into the data in Table 1, undoped $SrZrO_3$ gives the lowest PersL time of 100s with emission in the UV-A region [74], while longest PersL decay time is obtained for $LiYGeO_4$ with emission for 300 h [75] The excitation or charging step can be adopted by either choosing high energy for movement through the conduction band or by using low energy for more localized excitation. Thus, based upon these two aspects, the mechanisms of PersL materials can be divided into the delocalized mechanism [21,22] and the localized mechanism [20]. In the delocalized mechanism, the charge trapping–detrapping occurs via conduction band or valence band for electrons and holes; while in localized mechanism, the trapping–detrapping is similar to that of molecular systems [76].

Table 1. Different hosts emitting UV-A, UV-B and UV-C PersL emission. The information for the dopant ions, corresponding decay time and application domain is also provided.

Host	Dopant	Emission λ (nm)	PersL Duration	Application	Reference
UV-A Emission					
$LiScGeO_4$	Bi^{3+}	361	>12 h	information storage	[77]
$SrLaAlO_4$	Bi^{3+}	380	60 min	photodynamic therapy	[78]
$LiYGeO_4$	Bi^{3+}	350	72–300 h	biomedical, catalysis	[75]
CaB_2O_4	Ce^{3+}	365	15 h	UV Phototherapy	[79]
$Sr_2MgGe_2O_7$	Pb^{2+}	370	>12 h	anti-counterfeiting	[80]
$LiScGeO_4$	Bi^{3+}	365	120 h	photodynamic therapy	[81]
$NaLuGeO_4$	Bi^{3+}	400	63 h	photodynamic therapy	[82]
SrO	Pb^{2+}	390	>1 h	–	[83]
CaO	Pb^{2+}	360	>1 h	–	[83]
$MO–Al_2O_3–SiO_2$	Ce^{3+}	396	2 min	photocatalysis	[84]
$SrZrO_3$	undoped	395	100 s	information storage	[85]
$SrZrO_3$	Pr^{3+}	300–450	10 min	–	[74]

Table 1. Cont.

Host	Dopant	Emission λ (nm)	PersL Duration	Application	Reference
$CdSiO_3$	Bi^{3+}	360	<5 min	photocatalysis	[25]
$CdSiO_3$	Bi^{3+}	360	<10 min	disinfection	[25]
$CdSiO_3$	$Gd^{3+}-Bi^{3+}$	344	24 h	photocatalysis	[86]
Zn_2SiO_4	$Ga^{3+}-Bi^{3+}$	384, 374	4 h	photocatalysis	[87]
$LiLuGeO_4$	$Bi^{3+}-Yb^{3+}$	350	15 h	biophotonics	[88]
$CaAl_2O_4$	Ce^{3+}	400	>10 h	–	[89]
UV-B Emission					
$CaZnGeO_6$	Bi^{3+}	300–600	>12 h	photocatalysis	[57]
CYAS	Pr^{3+}	266/311	>12 h	Germ killing	[90]
Li_2CaGeO_3	Pr^{3+}	240–330	20 min	Sterilization	[91]
MLGB	Bi^{3+}	306	>12 h	multimode imaging	[92]
$(Y,Gd)_3Ga_5O_{12}$	Bi^{3+}	313	24 h	optical tagging	[93]
MLGO	Bi^{3+}	310–350	24 h	anticounterfeiting	[94]
LAGO	Pr^{3+}	302	60 h	optical tagging	[95]
$(Lu,Y)_3(Al,Ga)_5O_{12}$	Bi^{3+}	302–313	72 h	data encryption	[63]
YGG	Bi^{3+}	316	60 min	–	[96]
YAG	Bi^{3+}	303	60 min	–	[96]
BLAGSO	Pr^{3+}	301	3 h	photocatalysis	[97]
SYSO	Gd^{3+}	299	12 h	dermatology therapy	[98]
UV-C Emission					
Cs_2NaYF_6	Pr^{3+}	250	2 h	sensing/biomedicine	[75]
$LaPO_4$	Pr^{3+}	231	2 h	optoelectronic materials	[99]
SYSO	Pr^{3+}	266	12 h	dermatology therapy	[98]
YPO_4	Bi^{3+}	240	2 h	cancer therapy	[100]
Lu_5SiO_5	Pr^{3+}	200–280	12 h	optical tagging	[37]

6. Luminescence Mechanisms

The mechanism of charge trapping and detrapping is simple, yet complicated. In general, when a sample is excited using an appropriate excitation energy, the electrons move from valence band to the conduction band followed by their trapping at the defect sites. These charges upon trapping require some external stimulation such as heat or optical energy, to get out from such defects. There can be two different ways of charge trapping: (a) one related to electrons (electron trapping), and (b) related to holes (hole trapping), as shown in Figure 3. In both cases, the released charge carriers can recombine with their charge carrier counterpart at luminescence centers producing luminescence due to electron–hole recombination.

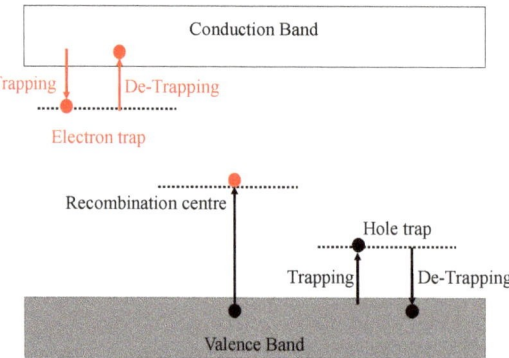

Figure 3. The basic mechanism of charge trapping and detrapping for electron and hole traps. The figure is adapted from Ref. [101].

6.1. Delocalized Mechanism

Among the most studied mechanisms of charge recombination (which produces luminescence) is delocalized mechanism. Matsuzawa was the first person to explain the mechanism in $MAl_2O_2:Eu^{2+}$ (M = Ca and Sr) material, as shown in Figure 4, assuming holes as the main charge carrier determined using photoconductivity studies [8]. The holes (or traps) are considered to be due to Sr^{2+} vacancies. When the incident photons excite Eu^{2+} ions, Eu^+ is formed due to escape of a hole. This hole is captured by Dy^{3+} converting to Dy^{4+}. The thermal energy due to room temperature is considered to be sufficient for these holes to detrap back to the valence band. This follows trapping of these hole back at Eu^+ converting to Eu^{2+} again due to electron–hole recombination at room temperature producing PersL.

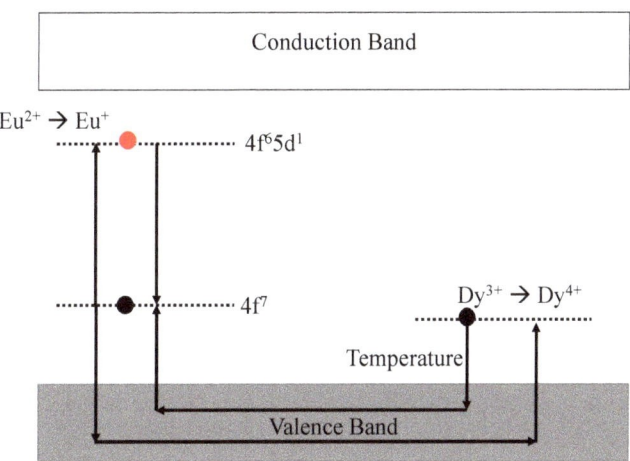

Figure 4. The mechanism of charge trapping and detrapping as proposed by Matsuzawa [8].

After the initial work of Matsuzawa, further work in this direction was performed by Aitasalo in the year 2003 [102]. In Aitasalo's mechanism, the electrons are directly trapped at defects and the holes are trapped at calcium vacancies. Matsuzawa model was rejected by Aitasalo as the PersL from the non-Dy^{3+} doped sample could not be explained by Matsuzawa. Later in the year 2005, Dorenbos and Clabau proposed two different models of charge trapping–detrapping followed by their recombination at room temperature [103,104]. In the explanation from Dorenbos, the higher states of Eu^{2+} are determined to be within conduction band. This implies that upon excitation the electrons are within conduction band changing Dy^{3+} to Dy^{2+} with a trap depth of 0.9 eV in both the cases. Upon increasing the temperature, the electrons stored in these 0.9 eV traps come out and recombine at the luminescent center. On the other hand, Clabau determined that the electron spin resonance (ESR) signal due to Eu^{2+} ions decreases as a function of excitation energy suggesting a pathway involving Eu^{2+} ions. Upon detrapping, the concentration of Eu^{2+} was found to increase further inferring that the trapping might be at the Eu^{2+} excited state. This situation (interpretation) contradicts models by Dorenbos and Aitasalo. However, in reality, in contrast to Dorenbos, based on the temperature dependent photoconductivity measurements, no direct migration of electrons through conduction band and the nature of traps upon Dy^{3+} doping, could be observed [105]. Both Dorenbos and Clabau used electron trapping model, while vacancies were considered to be due to missing oxygen in the lattice.

6.2. Localized Mechanism

In the localized mechanism, the origin of PersL is the presence of neighboring antisite defects which are close to the luminescent centre. The same can be explained using an example of Cr^{3+} substituting Ga^{3+} ion in spinel $ZnGa_2O_4$–matrix host. Antisite defects are the defects resulting from the exchange in site positions of A and B ions in the spinel structure AB_2O_4 [24,76,106–109]. An important point in this mechanism is that the luminescent centre liberates an electron-hole pair during excitation without changing its oxidation state [76]. The steps that are followed for charge trapping detrapping and recombination in such type of mechanism, for example in $ZnGa_2O_4$:Cr^{3+}, is shown in Figure 5. The steps followed are:

- **Step 1:** the excitation of the Cr_{N2} ions.
- **Step 2:** the excitation is dissociated by the local electric field into an electron and a hole.
- **Step 3:** The excitation is thus trapped in the vicinity of Cr^{3+} in the form of a pair of neutral defects, while Cr^{3+} returns to its $^4A_2$ ground state. Electrons and hole can then migrate far from Cr^{3+} ion, so that this storage mechanism can proceed many times with the same Cr_{N2} ion.
- **Step 4:** the reverse reaction (electron–hole release and capture by Cr^{3+}) is thermally activated followed by recombination or release of photons.

Such mechanism does not require movement of charges through the conduction band and is known as the localized mechanism. More details of this type of mechanism have been discussed in Refs. [19,20,76].

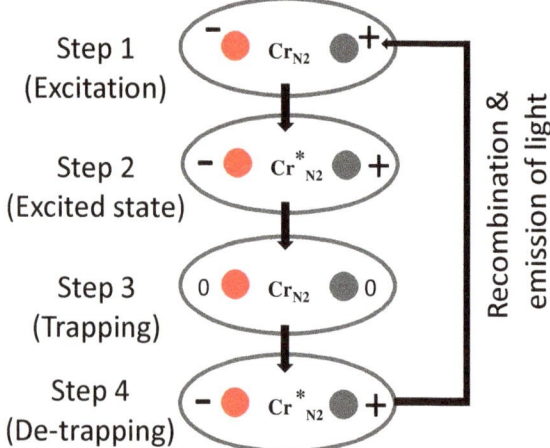

Figure 5. The localized mechanism of charge trapping and detrapping followed by recombination which produces near-IR light emission. The figure is adapted from Ref. [19].

6.3. New Mechanism by Dorenbos

Usually while discussing mechanism of PersL materials, electron trapping and release is preferred and discussed. More recently, in 2018, Dorenbos proposed another mechanism based on the hole-trapping model. Before this, the hole-trapping mechanism was used in the field of semiconductors, and was hardly used in the field of wide band gap inorganic compounds [110]. Such mechanism deals with the difficulties that arise from hole trapping, charge transfer luminescence and luminescence quenching of Eu^{3+} emission. In this mechanism, the hole ground state of a trivalent lanthanoid is placed at the same location as the electron ground state of the corresponding divalent lanthanoids. Quenching by hole ionization to the valence band then appears a mirror image to quenching by electron ionization to the conduction band. The excited hole state is given by the upside-down

Dieke diagrams, and the quenching is described by upside-down configuration coordinate diagrams. The reader is referred to Ref. [110] for further reading.

7. Future Direction

Despite several attempts to prepare and understand the UV-emitting PersL materials, the mechanism remains unclear and deserves due attention of the scientific community. The most controversial of all is Strontium Aluminate phosphor, whose mechanism has been under debate for the past 25 years now. Looking in to the role of the trapping centers for applications such as information storage and optical tagging, it is very important to elucidate the mechanism of charge storage, their release and re-trapping at room temperature and even at high temperatures. For example, in the case of Ce^{3+} doped materials, the lowest of 5d levels ($5d_1$) should be very close to the conduction band for efficient trapping and detrapping at the room temperature. Apart from physical efforts to synthesize new materials, more smart ways like preparing vacuum referred binding energy (VRBE) diagrams should be preferred before actual synthesis of the phosphors [111].

From the prior knowledge of host bandgap and corresponding energetic excitation that causes a maximum photocurrent upon Ce^{3+} doping, the information about lowest T_{2g} (for cubic environment) can be determined before the synthesis step. Based on these procedures, afterglow emission change from 2 min to maximum of 1200 min has been reported. Similarly, the duration of the PersL decay time depends upon the delocalization of Ce^{3+} 5d electrons and their separation from the top of the conduction band. Co-doping such phosphors with other suitable co-dopant ions (e.g., Mn^{2+}, $Eu^{2+/3+}$, Tb^{3+}) increases the PersL decay time and some new series of materials using energy-transfer phenomenon, can be prepared.

Overall, several compounds with excellent PersL decay emission have already been reported. However, the challenges like particles with small grain size, desired morphology, and emission window exhibiting higher efficiency, are still open. It is known that the PersL decreases with decreasing grain size of particles. However, if it is true for all type of hosts (organic and/or inorganic) or for only few, is still unknown. The pathways to solve such issues require immediate attention of the scientific community. Another area of immense interest for researchers' working on PersL, is to prepare good, efficient phosphors for security and surveillance purposes.

8. Conclusions

The recent developments on PersL materials especially those emitting UV radiation have found applications in photodynamic therapy, information storage, anticounterfeiting, photocatalysis, etc. To prepare UV-emitting materials, which can be divided into UV-A, UV-B and UV-C regions, one needs to follow smart techniques (calculations) rather than physical efforts. While VRBE gives us information about the exact location of rare earth ions in individual hosts, the options to prepare UV-emitting materials is limited as the transitions in lanthanoids are restricted because of their f–f nature (for Ln^{3+} ions). However, if one considers the Ln^{2+} ions, and an f–d transition, the emission can vary quite a lot. Only few of the rare earth ions, Gd^{3+}, Ce^{3+} and Pr^{3+}, are capable of emitting in the UV-region. The Gd^{3+} and Pr^{3+} emission is independent of the host crystal field due to shielding of the outermost electrons. The most promising among all is Ce^{3+} whose 5d shell is affected by the choice of host lattice and the emission can be tuned from UV to red region. Herein, we have summarized such materials and found it to be promising to work on these materials emitting UV–PersL.

Author Contributions: Conceptualization, S.K.S.; methodology, S.K.S., J.J. and S.K.G.; software, S.K.S.; formal analysis, S.K.S., J.J. and S.K.G.; data curation, S.K.S., J.J. and S.K.G.; writing—original draft preparation, S.K.S.; writing—review and editing, S.K.S., J.J., S.K.G. and S.H.; supervision, S.K.S.; project administration, S.K.S. and S.H.; funding acquisition, S.K.S. and S.H. All authors have read and agreed to the published version of the manuscript.

Funding: S.K.S. would like to acknowledge the UGC-DAE CSR long-term proposal (Ref:CRS/2021-22/04/616) for the financial support.

Institutional Review Board Statement: Not applicable.

Informed Consent Statement: Not applicable.

Data Availability Statement: All data are provided within this manuscript.

Acknowledgments: The authors would like to thank R.K.Kohli, Vice Chancellor, Amity University, Mohali-Punjab, India, for their continuous support and encouragement. The author S.K.S. thanks Dirk Poelman for the invitation to write the article and for full fee waiver to publish the article, open access. This work was partially carried out using the facilities of UGC–DAE CSR.

Conflicts of Interest: The authors declare no conflict of interest.

Abbreviations

The following abbreviations are used in this manuscript:

PersL	Persistent Luminescence
PL	Photoluminescence
TL	Thermoluminescence
OSL	Optically Stimulated luminescence
VRBE	Vacuum referred binding energy

References

1. Arpiarian, N. The Centenary of the Discovery of Luminescent Zinc Sulphide. In Proceedings of the International Conference on Luminescence, Budapest, Hungary, 1966; pp. 903–906.
2. Leverenz, H.W. *An Introduction to Luminescence of Solids*; Dover Publications: New York, NY, USA, 1968.
3. Durant, W. *Our Oriental Heritage: The Story of Civilization*; MJF Books: New York, NY, USA, 1966.
4. Yocom, P. Future requirements of display phosphors from an historical perspective. *J. Soc. Inf. Disp.* **1966**, *4*, 903–906. [CrossRef]
5. Harvey, E. *A History of Luminescence from the Earliest Times until 1900*; J.H. Furst Company: Baltimore, MD, USA, 1957.
6. Hoogenstraaten, W.; Klasens, H.A. Some Properties of Zinc Sulfide Activated with Copper and Cobalt. *J. Electrochem. Soc.* **1953**, *100*, 366. [CrossRef]
7. Abbruscato, V. Optical and Electrical Properties of $SrAl_2O_4:Eu^{2+}$. *J. Electrochem. Soc.* **1971**, *118*, 930. 1.2408226. [CrossRef]
8. Matsuzawa, T. A New Long Phosphorescent Phosphor with High Brightness, $SrAl_2O_4:Eu^{2+},Dy^{3+}$. *J. Electrochem. Soc.* **1996**, *143*, 2670. [CrossRef]
9. Takasaki, H.; Tanabe, S.; Hanada, T. Long-Lasting Afterglow Characteristics of Eu, Dy Codoped $SrO-Al_2O_3$ Phosphor. *J. Ceram. Soc. Jpn.* **1996**, *104*, 322–326. [CrossRef]
10. Katsumata, T.; Nabae, T.; Sasajima, K.; Komuro, S.; Morikawa, T. Effects of Composition on the Long Phosphorescent $SrAl_2O_4:Eu^{2+}$, Dy^{3+} Phosphor Crystals. *J. Electrochem. Soc.* **1997**, *144*, L243. [CrossRef]
11. Katsumata, T.; Nabae, T.; Sasajima, K.; Matsuzawa, T. Growth and characteristics of long persistent $SrAl_2O_4$- and $CaAl_2O_4$-based phosphor crystals by a floating zone technique. *J. Cryst. Growth* **1998**, *183*, 361–365. [CrossRef]
12. Hölsä, J.; Jungner, H.; Lastusaari, M.; Niittykoski, J. Persistent luminescence of Eu^{2+} doped alkaline earth aluminates, $MAl_2O_4:Eu^{2+}$. *J. Alloys Compd.* **2001**, *326–330*. [CrossRef]
13. Van der Heggen, D.; Joos, J.J.; Feng, A.; Fritz, V.; Delgado, T.; Gartmann, N.; Walfort, B.; Rytz, D.; Hagemann, H.; Poelman, D.; et al. Persistent Luminescence in Strontium Aluminate: A Roadmap to a Brighter Future. *Adv. Funct. Mater.* **2022**, *32*, 2208809.
14. Lin, Y.; Tang, Z.; Zhang, Z. Preparation of a new long afterglow blue-emitting $Sr_2MgSi_2O_7$-based photoluminescent phosphor. *J. Mater. Sci. Lett.* **2001**, *20*, 1505–1506. [CrossRef]
15. Xu, J.; Tanabe, S. Persistent luminescence instead of phosphorescence: History, mechanism, and perspective. *J. Lumin.* **2019**, *205*, 581–620. [CrossRef]
16. Yu, F.; Yang, Y.; Su, X.; Mi, C.; Seo, H.J. Novel long persistent luminescence phosphors: Yb^{2+} codoped MAl_2O_4 (M = Ba, Sr). *Opt. Mater. Express* **2015**, *5*, 585–595. [CrossRef]
17. le Masne de Chermont, Q.; Chanéac, C.; Seguin, J.; Pellé, F.; Maîtrejean, S.; Jolivet, J.P.; Gourier, D.; Bessodes, M.; Scherman, D. Nanoprobes with near-infrared persistent luminescence for in vivo imaging. *Proc. Natl. Acad. Sci. USA* **2007**, *104*, 9266–9271.
18. Kabe, R.; Adachi, C. Organic long persistent luminescence. *Nature* **2017**, *550*, 384–387. [CrossRef]
19. Bessiére, A.; Sharma, S.K.; Basavaraju, N.; Binet, L.; Viana, B.; Bos, A.J.J.; Maldiney, T.; Richard, C.; Scherman, D.; Gourier, D. Storage of visible light for long-lasting phosphorescence in chromium-doped zinc gallate. *Chem. Mater.* **2014**, *26*, 1365–1373. [CrossRef]

20. Maldiney, T.; Bessière, A.; Seguin, J.; Teston, E.; Sharma, S.K.; Viana, B.; Dorenbos, P.; Gourier, D.; Richard, C. The in vivo activation of persistent nanophosphors for optical imaging of vascularization, tumours and grafted cells. *Nat. Mater.* **2014**, *13*, 418–426. [CrossRef]
21. Furetta, C. *Handbook of Thermoluminescence*; World Scientific: Singapore, 2003.
22. Bøtter-Jensen, L.; McKeever, S.; Wintle, A. Chapter 4—Passive optically stimulated luminescence dosimetry. In *Optically Stimulated Luminescence Dosimetry*; Elsevier: Amsterdam, the Netherlands, 2003. [CrossRef]
23. Blasse, G. New luminescent materials. *Chem. Mater.* **1989**, *1*, 294–301.
24. Sharma, S.K.; Bessière, A.; Basavaraju, N.; Priolkar, K.R.; Binet, L.; Viana, B.; Gourier, D. Interplay between chromium content and lattice disorder on persistent luminescence of $ZnGa_2O_4:Cr^{3+}$ for in vivo imaging. *J. Lumin.* **2014**, *155*, 251–256. [CrossRef]
25. Qu, X.; Cao, L.; Liu, W.; Su, G.; Wang, P.; Schultz, I. Sol–gel synthesis of long-lasting phosphors $CdSiO_3: Mn^{2+}$, RE^{3+} (RE=Tb, Eu, Nd) and luminescence mechanism research. *Mater. Res. Bull.* **2012**, *47*, 1598–1603. [CrossRef]
26. Gupta, S.; Mohapatra, M.; Kaity, S.; Natarajan, V.; Godbole, S. Structure and site selective luminescence of sol–gel derived $Eu:Sr_2SiO_4$. *J. Lumin.* **2012**, *132*, 1329–1338. [CrossRef]
27. Hreniak, D.; Stręk, W.; Mazur, P.; Pazik, R.; Ząbkowska-Wacławek, M. Luminescence properties of $Tb^{3+}:Y_3Al_5O_{12}$ nanocrystallites prepared by the sol–gel method. *Opt. Mater.* **2004**, *26*, 117–121. [CrossRef]
28. Głuchowski, P.; Stręk, W.; Lastusaari, M.; Hölsä, J. Optically stimulated persistent luminescence of europium-doped $LaAlO_3$ nanocrystals. *Phys. Chem. Chem. Phys.* **2015**, *17*, 17246–17252. [CrossRef] [PubMed]
29. Kim, J.; Lee, C.K.; Kim, Y.J. Low temperature synthesis of $Lu_3Al_{5-x}Ga_xO_{12}:Ce^{3+}$, Cr^{3+} powders using a sol-gel combustion process and its persistent luminescence properties. *Opt. Mater.* **2020**, *104*, 109944. [CrossRef]
30. Sengar, P.; García-Tapia, K.; Can-Uc, B.; Juárez-Moreno, K.; Contreras-López, O.E.; Hirata, G.A. Simultaneous paramagnetic and persistence-luminescence in GAGG:Ce,Pr nanoparticles synthesized by sol-gel for biomedical applications. *J. Appl. Phys.* **2019**, *126*, 083107.
31. Freeda, M.; Subash, T. Photoluminescence investigations of Ytterbium doped Calcium Aluminate nanophosphor synthesized by sol-gel technique ($CaAl_2O_4$: Yb). *Mater. Today Proc.* **2020**, *24*, 2149–2156.
32. Homayoni, H.; Sahi, S.; Ma, L.; Zhang, J.; Mohapatra, J.; Liu, P.; Sotelo, A.P.; Macaluso, R.T.; Davis, T.; Chen, W. X-ray excited luminescence and persistent luminescence of $Sr_2MgSi_2O_7:Eu^{2+}$, Dy^{3+} and their associations with synthesis conditions. *J. Lumin.* **2018**, *198*, 132–137. [CrossRef]
33. Wei, M.; Feng, S.; Tian, X.; Ji, C.; Huang, Z.; Wen, J.; Liu, X.; Luo, F.; Li, C.; Li, J.; et al. Albumin assisted sol-gel synthesized $SrSnO_3: Pr^{3+}$ red persistent phosphors for temperature sensing. *J. Lumin.* **2021**, *239*, 118328. [CrossRef]
34. Keskin, İ.Ç. Radioluminescence results, thermoluminescence analysis and kinetic parameters of $Y_2O_3:Ln^{3+}$ (Ln: Dy, Nd, Sm) nanophosphors obtained by sol-gel method. *Ceram. Int.* **2022**, *48*, 20579–20590. [CrossRef]
35. Du, P.; Meng, Q.; Wang, X.; Zhu, Q.; Li, X.; Sun, X.; Li, J.G. Sol-gel processing of Eu^{3+} doped $Li_6CaLa_2Nb_2O_{12}$ garnet for efficient and thermally stable red luminescence under near-ultraviolet/blue light excitation. *Chem. Eng. J.* **2019**, *375*, 121937. [CrossRef]
36. Singh, V.; Tiwari, M.K. UV emitting Pb^{2+} doped $Ca_2La_8(SiO_4)6O_2$ phosphors prepared by sol-gel procedure. *Optik* **2020**, *206*, 163600. [CrossRef]
37. Yan, S.; Liang, Y.; Chen, Y.; Liu, J.; Chen, D.; Pan, Z. Ultraviolet-C persistent luminescence from the $Lu_2SiO_5:Pr^{3+}$ persistent phosphor for solar-blind optical tagging. *Dalton Trans.* **2021**, *50*, 8457–8466. [CrossRef] [PubMed]
38. Singh, V.; Tiwari, M.K. Pb^{2+} doped diopside $CaMgSi_2O_6$: New UV luminescent phosphor. *Optik* **2020**, *202*, 163542. [CrossRef]
39. Van den Eeckhout, K.; Smet, P.F.; Poelman, D. Persistent Luminescence in Eu^{2+}-Doped Compounds: A Review. *Materials* **2010**, *3*, 2536–2566. [CrossRef]
40. Cheng, B.; Zhang, Z.; Han, Z.; Xiao, Y.; Lei, S. $SrAl_xO_y:Eu^{2+}$, Dy^{3+} (x = 4) nanostructures: Structure and morphology transformations and long-lasting phosphorescence properties. *CrystEngComm* **2011**, *13*, 3545–3550. [CrossRef]
41. Li, Y.J.; Wang, M.W.; Zhang, L.D.; Gao, D.; Liu, S.X. Soft chemical synthesis and luminescence properties of red long-lasting phosphors $Y_2O_2S:Sm^{3+}$. *Int. J. Miner. Metall. Mater.* **2013**, *20*, 972–977. [CrossRef]
42. Liu, D.; Cui, C.; Huang, P.; Wang, L.; Jiang, G. Luminescent properties of red long-lasting phosphor $Y_2O_2S:Eu^{3+}$, M^{2+} (M=Mg, Ca, Sr, Ba), Ti^{4+} nanotubes via hydrothermal method. *J. Alloys Compd.* **2014**, *583*, 530–534. [CrossRef]
43. Xu, Y.F.; Ma, D.K.; Guan, M.L.; Chen, X.A.; Pan, Q.Q.; Huang, S.M. Controlled synthesis of single-crystal $SrAl_2O_4:Eu^{2+},Dy^{3+}$ nanosheets with long-lasting phosphorescence. *J. Alloys Compd.* **2010**, *502*, 38–42. [CrossRef]
44. Yu, N.; Liu, F.; Li, X.; Pan, Z. Near infrared long-persistent phosphorescence in $SrAl_2O_4:Eu^{2+},Dy^{3+},Er^{3+}$ phosphors based on persistent energy transfer. *Appl. Phys. Lett.* **2009**, *95*, 231110.
45. Pan, W.; Ning, G.; Zhang, X.; Wang, J.; Lin, Y.; Ye, J. Enhanced luminescent properties of long-persistent $Sr_2MgSi_2O_7:Eu^{2+}$, Dy^{3+} phosphor prepared by the co-precipitation method. *J. Lumin.* **2008**, *128*, 1975–1979. [CrossRef]
46. Cheng, B.; Liu, H.; Fang, M.; Xiao, Y.; Lei, S.; Zhang, L. Long-persistent phosphorescent $SrAl_2O_4:Eu^{2+}$, Dy^{3+} nanotubes. *Chem. Commun.* **2009**, 944–946. [CrossRef]
47. Liu, Y.; xiang Liu, S.; wen Wang, M.; jun Li, W.; Zhang, T.; Zhang, X. Synthesis and luminescence properties of Eu^{3+}, Sm^{3+} doped $(YxGd1-x)_2O_3:Si^{4+}$, Mg^{2+} long-lasting phosphor. *Int. J. Miner. Metall. Mater.* **2010**, *17*, 347–352. [CrossRef]
48. Yao, K.; Wang, M.; Liu, S.; Zhang, L.; Li, W. Effects of Host Doping on Spectral and Long-Lasting Properties of Sm^{3+}-Doped Y_2O_2S. *J. Rare Earths* **2006**, *24*, 524–528. [CrossRef]

49. Marcilly, C.; Courty, P.; Delmon, B. Preparation of Highly Dispersed Mixed Oxides and Oxide Solid Solutions by Pyrolysis of Amorphous Organic Precursors. *J. Am. Ceram. Soc.* **1970**, *53*, 56–57. [CrossRef]
50. Zhang, H.; Fu, X.; Niu, S.; Sun, G.; Xin, Q. Photoluminescence of YVO$_4$:Tm phosphor prepared by a polymerizable complex method. *Solid State Commun.* **2004**, *132*, 527–531. [CrossRef]
51. Zhang, H.; Fu, X.; Niu, S.; Xin, Q. Synthesis and photoluminescence properties of Eu^{3+}-doped AZrO$_3$ (A=Ca, Sr, Ba) perovskite. *J. Alloys Compd.* **2008**, *459*, 103–106. [CrossRef]
52. Lima, S.; Sigoli, F.; Davolos, M.; Jafelicci, M. Europium(III)-containing zinc oxide from Pechini method. *J. Alloys Compd.* **2002**, *344*, 280–284. [CrossRef]
53. Li, Y.; Gecevicius, M.; Qiu, J. Long persistent phosphors—From fundamentals to applications. *Chem. Soc. Rev.* **2016**, *45*, 2090–2136. [CrossRef]
54. Aitasalo, T.; Hassinen, J.; Hölsä, J.; Laamanen, T.; Lastusaari, M.; Malkamäki, M.; Niittykoski, J.; Novák, P. Synchrotron radiation investigations of the Sr$_2$MgSi$_2$O$_7$:Eu^{2+},R^{3+} persistent luminescence materials. *J. Rare Earths* **2009**, *27*, 529–538. [CrossRef]
55. Sortino, S. Photoactivated nanomaterials for biomedical release applications. *J. Mater. Chem.* **2012**, *22*, 301–318. [CrossRef]
56. Inoue, S.i.; Tamari, N.; Taniguchi, M. 150 mW deep-ultraviolet light-emitting diodes with large-area AlN nanophotonic light-extraction structure emitting at 265 nm. *Appl. Phys. Lett.* **2017**, *110*, 141106.
57. Dou, X.; Xiang, H.; Wei, P.; Zhang, S.; Ju, G.; Meng, Z.; Chen, L.; Hu, Y.; Li, Y. A novel phosphor CaZnGe$_2$O$_6$:Bi^{3+} with persistent luminescence and photo-stimulated luminescence. *Mater. Res. Bull.* **2018**, *105*, 226–230. [CrossRef]
58. Juzeniene, A.; Moan, J. Beneficial effects of UV radiation other than via vitamin D production. *Dermato-Endocrinology* **2012**, *4*, 109–117.
59. Hönigsmann, H. Erythema and pigmentation. *Photodermatol. Photoimmunol. Photomed.* **2002**, *18*, 75–81. [CrossRef] [PubMed]
60. Gandini, S.; Autier, P.; Boniol, M. Reviews on sun exposure and artificial light and melanoma. *Prog. Biophys. Mol. Biol.* **2011**, *107*, 362–366.
61. Matsumura, Y.; Ananthaswamy, H.N. Toxic effects of ultraviolet radiation on the skin. *Toxicol. Appl. Pharmacol.* **2004**, *195*, 298–308. [CrossRef]
62. Doré, J.F.; Chignol, M.C. Tanning salons and skin cancer. *Photochem. Photobiol. Sci.* **2012**, *11*, 30–37. C1PP05186E. [CrossRef]
63. Wang, C.; Jin, Y.; Zhang, R.; Yuan, L.; Li, Z.; Wu, H.; Chen, L.; Hu, Y. Tunable ultraviolet-B full-spectrum delayed luminescence of bismuth-activated phosphors for high-secure data encryption and decryption. *J. Alloys Compd.* **2022**, *902*, 163776. [CrossRef]
64. Xiong, P.; Peng, M. Recent advances in ultraviolet persistent phosphors. *Opt. Mater. X* **2019**, *2*, 100022. [CrossRef]
65. Sharma, S.K.; Behm, T.; Köhler, T.; Beyer, J.; Gloaguen, R.; Heitmann, J. Library of UV-Visible Absorption Spectra of Rare Earth Orthophosphates, LnPO$_4$ (Ln = La-Lu, except Pm). *Crystals* **2020**, *10*, 593. [CrossRef]
66. Blasse, G.; Grabmaier, B.C. *Luminescent Materials*; Springer: Berlin, Germany, 1994.
67. Dorenbos, P.; Andriessen, J.; Van Eijk, C.W. 4f^{n-1}5d centroid shift in lanthanides and relation with anion polarizability, covalency, and cation electronegativity. *J. Solid State Chem.* **2003**, *171*, 133–136. [CrossRef]
68. Dorenbos, P. Absolute location of lanthanide energy levels and the performance of phosphors. *J. Lumin.* **2007**, *122–123*, 315–317.
69. Kodama, N.; Takahashi, T.; Yamaga, M.; Tanii, Y.; Qiu, J.; Hirao, K. Long-lasting phosphorescence in Ce^{3+}-doped Ca$_2$Al$_2$SiO$_7$ and CaYAl$_3$O$_7$ crystals. *Appl. Phys. Lett.* **1999**, *75*, 1715–1717.
70. Kruk, A. Optical and structural properties of arc melted Ce or Pr –doped Y$_2$O$_3$ transparent ceramics. *Ceram. Int.* **2017**, *43*, 16909–16914. [CrossRef]
71. Qiu, J.; Kodama, N.; Yamaga, M.; Miura, K.; Mitsuyu, T.; Hirao, K. Infrared femtosecond laser pulse-induced three-dimensional bright and long-lasting phosphorescence in a Ce^{3+}-doped Ca$_2$Al$_2$SiO$_7$ crystal. *Appl. Opt.* **1999**, *38*, 7202–7205. [CrossRef] [PubMed]
72. Jia, D. Relocalization of Ce^{3+} 5d electrons from host conduction band. *J. Lumin.* **2006**, *117*, 170–178. [CrossRef]
73. Xu, X.; Wang, Y.; Yu, X.; Li, Y.; Gong, Y. Investigation of Ce–Mn Energy Transfer in SrAl$_2$O$_4$:Ce^{3+},Mn^{2+}. *J. Am. Ceram. Soc.* **2011**, *94*, 160–163. [CrossRef]
74. Jin, Y.; Hu, Y.; Chen, L.; Wang, X.; Ju, G.; Mou, Z. Luminescence Properties of Dual-Emission (UV/Visible) Long Afterglow Phosphor SrZrO$_3$: Pr^{3+}. *J. Am. Ceram. Soc.* **2013**, *96*, 3821–3827.
75. Shi, H.; An, Z. Ultraviolet afterglow. *Nat. Photonics* **2019**, *13*, 74–75. [CrossRef]
76. Gourier, D.; Bessière, A.; Sharma, S.; Binet, L.; Viana, B.; Basavaraju, N.; Priolkar, K.R. Origin of the visible light induced persistent luminescence of Cr^{3+}-doped zinc gallate. *J. Phys. Chem. Solids* **2014**, *75*, 826–837. [CrossRef]
77. Zhou, Z.; Xiong, P.; Liu, H.; Peng, M. Ultraviolet-A Persistent Luminescence of a Bi^{3+}-Activated LiScGeO$_4$ Material. *Inorg. Chem.* **2020**, *59*, 12920–12927. PMID: 32822162. [CrossRef]
78. Liu, B.M.; Gan, W.J.; Lou, S.Q.; Zou, R.; Tang, Q.; Wang, C.X.; Jiao, J.; Wang, J. X-ray-activated, UVA persistent luminescent materials based on Bi-doped SrLaAlO$_4$ for deep-Seated photodynamic activation. *J. Appl. Phys.* **2021**, *129*, 120901.
79. Sharma, S.K.; Bettinelli, M.; Carrasco, I.; Beyer, J.; Gloaguen, R.; Heitmann, J. Dynamics of Charges in Superlong Blacklight-Emitting CaB$_2$O$_4$:Ce^{3+} Persistent Phosphor. *J. Phys. Chem. C* **2019**, *123*, 14639–14646.
80. Liang, Y.; Liu, F.; Chen, Y.; Sun, K.; Pan, Z. Long persistent luminescence in the ultraviolet in Pb^{2+}-doped Sr$_2$MgGe$_2$O$_7$ persistent phosphor. *Dalton Trans.* **2016**, *45*, 1322–1326. [CrossRef] [PubMed]
81. Zhang, Y.; Chen, D.; Wang, W.; Yan, S.; Liu, J.; Liang, Y. Long-lasting ultraviolet-A persistent luminescence and photostimulated persistent luminescence in Bi^{3+}-doped LiScGeO$_4$ phosphor. *Inorg. Chem. Front.* **2020**, *7*, 3063–3071. [CrossRef]

82. Wang, W.; Sun, Z.; He, X.; Wei, Y.; Zou, Z.; Zhang, J.; Wang, Z.; Zhang, Z.; Wang, Y. How to design ultraviolet emitting persistent materials for potential multifunctional applications: A living example of a NaLuGeO$_4$:Bi^{3+},Eu^{3+} phosphor. *J. Mater. Chem. C* **2017**, *5*, 4310–4318. [CrossRef]
83. Fu, J. Orange- and Violet-Emitting Long-Lasting Phosphors. *J. Am. Ceram. Soc.* **2002**, *85*, 255–257. [CrossRef]
84. Gutiérrez-Martín, F.; Fernández-Martinez, F.; Díaz, P.; Colón, C.; Alonso-Medina, A. Persistent UV phosphors for application in photo catalysis. *J. Alloys Compd.* **2010**, *501*, 193–197. [CrossRef]
85. Wang, Z.; Zhang, J.; Zheng, G.; Peng, X.; Dai, H. Violet-blue afterglow luminescence properties of non-doped SrZrO$_3$ material. *J. Lumin.* **2013**, *144*, 30–33. [CrossRef]
86. Lai, S.; Yang, Z.; Liao, J.; Qiu, J.; Song, Z.; Yang, Y.; Zhou, D. Investigation of persistent luminescence property of Bi^{3+}, Dy^{3+} co-doped CdSiO$_3$ phosphor. *Mater. Res. Bull.* **2014**, *60*, 714–718. [CrossRef]
87. Mei, Y.; Xu, H.; Zhang, J.; Ci, Z.; Duan, M.; Peng, S.; Zhang, Z.; Tian, W.; Lu, Y.; Wang, Y. Design and spectral control of a novel ultraviolet emitting long lasting phosphor for assisting TiO2 photocatalysis: Zn$_2$SiO$_4$:Ga^{3+}, Bi^{3+}. *J. Alloys Compd.* **2015**, *622*, 908–912. [CrossRef]
88. Cai, H.; Song, Z.; Liu, Q. Infrared-photostimulable and long-persistent ultraviolet-emitting phosphor LiLuGeO$_4$:Bi^{3+},Yb^{3+} for biophotonic applications. *Mater. Chem. Front.* **2021**, *5*, 1468–1476. [CrossRef]
89. Jia, D.; Yen, W.M. Trapping Mechanism Associated with Electron Delocalization and Tunneling of CaAl$_2$O$_4$:Ce^{3+}, A Persistent Phosphor. *J. Electrochem. Soc.* **2003**, *150*, H61. [CrossRef]
90. Wang, X.; Mao, Y. Achieving Ultraviolet C and Ultraviolet B Dual-Band Persistent Luminescence by Manipulating the Garnet Structure. *Adv. Opt. Mater.* **2022**, *10*, 2102157.
91. Zhou, X.; Qiao, J.; Zhao, Y.; Han, K.; Xia, Z. Multi-responsive deep-ultraviolet emission in praseodymium-doped phosphors for microbial sterilization. *Sci. China Mater.* **2021**, *65*, 1103–1111. [CrossRef] [PubMed]
92. Liu, L.; Peng, S.; Lin, P.; Wang, R.; Zhong, H.; Sun, X.; Song, L.; Shi, J.; Zhang, Y. High-level information encryption based on optical nanomaterials with multi-mode luminescence and dual-mode reading. *Inorg. Chem. Front.* **2022**, *9*, 4433–4441. [CrossRef]
93. Liu, J.; Liang, Y.; Yan, S.; Chen, D.; Miao, S.; Wang, W. Narrowband ultraviolet-B persistent luminescence from (Y,Gd)$_3$Ga$_5$O$_{12}$:Bi^{3+} phosphors for optical tagging application. *Dalton Trans.* **2021**, *50*, 15413–15421. [CrossRef]
94. Liu, L.; Shi, J.; Li, Y.; Peng, S.; Zhong, H.; Song, L.; Zhang, Y. Disguise as fluorescent powder: Ultraviolet-B persistent luminescence material without visible light for advanced information encryption and anti-counterfeiting applications. *Chem. Eng. J.* **2022**, *430*, 132884. [CrossRef]
95. Yan, S.; Liang, Y.; Liu, J.; Chen, D.; Miao, S.; Bi, J.; Sun, K. Development of ultraviolet-B long-lived persistent phosphors in Pr^{3+}-doped garnets. *J. Mater. Chem. C* **2021**, *9*, 14730–14739. [CrossRef]
96. Sun, H.; Gao, Q.; Wang, A.; Liu, Y.; jun Wang, X.; Liu, F. Ultraviolet-B persistent luminescence and thermoluminescence of bismuth ion doped garnet phosphors. *Opt. Mater. Express* **2020**, *10*, 1296–1302. [CrossRef]
97. Yuan, W.; Tan, T.; Wu, H.; Pang, R.; Zhang, S.; Jiang, L.; Li, D.; Wu, Z.; Li, C.; Zhang, H. Intense UV long persistent luminescence benefiting from the coexistence of Pr^{3+}/Pr^{4+} in a praseodymium-doped BaLu$_2$Al$_2$Ga$_2$SiO$_{12}$ phosphor. *J. Mater. Chem. C* **2021**, *9*, 5206–5216. [CrossRef]
98. Wang, X.; Chen, Y.; Kner, P.A.; Pan, Z. Gd^{3+}-activated narrowband ultraviolet-B persistent luminescence through persistent energy transfer. *Dalton Trans.* **2021**, *50*, 3499–3505. [CrossRef] [PubMed]
99. Li, H.; Liu, Q.; Ma, J.P.; Feng, Z.Y.; Liu, J.D.; Zhao, Q.; Kuroiwa, Y.; Moriyoshi, C.; Ye, B.J.; Zhang, J.Y.; et al. Theory-Guided Defect Tuning through Topochemical Reactions for Accelerated Discovery of UVC Persistent Phosphors. *Adv. Opt. Mater.* **2020**, *8*, 1901727.
100. Liu, Q.; Feng, Z.Y.; Li, H.; Zhao, Q.; Shirahata, N.; Kuroiwa, Y.; Moriyoshi, C.; Duan, C.K.; Sun, H.T. Non-Rare-Earth UVC Persistent Phosphors Enabled by Bismuth Doping. *Adv. Opt. Mater.* **2021**, *9*, 2002065.
101. Saadatkia, P.; Varney, C.; Selim, F. Trap Level Measurements in Wide Band Gap Materials by Thermoluminescence. In *Luminescence*; Thirumalai, J., Ed.; IntechOpen: Rijeka, Rijeka, 2016; Chapter 10. [CrossRef]
102. Aitasalo, T.; Dereń, P.; Hölsä, J.; Jungner, H.; Krupa, J.C.; Lastusaari, M.; Legendziewicz, J.; Niittykoski, J.; Stręk, W. Persistent luminescence phenomena in materials doped with rare earth ions. *J. Solid State Chem.* **2003**, *171*, 114–122.
103. Dorenbos, P. Mechanism of persistent luminescence in Sr$_2$MgSi$_2$O$_7$:Eu^{2+}; Dy^{3+}. *Phys. Status Solidi B* **2005**, *242*, R7–R9.
104. Clabau, F.; Rocquefelte, X.; Jobic, S.; Deniard, P.; Whangbo, M.H.; Garcia, A.; Le Mercier, T. Mechanism of Phosphorescence Appropriate for the Long-Lasting Phosphors Eu^{2+}-Doped SrAl$_2$O$_4$ with Codopants Dy^{3+} and B^{3+}. *Chem. Mater.* **2005**, *17*, 3904–3912.
105. Sharma, S.K. Persistent luminescence: Cerium-doped phosphors. In *Phosphors Synthesis and Applications*; Jenny Stanford Publishing: Dubai, United Arab Emirates, 2018; p. 49.
106. Binet, L.; Sharma, S.K.; Gourier, D. Interaction of Cr^{3+} with valence and conduction bands in the long persistent phosphor ZnGa$_2$O$_4$:Cr^{3+}, studied by ENDOR spectroscopy. *J. Phys. Condens. Matter* **2016**, *28*, 385501. [CrossRef]
107. Nie, Y.; Michel-Calendini, F.; Linarès, C.; Boulon, G.; Daul, C. New results on optical properties and term-energy calculations in Cr^{3+}-doped ZnAl$_2$O$_4$. *J. Lumin.* **1990**, *46*, 177–190. [CrossRef]
108. Mikenda, W.; Preisinger, A. N-lines in the luminescence spectra of Cr^{3+}-doped spinels (II) origins of N-lines. *J. Lumin.* **1981**, *26*, 67–83. [CrossRef]

109. Zhang, W.; Zhang, J.; Chen, Z.; Wang, T.; Zheng, S. Spectrum designation and effect of Al substitution on the luminescence of Cr^{3+} doped $ZnGa_2O_4$ nano-sized phosphors. *J. Lumin.* **2010**, *130*, 1738–1743. [CrossRef]
110. Dorenbos, P. The hole picture as alternative for the common electron picture to describe hole trapping and luminescence quenching. *J. Lumin.* **2018**, *197*, 62–65. [CrossRef]
111. Dorenbos, P. Modeling the chemical shift of lanthanide 4f electron binding energies. *Phys. Rev. B—Condens. Matter Mater. Phys.* **2012**, *85*, 165107. [CrossRef]

Disclaimer/Publisher's Note: The statements, opinions and data contained in all publications are solely those of the individual author(s) and contributor(s) and not of MDPI and/or the editor(s). MDPI and/or the editor(s) disclaim responsibility for any injury to people or property resulting from any ideas, methods, instructions or products referred to in the content.

Article

Efficient Sensitized Photoluminescence from Erbium Chloride Silicate via Interparticle Energy Transfer

Hao Shen [1,2,†], Huabao Shang [1,†], Yuhan Gao [1], Deren Yang [1] and Dongsheng Li [1,*]

1. State Key Laboratory of Silicon Materials, School of Materials Science and Engineering, Zhejiang University, Hangzhou 310027, China; 21426020@zju.edu.cn (H.S.); 12026010@zju.edu.cn (H.S.); 11326024@zju.edu.cn (Y.G.); mseyang@zju.edu.cn (D.Y.)
2. Institute of Fluid Physics, China Academy of Engineering Physics, Mianshan Road 64#, Mianyang 621900, China
* Correspondence: mselds@zju.edu.cn
† These authors contributed equally to this paper.

Abstract: In this study, we prepare Erbium compound nanocrystals and Si nanocrystal (Si NC) co-embedded silica film by the sol-gel method. Dual phases of Si and Er chloride silicate (ECS) nanocrystals were coprecipitated within amorphous silica. Effective sensitized emission of Er chloride silicate nanocrystals was realized via interparticle energy transfer between silicon nanocrystal and Er chloride silicate nanocrystals. The influence of density and the distribution of sensitizers and Er compounds on interparticle energy transfer efficiency was discussed. The interparticle energy transfer between the semiconductor and erbium compound nanocrystals offers some important insights into the realization of efficient light emission for silicon-based integrated photonics.

Keywords: interparticle energy transfer; erbium chloride silicate; sensitized emission

1. Introduction

Rare earth ion (RE^{3+})-containing luminescent materials, which possess abundant and sharp emissions via intra-4f transitions, have been gaining considerable interest in many areas such as optical amplifiers for telecommunications [1], solid-state light sources [2], bio-sensors [3], solar cells [4], etc. Among these RE^{3+} ions, erbium ions have been already widely employed in long-haul telecommunications due to their transition from first excited state to ground state emitting a photon at 1.5 µm that coincides with the minimum loss window of silica optical fiber [5]. Since the mid-1980s, silicon-based erbium-doped materials (including Si [6], SiO_x [7] SiN_x [8], etc.) were considered as a very promising platform for on-chip integrated photonics applications. The optical gain of erbium-containing materials is proportional to the erbium concentration according to a simple two-level model [9]. In order to satisfy the requirement for chip-scale integration application, it is necessary that there should be a much higher erbium doping density compared with Er-doped fiber amplifiers [10]. However, increasing the erbium concentration is limited by low Er solubility in many hosts (for instance, ~10^{20} cm^{-3} in silica [11]). The concentration quenching effect would be severe owing to the formation of optically inactive erbium precipitates and clusters inside the host when the doping density exceeds solid solubility [12]. Crystalline erbium compounds, such as $Er_2Si_2O_7$ [13] and erbium chloride silicate (ECS) [14], have been expected to overcome the solubility limit because erbium ions are periodically arranged at lattice sites. For example, giant net optical gain over 100 dB/cm has been realized in single-crystal ECS nanowires [15].

Although erbium compounds with high erbium content have great potential in chip-scale integration, they also suffer from very low excitation cross section owing to the parity-forbidden intra-4f transitions of Er^{3+} [16]. In general, silicon nanocrystals (Si NCs) that absorb broadband excitation light and then transfer energy to a nearby Er^{3+} can be introduced conveniently inside Er-doped materials as sensitizers [17]. Experimental data that the effective cross section of film embedded with Si NCs receive a nearly five of orders magnitude enhancement, confirming the high efficiency of this energy transfer process [18]. However, the strategy of co-doping Si NC cannot work in crystalline erbium compounds in view of the physical separation of Si NCs and erbium compounds [19,20]. The distance between Si NC and Er^{3+} is lower than the energy transfer distance because the Er^{3+} ions in the crystalline compound are fixed in the lattice position rather than distributed in the amorphous matrix. To overcome the separation, researchers attempted to constrain sensitizers and active centers into a core–shell nanostructure [21,22] or synthesize free-standing nanoparticles followed by incorporating them into a colloidal mixture for interparticle energy transfer (IPET) [23,24]. Although some exciting results have been reported based on energy transfer in core–shell or interparticle nanostructure, there are some limitations that prevent these methods from practical application, such as complex fabrication procedures and quenching effects induced by surface state [25]. Simultaneously tailoring sensitizers and active center phase precipitation in a solid matrix may be a possible solution to the abovementioned problems, but the lattice and dimension mismatch between Si NC and Er compounds restrict the IPET process inside host materials [19]. To the best of our knowledge, there is no report of sensitized crystalline Er compound emissions via IPET inside the matrix. Moreover, it is unknown how the density and distribution of sensitizers and Er compounds in host material affect the IPET efficiency. Here, we fabricate the amorphous silica film with Si NCs and ECS nanocrystals embedded using a sol-gel method. We demonstrate that the efficient sensitized emission of crystalline ECS can be achieved via interparticle energy transfer from Si NCs. Additionally, the influence of density and the distribution of sensitizers and Er compounds on IPET efficiency were discussed based on different Er^{3+} concentrations.

2. Materials and Methods

The Er compound nanocrystal and Si NC co-embedded silica film was prepared by the sol-gel process followed by spin-on method. The Si NC was obtained from thermal dissociation of Si-H bonds generated by the hydrolysis–condensation reactions of triethoxysilane (H-Si(OC$_2$H$_5$)$_3$, TES) in an annealing process. At first, TES was dissolved in ethanol and de-ionized water to form silica sol with excess silicon. Then, different amounts of erbium chloride solution with 0–40% molar ratio of erbium to silicon were added to the silica sol as a precursor of Er compound nanocrystals. The diluted HCl was added dropwise to the mixture solution under vigorous stirring for adjusting the PH to 5. After a complex hydrolysis–condensation reaction, the uniform pink sol containing Er^{3+} and excess silicon was formed. Subsequently, the as-prepared sol was spin-coated on a clean p-Si substrate, followed by a careful drying process for liquid solvent removal at 80 °C. Finally, all the samples were annealed for one hour at 1000 °C in a tube furnace under Ar atmosphere. The reference sample with no excess silicon, which was prepared by substituting precursor TES with tetraethyl orthosilicate (Si(OC$_2$H$_5$)$_4$, TEOS), was treated with the same annealing procedure and labeled as 10% TEOS.

To identify the structures and crystallinity of the Er-related phase, X-ray diffraction (XRD) data were collected on an X-ray diffractometer (Rigaku D/max-2550pc, Rigaku Company, Tokyo, Japan) equipped with a Cu $K\alpha$ radiation source (λ = 0.154 nm). Cross-section samples prepared were observed under an electron microscope to study the microstructure and crystalline size. The specific preparation method is as follows: the samples were bonded with M-bond 610 and double-sided thinned to reduce the thickness to less than 20 microns. The samples were made into a thin area using a Gatan 691 ion-thinning instrument. Then, the cross-section samples were studied by using a TECHNAI-F20G2 transmission electron microscope (TEM, FEI Company, Hillsboro, OR, USA) under 200 keV accelerating voltage. The line resolution of the TEM is 0.102 nm, and the point resolution is 0.24 nm. Steady-state PL in the visible and near-infrared range is detected using a charge-coupled device (PIXIS:100BR, Princeton Instruments company, America) and InGaAs photomultiplier tube (PMT, R5509, Hamamatsu Company, Hamamatsu, Japan), respectively. Two lasers with wavelengths of 473 nm and 980 nm were used as a light source for distinguishing sensitized Er^{3+} emission. Time-resolved visible PL spectra were recorded on an Photonics FLS spectrophotometer (Edinburgh Instruments company, Britain), equipped with a µF 920 adjustable microsecond lamp as the excitation source.

3. Results and Discussion

XRD patterns of the reference sample and samples with different Er^{3+} concentrations are provided in Figure 1. XRD analysis illustrated that crystalline structures exist in all samples. It is apparent that the crystallization peaks shall be assigned to the Er-related phase since the diffraction peak intensity increases with Er^{3+} concentration. Here, the observed stronger diffraction peaks, which occurred in higher Er^{3+} concentration samples, indicate increased growth and crystallinity of matrix-embedded Er-related crystals. In other words, the diameter of Er-related crystal shall increase with Er^{3+} concentration, which is consistent with the other system of crystal embedded in amorphous matrix [26]. For samples with a higher Er:Si ratio, the diffraction peaks match well with erbium chloride silicate (ECS) (JCPDS No. 042-0365, Pnma), which exhibited excellent optical properties in the form of nanowire and nanosheet [27,28]. The main peaks of the ECS phase are marked with asterisks and crystal plane indices.

It is noteworthy that the reference sample and samples with a lower Er^{3+} concentration show some new diffraction peaks in addition to ECS crystals, which were marked with diamonds shown in Figure 1. These peaks are supposed to arise from erbium oxide or silicate, but there is no known Er-related phase in the JCPDS reference database fully corresponding to the XRD pattern. Other RE^{3+}-related phases, especially those with a close ionic radius to Er^{3+}, may be retrieved for reference as the crystal structures of rare earth oxide and silicate are dependent on ionic radius. This method was proved to be appropriate in previous studies [29,30]. According to the fact that the ionic radii of Er^{3+} and Y^{3+} are nearly the same, it is reasonable to attribute these new diffraction peaks to y-$Er_2Si_2O_7$ in the context of the large similarity between y-$Y_2Si_2O_7$ (JCPDS No. 74-1994) and observed peaks [29]. The appearance of y-$Er_2Si_2O_7$ phases in samples with lower Er^{3+} concentration is likely to be related to the local Er:Si molar ratio. When the Er^{3+} concentration is too low, local Er:Si molar ratio is not large enough for the formation of pure ECS with a 3:2 Er:Si ratio, as reported in Er-doped silica film deposited by magnetron sputtering [31]. Therefore, the reference sample and samples with a lower Er^{3+} concentration are composed of a mixture of ECS and y-$Er_2Si_2O_7$, and the reference sample contains more y-$Er_2Si_2O_7$ than ECS due to a fairly strong diffraction peak of y-$Er_2Si_2O_7$.

In order to further analyze the microstructure and crystallization phase, we performed TEM and HRTEM, including fast Fourier transform (FFT) of HRTEM. As shown in Figure 2a,d,g, TEM images of samples with different Er^{3+} concentrations display that a large number of dark nanoparticles are precipitated within the light matrix, while the distributions of nanoparticles are not very homogenous. It is apparent that the average size of these nanoparticles increases with increasing Er^{3+} concentrations, that is, the average diameter of these nanoparticles increases from 3.6 nm to 8.6 nm with the Er:Si ratio increasing from 10% to 40%.

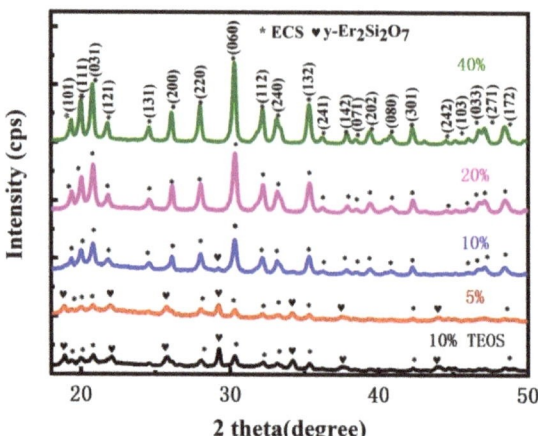

Figure 1. XRD patterns of reference sample with TEOS as precursor and samples with different Er^{3+} concentrations. The asterisks and diamonds represent the diffraction peaks of ECS and y-$Er_2Si_2O_7$, respectively.

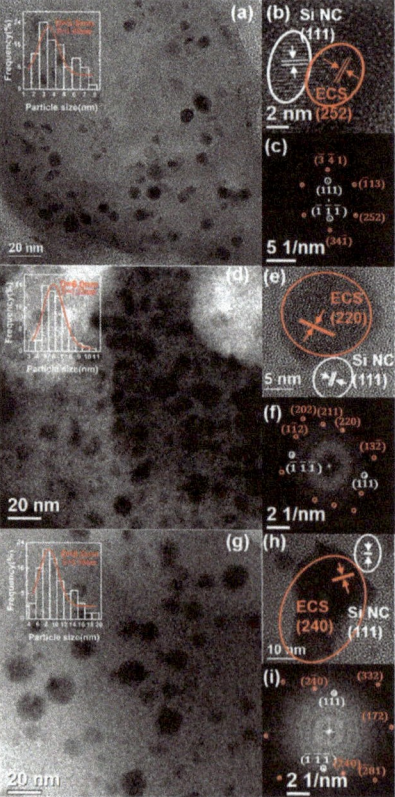

Figure 2. TEM images (inset is the nanoparticle size distribution histogram), HRTEM images of two nearby nanoparticle and the corresponding FFT patterns of the sample with different Er^{3+} concentrations: (**a**–**c**) 10% Er:Si ratio; (**d**–**f**) 20% Er:Si ratio; (**g**–**i**) 40% Er:Si ratio.

The HRTEM images as shown in Figure 2b,e,h demonstrate that there is another type of nanocrystal near to some dark nanoparticles and confirm their crystalline nature with high crystallinity. The measured interplanar spacings of dark nanoparticles are 1.92, 3.18 and 2.70 Å, consistent with the (252), (220) and (240) crystalline planes of orthorhombic structure of ECS, respectively. On the other hand, all the lighter nanocrystals present a lattice spacing of about 3.1 Å that is well matched with the (111) interplane distance of the diamond structure of Si. Although the measured interplane spacings of these nanocrystals are in good agreement with the designated crystalline planes, the confirmation of the phase structure would require a more in-depth analysis owing to the close interplane spacing within different Er-related phases and silicon. FFT patterns of corresponding HRTEM images further demonstrate that these two types of nanoparticles are attributed to be ECS and Si NC, and the determined Miller indices of crystalline planes were marked in Figure 2c,f,i.

The indirect excitation of Er^{3+} in Er-compound crystalline is difficult to achieve because it is an intriguing challenge to simultaneously control the crystallization of two types of phases in the neighborhood within a solid matrix. For example, it was found that the crystallization of Er silicates annealed at higher temperatures lead to the disappearance of Si NC formed at low annealing temperatures in Er-doped Si-rich silica film [19]. In our system, samples with different erbium concentrations have obtained both Si nanoparticle and ECS nanoparticle crystalline structure, and the distance between the particles is very close. The main reason is that the size difference between Si nanoparticles and ECS nanoparticles in our system is not too large, so the interface of the particles has relatively little effect on the crystallization of the particles. Secondly, the lattice mismatch of ECS relative to erbium silicate and Si is smaller. It seems possible that the Si NCs observed in our system act as sensitizers for Er^{3+} emission, as analogous to those Er-doped materials [32].

In order to identify the spectral overlap between sensitizers and ECS as well as the concentration of sensitizers, steady-state and time-resolved PL in visible range were carried out. Figure 3a shows the PL spectra of reference sample and samples with a different Er:Si ratio. In the spectra from samples with a different Er:Si ratio, the broad PL spectra peaks at about 800 nm are observed. This emission band may be attributed to confined exciton recombination from Si NC, which is in line with previous studies [16,33]. On the other hand, no PL signal is found from the reference sample due to no excess Si in the film. At first glance, there is no obvious shift of PL peak as the Er:Si ratio increases. The normalized PL spectra shown in Figure 3b further revealed that the emission peaks are almost independent of Er^{3+} concentration, which can be explained as the nearly constant average sizes of Si NCs, whereas the spectral bandwidths of Si NC emissions slightly increase with increasing the Er:Si ratio. A possible explanation for this might be that the Si NC size distribution in samples with a higher Er:Si ratio is broader than those with a lower Er:Si ratio owing to the interaction between Si NC and Er^{3+} ions, since the luminescence energy as well as the PL peak are determined by the size of Si NC [16].

In addition to emission peak and spectral bandwidth, it is also apparent from Figure 3a that the signal from Si NC at about 800 nm decreases with increasing the Er^{3+} concentration. This result may partly be explained by the reduction in Si NC concentration within amorphous silica. As the Er^{3+} concentration increases, more and more ECS were generated by reaction with silicon and oxygen during thermal treatment, leading to a slight reduction in excess silicon for Si NC precipitation. More importantly, the decrease in luminescence from Si NC may be the result of the fact that a growing number of excitons confined in the Si NCs transfer energy to ECS and no longer participate in the visible luminescence due to the increased amount of ECS with increasing the Er^{3+} concentration. If the energy transfer from Si NC to ECS occurs as expected, the total de-excitation rate of Si NC will increase after Er^{3+} doping because another non-radiative recombination pathway was introduced. It can thus be suggested that not only the emissions of Si NCs decrease, but also its decay time as the Er^{3+} concentration increases.

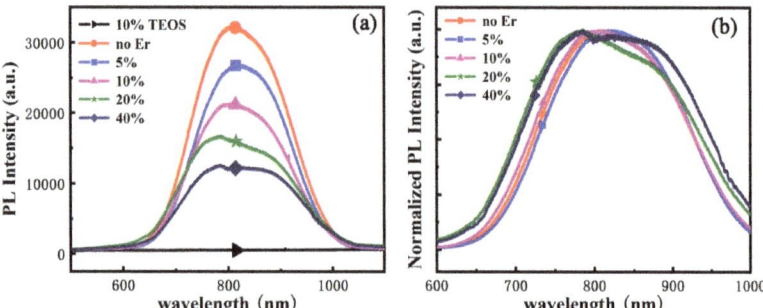

Figure 3. (a) The visible range PL spectra of reference sample and samples with different Er:Si ratio (0–40%), (b) normalized PL spectra of (a) excluding reference sample.

In order to confirm the aforementioned energy transfer process, time-resolved PL measurements of the Si NC emission were performed, and a typical decay curve was displayed in Figure 4a. The PL decay curve is not a single exponential and can be well described by a stretched exponential function like those Si NC embedded in silica [34]:

$$I(t) = I_0 \exp[-(t/\tau)^\beta] \qquad (1)$$

where $I(t)$ and I_0 are the PL intensity as a function of time and at $t = 0$, τ represents the mean luminescence lifetime, and β ($0 < \beta \leq 1$) is a dispersion factor that describes the distribution of lifetimes which becomes broader as β decreases to 0. Figure 4b provides the lifetimes and distribution factors fitted with stretched exponential as a function of Er:Si ratio. The lifetime decreases from 72.1 to 41.8 μs and the dispersion factor also decreases from 0.69 to 0.62 as the Er:Si ratio increases from 0 to 40%. The values of tens of μs are in accordance with the lifetimes observed from Si NC [34]. Again, the trend observed in the lifetime identifies an additional recombination channel that most likely is the energy transfer from Si NC to ECS, as reported in other sensitization Er^{3+} emission system [35,36]. The energy transfer efficiency (ETE) is the fraction of donors that are depopulated by energy transfer process to acceptors over the total number of donors being excited, which can be expressed as a function of the PL decay lifetimes, shown below:

$$\eta_{Er} = 1 - \frac{\tau_{Er}}{\tau_{Er-free}} \qquad (2)$$

The ETE is calculated as 34.4%, 35.1%, 38.3%, and 42.0% for Er:Si ratios of 5%, 10%, 20%, and 40% samples, respectively. Note that the slight decrease in dispersion factor with increasing Er:Si ratio is clear-cut evidence of inhomogeneous distribution of the lifetime caused by broad Si NC size distribution.

The occurrence of an energy-transfer process from Si NC to ECS can be investigated by using both resonate and non-resonate wavelength lasers as excitation sources. Figure 5a presents the PL spectra of the reference sample and samples with different Er:Si ratios under resonant wavelength (980 nm) excitation. Several sharp peaks with the main peak at 1533 nm that share almost the same spectrum shape are observed in all samples, similar to the spectrum of single-crystal ECS nanowire in a previous study [14]. These peaks arise from the transition between the Stark levels of ground and first state, indicating that the optically active Er^{3+} ions are mainly in a crystalline environment. For the reference sample and samples with a 5% Er:Si ratio, some weak peaks at 1536, 1540 and 1556 nm ascribed to the emission of y-$Er_2Si_2O_7$ can be seen [31], which is consistent with the XRD results. It is clear that the PL intensity of the samples increases with the increasing Er:Si ratio, which could be attributed to the increased optically active Er^{3+} concentration. It is noticed that the PL intensity of reference sample is even weaker than samples with 5% Er:Si ratio. This result demonstrates to some extent that no extra non-radiative recombination pathway will form after the introduction of Si NC via precursor TES. On the other hand, it is also suggested that ECS yields much stronger luminescence than y-$Er_2Si_2O_7$ because the PL spectrum of the reference sample agrees well with PL peaks from ECS in the context of a much larger proportion of y-$Er_2Si_2O_7$ than ECS phase [31].

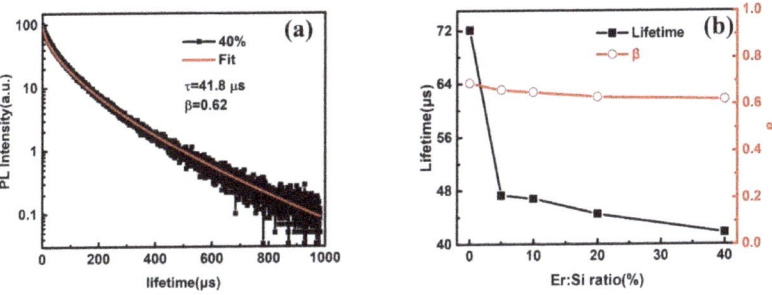

Figure 4. (**a**) PL decay curve of Si NC-related luminescence at 800 nm of sample with 10% Er:Si ratio and corresponding fit curve by stretched exponential function, (**b**) lifetimes and distribution factors (β) extracted from stretched exponential function of samples with different Er:Si ratio.

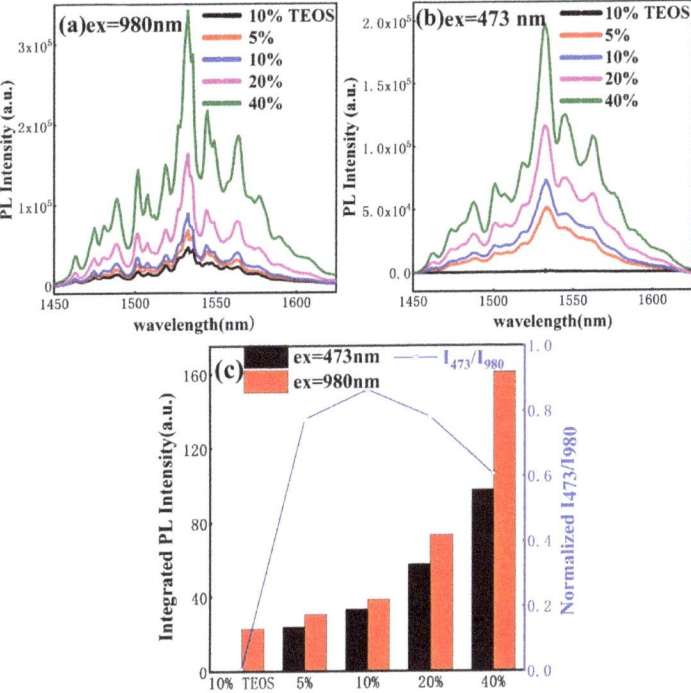

Figure 5. Near-infrared range PL spectra of reference sample and samples with different Er:Si ratio excited by (**a**) 980 nm laser and (**b**) 473 nm laser; (**c**) Integrated PL intensity and the ratio of integrated intensity excited by 473 nm to that by 980 nm laser (I_{473}/I_{980}).

Figure 5b shows the PL spectra of the reference sample and samples with different Er:Si ratios obtained using a 473 nm line laser. It is apparent that no PL signal was observed from the reference sample, while all samples with a different Er:Si ratio exhibited intense emission peaks characteristic of Er^{3+} ions. Note that the excitation wavelength of 473 nm does not coincide with any absorption band of Er^{3+}, and hence the Er^{3+} can only be excited through a carrier-mediated process. Accordingly, the Er^{3+} in the reference sample cannot be excited under 473 nm pumping owing to the lack of photo-injected carriers, as shown in Figure 3a. In contrast to the reference sample, samples with different Er:Si ratios all contain Si NCs that are able to act as sensitizers for Er^{3+}. In addition, there is a competition between Er^{3+} and Si NC luminescence, that is, Er^{3+} emission increases with increasing Er:Si ratio accompanying a quenching of Si NC emission discussed above. These results provide strong evidence of an effective energy transfer from the excitons confined in Si NC to the Er^{3+}. Energy transfer from silicon nanocrystals to ECS nanocrystals belongs to that from semiconductor nanocrystals to dielectric nanocrystals, the emission spectrum of the donor is band-shaped and the absorption spectrum of the acceptor is line-shaped, meeting spectral overlap requirements. Under 473 nm optical pumping, electrons and holes confined in silicon nanocrystals are formed, and the peak of released energy by exciton recombination (~1.55 eV in our system) overlaps with the resonance absorption peak of erbium ions (1.53 eV). In this way, erbium ions are excited from the ground excitation state to the third excitation state ($^4I_{9/2}$), and finally show 1.55 μm luminescence after multiple transition processes [37]. It is important to note that the spectrum shape observed under 473 nm pumping is identical to that obtained with the 980 nm pump beam, indicating that the environment of excited Er^{3+} with resonant or non-resonant wavelength pumping is the same. In other words, the sensitized emission of crystalline ECS was acquired based on the interparticle energy transfer between Si NC and ECS.

As shown in Figure 5c, integrated PL intensitis excited by 473 and 980 nm all increase with increasing Er:Si ratio monotonically, whereas the ratio of them (I_{473}/I_{980}) displays a slight increase followed by a decline. Integrated PL intensity under 473 and 980 nm pumping are proportional to the Er^{3+} density excited by IPET from Si NC and the total optically active Er^{3+} density, respectively. Therefore, the term I_{473}/I_{980} stands for the relative proportion of sensitized Er^{3+} ions and can be used for comparing the sensitized emission between different samples. The I_{473}/I_{980} of the sample with a 10% Er:Si ratio is highest among these samples because it owns an optimal balance, that is, IPET from Si NC to ECS in samples with lower Er:Si ratio may suffer from low encounter probability of Si NC and ECS caused by low ECS concentration, while for samples with a higher Er:Si ratio, a possible explanation is that the relative interparticle distance between Si NC and ECS increases due to the increased size of ECS. Effective sensitized emission of Er chloride silicate nanocrystals was realized through interparticle energy transfer between silicon nanocrystals and Er chloride silicate nanocrystals co-embedded in amorphous silicon oxide. In addition, the influence of density and distribution of sensitizers and the Er compound on interparticle energy transfer efficiency was studied. This provides some insights for the sensitization of erbium ions through the particle energy transfer co-embedded in the matrix.

4. Conclusions

In conclusion, silicon-rich silicon oxide films doped with a high Er^{3+} concentration have been prepared using the sol-gel method with TES as precursor. Dual phases of Si and Er chloride silicate (ECS) nanocrystals were coprecipitated within amorphous silicon oxide during thermal treatment, and ECS was localized at the vicinity of Si NC. The efficient sensitized luminescence of crystalline (ECS) was achieved via an interparticle energy transfer (IPET) from Si NC to ECS. PL spectra under different excitations have confirmed that samples with a medium Er^{3+} concentration, e.g., 10% Er:Si ratio, holds the optimal sensitization that benefited from the efficient IPET process mainly as a result of the fine distribution of dual phase nanocrystals.

Author Contributions: Conceptualization, D.L., H.S. (Hao Shen) and D.Y.; methodology, H.S. (Hao Shen) and D.L.; data curation, H.S. (Hao Shen), H.S. (Huabao Shang) and Y.G.; writing—original draft preparation, H.S. (Hao Shen), H.S. (Huabao Shang) and D.L.; writing—review and editing, D.L. and D.Y.; project administration, D.L. and D.Y.; funding acquisition, D.L. and D.Y. All authors have read and agreed to the published version of the manuscript.

Funding: This research was funded by the National Key R&D Program of China [2018YFB2200102] and the Natural Science Foundation of China [61874095 and 61721005].

Institutional Review Board Statement: Not applicable.

Informed Consent Statement: Not applicable.

Data Availability Statement: All experimental data to support the findings of this study are available upon request by contacting the corresponding authors.

Conflicts of Interest: The authors declare no conflict of interest.

References

1. Chen, X.; Sun, T.; Wang, F. Lanthanide-Based Luminescent Materials for Waveguide and Lasing. *Chem.-Asian J.* **2020**, *15*, 21–33. [CrossRef] [PubMed]
2. Sen, P.; Kar, D.; Laha, R.; Balasubrahmaniyam, M.; Kasiviswanathan, S. Hot electron mediated enhancement in the decay rates of persistent photocurrent in gold nanoparticles embedded indium oxide films. *Appl. Phys. Lett.* **2019**, *114*, 211103. [CrossRef]
3. Achatz, D.E.; Ali, R.; Wolfbeis, O.S. Luminescent Chemical Sensing, Biosensing, and Screening Using Upconverting Nanoparticles. *Lumin. Appl. Sens. Sci.* **2011**, *300*, 29–50.
4. Lo Savio, R.; Miritello, M.; Shakoor, A.; Cardile, P.; Welna, K.; Andreani, L.C.; Gerace, D.; Krauss, T.F.; O'Faolain, L.; Priolo, F.; et al. Enhanced 1.54 mu m emission in Y-Er disilicate thin films on silicon photonic crystal cavities. *Opt. Express* **2013**, *21*, 10278–10288. [CrossRef] [PubMed]
5. Mears, R.J.; Reekie, L.; Jauncey, I.M.; Payne, D.N. Low-noise erbium-doped fiber amplifier operating at 1.54-mu-m. *Electron. Lett.* **1987**, *23*, 1026–1028. [CrossRef]
6. Ennen, H.; Pomrenke, G.; Axmann, A.; Eisele, K.; Haydl, W.; Schneider, J. 1.54-mu-m electroluminescence of erbium-doped silicon grown by molecular-beam epitaxy. *Appl. Phys. Lett.* **1985**, *46*, 381–383. [CrossRef]
7. Fujii, M.; Hayashi, S.; Yamamoto, K. Excitation of intra-4f shell luminescence of Yb3+ by energy transfer from Si nanocrystals. *Appl. Phys. Lett.* **1998**, *73*, 3108–3110. [CrossRef]
8. Gong, Y.; Yerci, S.; Li, R.; Dal Negro, L.; Vuckovic, J. Enhanced Light Emission from Erbium Doped Silicon Nitride in Plasmonic Metal-Insulator-Metal Structures. *Opt. Express* **2009**, *17*, 20642–20650. [CrossRef]
9. Isshiki, H.; Jing, F.; Sato, T.; Nakajima, T.; Kimura, T. Rare earth silicates as gain media for silicon photonics Invited. *Photonics Res.* **2014**, *2*, A45–A55. [CrossRef]
10. Wang, X.; Zhou, P.; He, Y.; Zhou, Z. Erbium silicate compound optical waveguide amplifier and laser Invited. *Opt. Mater. Express* **2018**, *8*, 2970–2990. [CrossRef]
11. Isshiki, H.; Kimura, T. Toward small size waveguide amplifiers based on erbium silicate for silicon photonics. *IEICE Trans. Electron.* **2008**, *E91C*, 138–144. [CrossRef]
12. Polman, A.; Jacobson, D.C.; Eaglesham, D.J.; Kistler, R.C.; Poate, J.M. Optical doping of wave-guide materials by mev er implantation. *J. Appl. Phys.* **1991**, *70*, 3778–3784. [CrossRef]
13. Yin, Y.; Sun, K.; Xu, W.J.; Ran, G.Z.; Qin, G.G.; Wang, S.M.; Wang, C.Q. 1.53 mu m photo- and electroluminescence from Er3+ in erbium silicate. *J. Phys.-Condens. Matter* **2009**, *21*, 012204. [CrossRef] [PubMed]
14. Pan, A.; Yin, L.; Liu, Z.; Sun, M.; Liu, R.; Nichols, P.L.; Wang, Y.; Ning, C.Z. Single-crystal erbium chloride silicate nanowires as a Si-compatible light emission material in communication wavelength. *Opt. Mater. Express* **2011**, *1*, 1202–1209. [CrossRef]
15. Sun, H.; Yin, L.; Liu, Z.; Zheng, Y.; Fan, F.; Zhao, S.; Feng, X.; Li, Y.; Ning, C.Z. Giant optical gain in a single-crystal erbium chloride silicate nanowire. *Nat. Photonics* **2017**, *11*, 589–593. [CrossRef]
16. Kik, P.G.; Polman, A. Exciton-erbium interactions in Si nanocrystal-doped SiO$_2$. *J. Appl. Phys.* **2000**, *88*, 1992–1998. [CrossRef]
17. Franzo, G.; Boninelli, S.; Pacifici, D.; Priolo, F.; Iacona, F.; Bongiorno, C. Sensitizing properties of amorphous Si clusters on the 1.54-mu m luminescence of Er in Si-rich SiO$_2$. *Appl. Phys. Lett.* **2003**, *82*, 3871–3873. [CrossRef]
18. Franzo, G.; Vinciguerra, V.; Priolo, F. The excitation mechanism of rare-earth ions in silicon nanocrystals. *Appl. Phys. A-Mater. Sci. Process.* **1999**, *69*, 3–12. [CrossRef]
19. Gao, Y.; Shen, H.; Li, D.; Yang, D. Efficient sensitized photoluminescence of Er silicate in silicon oxide films embedded with amorphous silicon clusters, part I: Fabrication. *Opt. Mater. Express* **2019**, *9*, 4329–4338. [CrossRef]
20. Yin, Y.; Xu, W.J.; Wei, F.; Ran, G.Z.; Qin, G.G.; Shi, Y.F.; Yao, Q.G.; Yao, S.D. Room temperature Er3+ 1.54 mu m electroluminescence from Si-rich erbium silicate deposited by magnetron sputtering. *J. Phys. D-Appl. Phys.* **2010**, *43*, 335102. [CrossRef]
21. Choi, H.J.; Shin, J.H.; Suh, K.; Seong, H.K.; Han, H.C.; Lee, J.C. Self-organized growth of Si/Silica/Er2Si2O7 core-shell nanowire heterostructures and their luminescence. *Nano Lett.* **2005**, *5*, 2432–2437. [CrossRef] [PubMed]

22. Shen, H.; Xu, L.; Li, D.; Yang, D. Sensitized photoluminescence of erbium silicate synthesized on porous silicon framework. *J. Appl. Phys.* **2017**, *122*, 113103. [CrossRef]
23. Lee, J.; Govorov, A.O.; Kotov, N.A. Bioconjugated superstructures of CdTe nanowires and nanoparticles: Multistep cascade forster resonance energy transfer and energy channeling. *Nano Lett.* **2005**, *5*, 2063–2069. [CrossRef] [PubMed]
24. Sarkar, S.; Meesaragandla, B.; Hazra, C.; Mahalingam, V. Sub-5 nm Ln(3+)-doped BaLuF5 Nanocrystals: A Platform to Realize Upconversion via Interparticle Energy Transfer (IPET). *Adv. Mater.* **2013**, *25*, 856–860. [CrossRef]
25. Johnson, N.J.J.; He, S.; Diao, S.; Chan, E.M.; Dai, H.; Almutairi, A. Direct Evidence for Coupled Surface and Concentration Quenching Dynamics in Lanthanide-Doped Nanocrystals. *J. Am. Chem. Soc.* **2017**, *139*, 3275–3282. [CrossRef]
26. Zhang, X.; Lin, T.; Zhang, P.; Xu, J.; Lin, S.; Xu, L.; Chen, K. Highly efficient near-infrared emission in Er3+ doped silica films containing size-tunable SnO2 nanocrystals. *Opt. Express* **2014**, *22*, 369–376. [CrossRef]
27. Yin, L.; Ning, H.; Turkdogan, S.; Liu, Z.; Nichols, P.L.; Ning, C.Z. Long lifetime, high density single-crystal erbium compound nanowires as a high optical gain material. *Appl. Phys. Lett.* **2012**, *100*, 241905. [CrossRef]
28. Zhang, X.; Yang, S.; Zhou, H.; Liang, J.; Liu, H.; Xia, H.; Zhu, X.; Jiang, Y.; Zhang, Q.; Hu, W.; et al. Perovskite-Erbium Silicate Nanosheet Hybrid Waveguide Photodetectors at the Near-Infrared Telecommunication Band. *Adv. Mater.* **2017**, *29*, 1604031. [CrossRef]
29. Gao, Y.; Fu, Q.; Shen, H.; Li, D.; Yang, D. Correlation of efficient luminescence with crystal structures of y-Er2Si2O7 and alpha-Er2Si2O7 in Er-doped silicon oxide films. *J. Mater. Sci.* **2019**, *54*, 12668–12675. [CrossRef]
30. Lo Savio, R.; Miritello, M.; Piro, A.M.; Priolo, F.; Iacona, F. The influence of stoichiometry on the structural stability and on the optical emission of erbium silicate thin films. *Appl. Phys. Lett.* **2008**, *93*, 021919. [CrossRef]
31. Gao, Y.; Shen, H.; Cao, J.; Li, D.; Yang, D. Control of the formation and luminescent properties of polymorphic erbium silicates on silicon. *Opt. Mater. Express* **2019**, *9*, 1716–1727. [CrossRef]
32. Kim, I.Y.; Shin, J.H.; Kim, K.J. Extending the nanocluster-Si/erbium sensitization distance in Er-doped silicon nitride: The role of Er-Er energy migration. *Appl. Phys. Lett.* **2009**, *95*, 221101. [CrossRef]
33. Soraru, G.D.; Modena, S.; Bettotti, P.; Das, G.; Mariotto, G.; Pavesi, L. Si nanocrystals obtained through polymer pyrolysis. *Appl. Phys. Lett.* **2003**, *83*, 749–751. [CrossRef]
34. Linnros, J.; Lalic, N.; Galeckas, A.; Grivickas, V. Analysis of the stretched exponential photoluminescence decay from nanometer-sized silicon crystals in SiO2. *J. Appl. Phys.* **1999**, *86*, 6128–6134. [CrossRef]
35. Krzylanowska, H.; Fu, Y.; Ni, K.S.; Fauchet, P.M. Efficient Energy Transfer between Si Nanostructures and Er Located at a Controlled Distance. *ACS Photonics* **2016**, *3*, 564–570. [CrossRef]
36. Zhang, X.W.; Lin, T.; Zhang, P.; Song, H.C.; Jin, H.; Xu, J.; Xu, J.; Wang, P.J.; Niu, K.Y.; Chen, K.J. Tunable quantum dot arrays as efficient sensitizers for enhanced near-infrared electroluminescence of erbium ions. *Nanoscale* **2018**, *10*, 4138–4146. [CrossRef] [PubMed]
37. Pacifici, D.; Irrera, A.; Franzo, G.; Miritello, M.; Iacona, F.; Priolo, F. Erbium-doped Si nanocrystals: Optical properties and electroluminescent devices. *Phys. E-Low-Dimens. Syst. Nanostruct.* **2003**, *16*, 331–340. [CrossRef]

Article

Exploring the Impact of Structure-Sensitivity Factors on Thermographic Properties of Dy^{3+}-Doped Oxide Crystals

Radosław Lisiecki [1,*], Jarosław Komar [1], Bogusław Macalik [1], Michał Głowacki [2], Marek Berkowski [2] and Witold Ryba-Romanowski [1]

[1] Institute of Low Temperature and Structure Research, Polish Academy of Sciences, ul. Okólna 2, 50-422 Wrocław, Poland; j.komar@intibs.pl (J.K.); b.macalik@intibs.pl (B.M.); w.ryba-romanowski@intibs.pl (W.R.-R.)
[2] Institute of Physics, Polish Academy of Sciences, Al. Lotnikow 32/46, 02-668 Warsaw, Poland; glowacki@ifpan.edu.pl (M.G.); berko@ifpan.edu.pl (M.B.)
* Correspondence: r.lisiecki@intibs.pl; Tel.: +48-713-954-182

Citation: Lisiecki, R.; Komar, J.; Macalik, B.; Głowacki, M.; Berkowski, M.; Ryba-Romanowski, W. Exploring the Impact of Structure-Sensitivity Factors on Thermographic Properties of Dy^{3+}-Doped Oxide Crystals. *Materials* **2021**, *14*, 2370. https://doi.org/10.3390/ma14092370

Academic Editor: Dirk Poelman

Received: 15 March 2021
Accepted: 30 April 2021
Published: 2 May 2021

Publisher's Note: MDPI stays neutral with regard to jurisdictional claims in published maps and institutional affiliations.

Copyright: © 2021 by the authors. Licensee MDPI, Basel, Switzerland. This article is an open access article distributed under the terms and conditions of the Creative Commons Attribution (CC BY) license (https:// creativecommons.org/licenses/by/ 4.0/).

Abstract: Optical absorption spectra and luminescence spectra were recorded as a function of temperature between 295 K and 800 K for single crystal samples of $Gd_2SiO_5:Dy^{3+}$, $Lu_2SiO_5:Dy^{3+}$, $LiNbO_3:Dy^{3+}$, and $Gd_3Ga_3Al_2O_{12}:Dy^{3+}$ fabricated by the Czochralski method and of $YAl_3(BO_3)_4:Dy^{3+}$ fabricated by the top-seeded high temperature solution method. A thermally induced change of fluorescence intensity ratio (FIR) between the $^4I_{15/2} \to {}^6H_{15/2}$ and $^4F_{9/2} \to {}^6H_{15/2}$ emission bands of Dy^{3+} was inferred from experimental data. It was found that relative thermal sensitivities S_R at 350 K are higher for $YAl_3(BO_3)_4:Dy^{3+}$ and $Lu_2SiO_5:Dy^{3+}$ than those for the remaining systems studied. Based on detailed examination of the structural peculiarities of the crystals it was ascertained that the observed difference between thermosensitive features cannot be attributed directly to the dissimilarity of structural factors consisting of the geometry and symmetry of Dy^{3+} sites, the number of non-equivalent Dy^{3+} sites, and the host anisotropy. Instead, it was found that a meaningful correlation between relative thermal sensitivity S_R and rates of radiative transitions of Dy^{3+} inferred from the Judd–Ofelt treatment exists. It was concluded that generalization based on the Judd–Ofelt parameters and luminescence branching ratio analysis may be useful during a preliminary assessment of thermosensitive properties of new phosphor materials.

Keywords: luminescence; Dy-doped crystals; optical temperature sensors

1. Introduction

The remote temperature readout is a useful and meaningful method, and consequently, great attention has been addressed towards distinct advanced luminescence thermometers. For this purpose, various sophisticated luminescence systems and temperature sensor techniques have been proposed and elaborated on within the last decade. The luminescence sensors, in the form of lanthanide-doped optical systems, quantum dots, organic fluorophores, or biomolecules, may be applied as potential luminescence thermometers and their temperature-dependent spectroscopic peculiarities and sensing capabilities have been reported and compared in the comprehensive review papers, e.g., [1–5].

A temperature readout above a thousand degrees is possible for limited luminescence systems, but inorganic amorphous materials or lanthanide-doped crystals showing efficient emission within wide UV-Vis-NIR spectral regions can be satisfactorily utilized there, in contrast to fluorophores or bio-molecules, which are susceptible to destruction [6–15].

In the present work we deal with the thermosensitive properties of Dy^{3+}-doped oxide crystals. Their advantage over other rare-earth-doped phosphors stems from a specific energy level scheme of the Dy^{3+} ions, in which the energy separation between the $^4F_{9/2}$ luminescent level and the next lower-energy dysprosium excited state is considerable, approaching 7000 cm^{-1}. As a consequence, the contribution of adverse multiphonon relaxation is substantially suppressed and the quantum efficiency of the $^4F_{9/2}$ luminescence is significant and weakly affected by the temperature for the most optical materials doped with dysprosium when the luminescence admixture concentration is adequately restricted [16]. With respect to the structural and optical properties of the host, the visible emission of the dysprosium can be differently distributed within blue, green, and red spectral regions, and, consequently, these individual materials' spectroscopic characteristics influence the resulting phosphor color. The diverse phosphor materials containing Dy^{3+} ions, e.g., single-doped with Dy^{3+} [17], double-doped with Dy + Mn [18], Dy + Eu [19], or triple-doped with Dy + Eu + Tb [20], have been described in numerous recent papers pointing out their utility for the design of novel lighting devices. The intensive development of UV and blue-emitting diode lasers, which can be applied as effective pumping sources, is extremely favorable for potential Dy^{3+}-doped laser materials. Recent deficiency of these efficient excitation sources significantly affected the progress of visible solid state lasers utilizing dysprosium-doped crystals. Fortunately, this inconvenience has been recently overcome, and, for instance, the laser performance of YAG:Dy^{3+} garnet crystals has been documented [21], describing a 12% slope efficiency of visible laser operation that was attained by applying a GaN laser diode as the optical pumping source.

Furthermore, dysprosium-doped crystals and glasses can be considered as potential optical temperature sensors and, as a result, several papers have been devoted to verify these possibilities [22–27]. The majority of these works were devoted to the preparation and assessment of the thermographic performance of new materials. There are, however, recent papers reporting more in-depth considerations, including the analysis of structure-sensitive factors. E. Hertle et al. [28] have investigated temperature-dependent emission qualities of Dy^{3+} in YAG, YAP, YSO, YSZ, and CASO, examining the impact of the host, and Er^{3+}-Pr^{3+} sensitizers' incorporation. In another paper, E. Hertle et al. [29] reported the in-depth investigation of (Gd,Lu)AlO_3:Dy^{3+} and (Gd,Lu)Al_5O_{12}:Dy^{3+}, unraveling the effect of substituting Gd^{3+} by Lu^{3+} ions on the garnet structure durability and the spectroscopic features of these luminescent materials. Perera and Rabufetti [30] reported the investigation of the thermosensitive properties of polycrystalline $NaLa_{1-x}Dy_x(MO_4)_2$ and $Na_5La_{1-x}Dy_x(MO_4)_4$ (M = Mo, W) materials, with special attention paid to the structural implication and the effect of Dy^{3+} concentration. All Dy^{3+}-doped systems mentioned above were prepared in a polycrystalline form by a high temperature solid state reaction. This method of material synthesis is cheap and time saving. In this way, series of samples differing in the concentration of luminescent rare-earth ions or in the substitution of cations in the host structure can be fabricated easily.

Our work deals with the examination of single crystals of Gd_2SiO_5:5at.%Dy^{3+} (GSO), Lu_2SiO_5:5at.%Dy^{3+} (LSO), $LiNbO_3$:1.94at.%Dy^{3+} (LNO), and $Gd_3Ga_3Al_2O_{12}$:1at.%Dy^{3+} (GGAG) fabricated by the Czochralski method, and of $YAl_3(BO_3)_4$:4at.%Dy^{3+} (YAB) fabricated by the top-seeded high temperature solution method. The choice of the crystals represents a trade-off between an intention to gather a set of samples showing inherent structural dissimilarity on one hand, and the availability of samples with the highest possible quality to warrant reliability of results on the other hand. The samples listed above comply with these requirements. The technology of their crystal growth has been mastered previously during works aiming at the design of visible lasers [31–35]. It will be shown in the following that the availability of single crystal samples is very relevant for comprehensive and reliable spectroscopic study. The intention of our investigation is to determine thermosensitive properties not yet reported for the systems studied and to correlate the obtained results with structural implications, attempting to establish a generalization regarding the effect of structure-sensitivity factors on luminescence thermometric qualities of Dy^{3+}-doped oxide crystals.

2. Materials and Methods

A Varian 5E UV-Vis-NIR spectrophotometer (Agilent, 5301 Stevens Creek Blvd, Santa Clara, CA 95051, USA) was applied to record the optical absorption spectra and 0.1 nm instrumental spectral bandwidth was then established. To determine crystal field splitting of Dy^{3+} excited multiplets, the absorption spectra were measured at a low temperature between 5 K and 10 K. For these low-temperature experiments, the crystals were mounted into an Oxford Model CF 1204 cryostat containing a liquid helium flow system and an adequate temperature controller. To record absorption spectra at different temperatures between 295 K and 800 K, the samples were placed into a chamber furnace. An Edinburgh Instruments FLS980 fluorescence spectrophotometer (Edinburgh Instruments Ltd. 2 Bain Square, Kirkton Campus, EH54 7DQ, UK) was utilized to measure the survey luminescence spectra and excitation spectra. A 450 W xenon lamp was utilized as an excitation source, and a Hamamatsu 928 PMT photomultiplier (Hamamatsu, 430-0852 2-25-7 Ryoke, Naka-ku, Japan) was used as the photon-sensitive detector. The acquired spectra were corrected on the experimental response of the used apparatus, employing their adequate sensitivity and spectral ranges. For measurements performed at a higher temperature, within 295–800 K, the samples were placed into a chamber furnace. The appropriate thermocouple was applied to temperature detection, and measurement accuracy was verified by a proportional-integral-derivative (PID) Omron E5CK controller. The samples were excited at 355 nm by a light beam consisting of a spectral band with 15 nm FWHM provided by the filtered output of a xenon lamp. The emission spectra were measured as a function of temperature within 295–800 K utilizing an Optron DM711 monochromator (DongWoo Optron Co. Ltd., Kyungg-do, Korea) with a 750 mm focal length. The resulting luminescence signal was detected applying a R3896 photomultiplier (Hamamatsu, 430-0852 2-25-7 Ryoke, Naka-ku, Japan).

3. Results and Discussion

Experimental data will be interpreted referring to the fundamental structural and optical data of the host crystals gathered in Table 1 and the energy level scheme for Dy^{3+} depicted in Figure 1. To construct this figure, the energy values for excited states determined in the past for Dy^{3+} (aquo) were taken from [36].

Table 1. Selected structural and optical features of the crystals under study.

	$Gd_3Ga_3Al_2O_{12}$ (GGAG)	$YAl_3(BO_3)_4$ (YAB)	$LiNbO_3$ (LNO)	Gd_2SiO_5 (GSO)	Lu_2SiO_5 (LSO)
Crystallographic system/ space group	Cubic/Ia-3d	Trigonal/R3c	Trigonal/rhombohedral/R3c	monoclinic/P2$_1$/c	monoclinic/C2/c
Unit cell (Å)	[31] A = 12.231	[32] a = 9.286 c = 7.231	[33] a = 5.15 c = 13.86	[34] a = 9.1105 b = 69783 c = 6.8544 β = 107.1411	[35] a = 14.277 (4) b = 6.6398 (4) c = 10.224 (6) β = 122.224 (1)°
Cut-off phonon energy [cm^{-1}]	808	1200	630	900	900
Bandgap (eV)	5.9	5.7	4.2	5.95	5.95
Dy^{3+} sites (Coordination No); Site symmetry	Dy (CN = 8); D_2	Dy (CN = 6); D_3	Dy (CN = 6); Close to C_3	Dy1 (CN = 9); Dy2 (CN = 7); C_{3v}, C_s	Dy1 (CN = 7); Dy2 (CN = 6); C_i for both sites

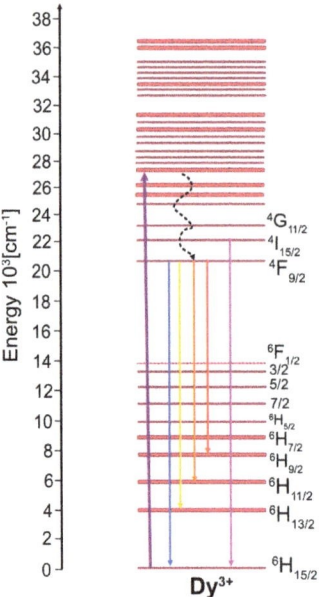

Figure 1. Energy levels scheme of Dy^{3+}.

The levels involved in luminescent transitions considered here are labelled with the symbols $^{2S+1}L_J$ of corresponding multiplets. Actually, for an ion imbedded in a crystalline host, each multiplet is split by the crystal field into crystal field components. Their number depends on the strength and symmetry of the crystal field, and, hence, on the structural features of the host crystal. In principle, low temperature absorption and luminescence spectra are able to offer detailed information regarding the number and nature of energy levels of rare-earth ions in crystals. Therefore, the interpretation of the observed luminescence phenomena refers to energy levels inferred from low temperature optical spectra for each Dy^{3+}-doped system under study. It follows from Figure 1 that excited multiplets created by the spin orbit splitting of the sextet ^6H and ^6F terms form a group of low energy levels located below about 14,300 cm^{-1}. A second group consists of high energy levels above about 20,400 cm^{-1}, related to closely spaced multiplets derived from the ^4F, ^4G, ^4H, ^4I, ^4K, ^4L, and ^4M quartet and ^6P sextet terms. It is worth noticing here that the $4f^9$ configuration of Dy^{3+} contains levels actually located at higher energies than those depicted in Figure 1. They have been omitted for the sake of clarity. Energy separation ΔE between neighboring excited levels of rare-earth ions in solids is a governing factor that determines the competition between radiative decay and nonradiative multiphonon relaxation. The latter process involves the simultaneous emission of the highest energy phonons available in the host, and the rate W_{mph} of this process depends on ΔE according to the energy gap law $W_{mph} = C\exp(-\alpha \Delta E)$, where C and α are host-dependent parameters. In the crystals studied, the high energy excited levels of Dy^{3+} ions relax nonradiatively, feeding the $^4F_{9/2}$ luminescent level. Its decay is governed by radiative transitions, mainly because the energy separation ΔE of ~7000 cm^{-1} between the $^4F_{9/2}$ level and the lower energy $^6F_{1/2}$ level is large when compared to the phonon energies listed in Table 1. The $^4F_{9/2}$ luminescence is related to the radiative transitions that terminate on multiplets derived from the ^6H and ^6F sextet terms. Transitions in the visible region are assigned and indicated by solid downward arrows in Figure 1. Transitions to remaining terminal levels are in the near infrared region and their intensities are small when compared to those in the visible region for virtually all Dy^{3+}-doped hosts. Figure 2 compares survey spectra of visible luminescence recorded at room temperature for the systems studied. The spectra shown deserve some comments to make the comparison meaningful. First, it follows from Table 1 that, except for cubic GGAG:Dy, the remaining crystals are anisotropic, i.e., GSO:Dy and LSO:Dy are optically biaxial whereas LNO:Dy and YAB:Dy are uniaxial. Their anisotropy was determined based on polarized optical spectra and has been reported in the past. Optical anisotropy is not relevant to our study; accordingly, the spectra in Figure 2 and all other spectra shown later on were recorded with unpolarized light. Second, instrumental spectral bandwidths for our measurement were carefully checked to avoid instrumental line broadening. With these points clarified, the impact of structural peculiarities listed in Table 1 on the spectral features of the luminescence bands becomes easier to see. Dy^{3+} ions substitute Gd^{3+} in GSO, and Lu^{3+} in LSO. They reside in two nonequivalent sites differing in the coordination number (CN), namely 9 and 7 for GSO [37] or 7 and 6 for LSO [38]. In the crystal structure of GSO, the two sites differ also in their local symmetry. Luminescence bands for GSO:Dy and LSO:Dy presented in Figure 2 show large overall widths and reach structures that stem from partly overlapping transitions between crystal field levels of two kinds of Dy^{3+} ions having dissimilar energies. In LNO, Dy^{3+} ions substitute in principle Li^+ ions entering sites characterized by CN = 6 and local symmetry close to C_3 [39]. However, observed spectra of LNO:Dy luminescence show large spectral width and poor band structure pointing at strong inhomogeneous broadening of spectral lines.

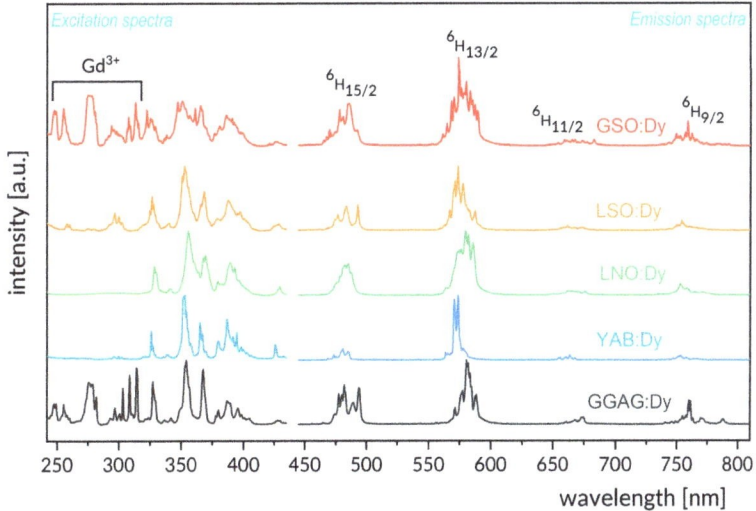

Figure 2. Survey spectra of visible luminescence (**right**) and survey excitation spectra of luminescence monitored at 575 nm (**left**) recorded at room temperature for the systems studied.

This effect is induced by the inherent structural disorder in a congruent LNO host, combined with problems with charge compensation in doped samples. In YAB, the Dy^{3+} ions substitute Y^{3+} ions entering one kind of well-defined site with CN = 6 and C_3 local symmetry [40]. As a consequence, observed spectral bands are relatively narrow and show some structure. In GGAG the Dy^{3+} ions substitute Gd^{3+} ions entering sites with CN = 8 and D_2 local symmetry [41]. GGAG host shows the structural disorder inherent for solid state solution crystals. Partial substitution of gallium ions by aluminum ions in this host brings about a dissimilarity of the crystal field acting on Dy^{3+} ions in different sites, inducing inhomogeneous spectral broadening, which, in contrast to LNO:Dy, is intentional. It can be seen in Figure 2 that the host crystal studied also affects the spectral distribution of luminescence intensity of incorporated Dy^{3+} ions, although the $^4F_{9/2} \rightarrow {}^6H_{13/2}$ band invariably dominates the spectra. Quantitative assessment of the distribution of luminescence intensity among spectral bands is commonly expressed in terms of luminescence branching ratios β, defined as the ratio of radiative transition rate for a particular transition from a luminescent level to the sum of rates of radiative transitions to all terminal levels. Experimental β_{exp} values can be evaluated by the numerical integration of bands in luminescence spectra. Table 2 compares percent values of β_{exp} determined by the numerical integration of spectra in Figure 2. It should be noticed that the sums of β_{exp} for four visible transitions equal to 100% because the contribution of weak infrared transition was neglected. Differences in the color of emitted light resulting from the dissimilarity of branching ratio values can be revealed based on the CIE chromaticity diagram shown in Figure 3 and the color coordinates gathered in the lowest part of Table 2.

Table 2. Experimental luminescence branching ratios β of the Dy^{3+} $^4F_{9/2} \rightarrow {}^{(2S+1)}L_J$ transitions and determined CIE values.

	Luminescence Branching Ratios β in %								
	GAGG		YAB		LNO		GSO		LSO
$^4F_{9/2} \rightarrow$ $^{(2S+1)}L_J$	β_{exp}		β_{exp}		β_{exp}		β_{exp}		β_{exp}
$^6H_{9/2}, ^6F_{11/2}$	4		7		4		2		6
$^6H_{11/2}$	8		11		9		5		5
$^6H_{13/2}$	49		65		54		58		55
$^6H_{15/2}$	42		17		33		35		34
CIE	x 0.393	y 0.396	x 0.405	y 0.456	x 0.431	y 0.426	x 0.403	y 0.412	x 0.401 y 0.432

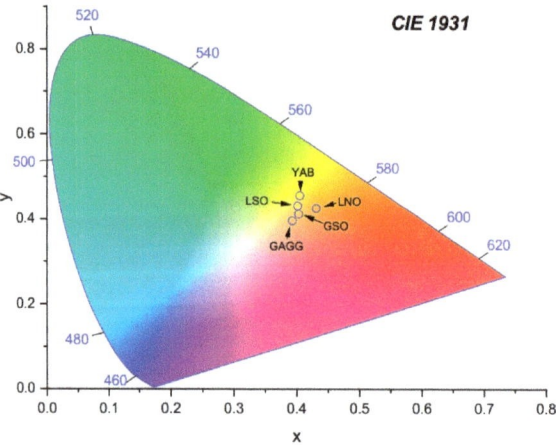

Figure 3. CIE chromaticity diagram constructed based on the color coordinates gathered in the bottom of Table 2.

In excitation spectra shown in Figure 2, the complex structure of bands is due to transitions within the $4f^9$ configuration of incorporated Dy^{3+} ions, except for strong Gd^{3+} bands located at around 250 nm and 310 nm in GSO:Dy and GGAG:Dy. The band located between about 340 nm and 360 nm is the most prominent. Its high intensity is due essentially to the $^6H_{15/2} \rightarrow {}^6P_{7/2}$ transition, although those ending on $(^4P, ^4D)_{3/2}$, $^6P_{5/2}$, $^4I_{11/2}$, $(^4M, ^4I)_{15/2}$, $(^4F, ^4D)_{5/2}$, and $^4I_{9/2}$ levels are also involved. These spectra imply that the intensity of Dy^{3+} luminescence depends critically on the wavelength of the incident excitation light. This shortcoming may not be encountered at higher temperatures because of thermal effects. Optical absorption and emission spectra of rare-earth ions located in non-centrosymmetric sites are related to pure electric dipole transitions, except for ions from the beginning or the end of rare-earth series, which show the contribution of vibronic transitions.

Thermally induced changes of the spectral bands of electric dipole transitions between multiplets of rare-earth ions in solids result from several factors. The governing factor follows from Boltzmann statistics, which determine the relative population of crystal field levels within multiplets, revealing, thereby, the number of band components and their intensity contribution as a function of the temperature. Other important factors relevant to narrow lines and related to transitions between individual crystal field levels are as follows: (i) thermal line broadening, a mechanism consisting of the Raman scattering of phonons by an ion in an excited state and (ii) thermal line shift, which determines the change of transition energy due to the temperature-induced displacement of levels involved in the transition. It is worth noticing here that the factors mentioned above affect the shapes of spectral bands and do not change the rates of the radiative transitions involved. Figure 4 compares optical absorption spectra in the UV-blue region recorded at several different temperatures between 300 K and 775 K for the systems under study. For the sake of clarity, the spectral region was restricted to 330–400 nm, where the most intense bands of interest for excitation purposes were located. In all spectra shown, the contribution of intense narrow lines and of local maxima diminishes with growing temperature, and eventually, above about 600 K, the spectra consist of a few broad and structureless bands. Spectra of GSO:Dy^{3+} provide a spectacular example of such an evolution, but the change of those for LSO:Dy^{3+} is less impressive. It follows from data in Table 1 that these orthosilicate hosts have ordered structures offering two different sites for Dy^{3+} ions. For each Dy^{3+} site, the crystal field splits the $^6H_{15/2}$ ground multiplet into eight components. As a consequence, partly overlapping homogeneously broadened lines related to transitions from 16 initial crystal field components contribute to the absorption bands of LSO:Dy and GSO:Dy. Low temperature luminescence spectra provided the overall ground state splitting of 933 cm^{-1} for Dy1 and Dy2 sites in LSO [35]. The overall ground state splitting of 922 cm^{-1} for the low symmetry Dy2 site and of 598 cm^{-1} for the high symmetry Dy1 site have been determined for GSO [34]. Different site symmetries combined with different ground state splitting results in the dissimilarity of LSO:Dy and GSO:Dy absorption spectra observed at room temperature. It can be seen in Figure 4 that this dissimilarity disappears gradually with increasing temperature. This is due to the increasing contribution of lines from higher energy crystal field components of the initial multiplet combined with thermal line broadening and thermal line shift. Unlike LSO:Dy and GSO:Dy, the Dy^{3+} ions are located in one kind of sites in a disordered structure of GGAG. As a consequence, their absorption bands consist of a superposition of lines related to transitions from eight crystal field components of the $^6H_{15/2}$ ground state, which shows an overall crystal field splitting of 674 cm^{-1} [31]. Owing to inhomogeneous line broadening, the spectral linewidths depend weakly on the temperature. Nevertheless, large inherent linewidths of several tens of nanometers combine with the increasing contribution of lines from higher energy crystal field components of the initial multiplet, contributing, thereby, to the thermally-induced broadening of the absorption bands. It is worth noticing that the spectra commented above do not contain bands of broad UV-blue absorption, indicating, thereby, that samples are free from point (color) defects. In the ordered structure of YAB, the Dy^{3+} ions substitute yttrium ions, and are located in one kind of site with CN equal to six and local symmetry D_3. In principle, their absorption bands should consist of a superposition of narrow lines related to transitions from eight crystal field components of the $^6H_{15/2}$ ground state, which shows an overall crystal field splitting of 468 cm^{-1} [32].

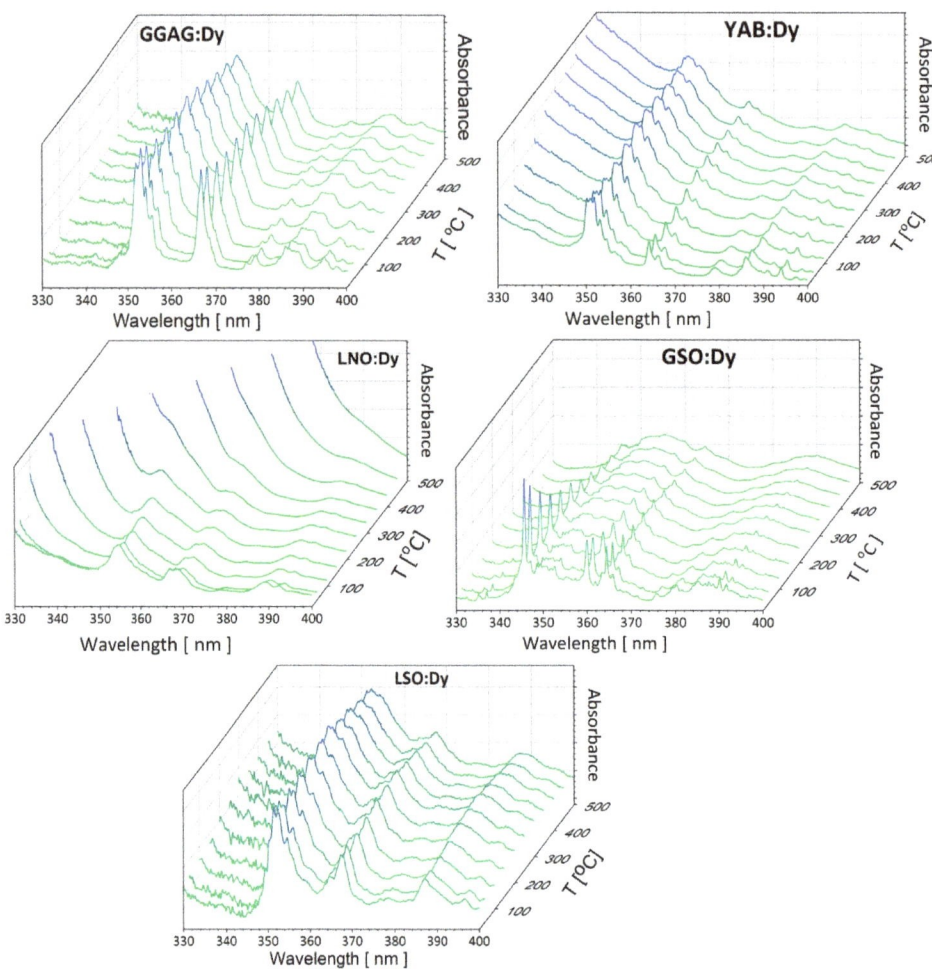

Figure 4. Optical absorption spectra in the UV-blue region recorded at several different temperatures between 295 K and 775 K for the systems under study. For the sake of clarity, the spectral region was restricted to 330–400 nm.

However, it can be seen in Figure 4 that, at 300 K, the baseline of the YAB:Dy spectra rises gently with decreasing wavelengths, but suffers from an upward shift at higher temperatures. This behavior indicates that the crystal structure of our YAB sample contains point defects, which show a thermally induced increase of absorption intensity. Occurrence of point defects gives rise to some inhomogeneous broadening of narrow band components, whereas a resulting parasitic absorption may adversely affect the efficiency of the Dy^{3+} excitation. It follows also from Figure 4 that these shortcomings are crucially relevant to the LNO:Dy system. Owing to a strong thermally induced increase of absorption intensity, which we interpret in terms of temperature-dependent charge transfer (CT) transition [42], the absorption bands of Dy^{3+} in the UV-blue region disappear in spectra recorded above about 500 K.

Recorded absorption spectra make it possible to determine quantitatively the effect of the sample temperature on Dy^{3+} luminescence intensity. For each system studied, the overall Dy^{3+} luminescence spectra recorded at different temperatures between 295 K and 725 K were numerically integrated within the 425–800 nm region. Next, the integrated luminescence intensities were normalized to unity at 295 K. Figure 5 compares the results obtained when exciting the samples at 355 nm with light consisting of a spectral band 15 nm FWHM provided by a filtered output of a xenon lamp. It can be seen in Figure 5 that, for the samples studied, the Dy^{3+} luminescence intensity excited at about 355 nm depends weakly on the temperature, except for the LNO:Dy crystal. It can be noticed also that, beginning at about 600 K, the YAB:Dy luminescence intensity is the lowest, likely because of the adverse contribution of defect centers commented on above.

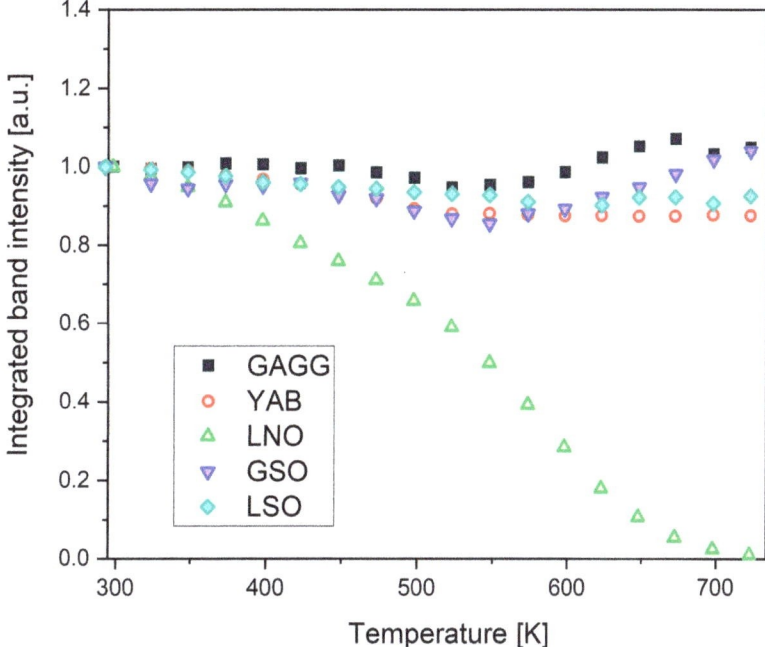

Figure 5. Integrated Dy^{3+} luminescence intensity within the 425–800 nm region at different temperatures between 295 K and 725 K. Results were normalized to unity at 295 K.

In the following, we examine luminescence phenomena related to transitions from the $^4I_{15/2}$, $^4F_{9/2}$ excited levels that are separated by about 1000 cm^{-1}, and whose populations are therefore governed by Boltzmann statistics. Accordingly, a thermally-induced change of fluorescence intensity ratio (FIR) between the $^4I_{15/2} \rightarrow {^6H_{15/2}}$ and $^4F_{9/2} \rightarrow {^6H_{15/2}}$ emission bands is a temperature-dependent parameter that can serve for temperature sensing. At 300 K, the Dy^{3+} luminescence spectrum consists essentially of the $^4F_{9/2} \rightarrow {^6H_{15/2}}$ band in the 465–500 nm region. With increasing temperature, the $^4I_{15/2}$ emission intensity between 450 nm and 465 nm, grows at the expense of the $^4F_{9/2}$ emission intensity.

Accordingly, the luminescence intensities are proportional to the population of the involved energy levels, and the FIR of two thermally coupled levels can be defined by the following equation [43]:

$$\text{FIR} = \frac{I_{(^4I_{15/2})}}{I_{(^4F_{9/2})}} = B \exp\left(-\frac{\Delta E}{kT}\right) \quad (1)$$

where B is the temperature-independent constant, ΔE is the energy gap between the two thermally coupled levels, and k is the Boltzmann constant. An optical thermometer may be quantitatively characterized with the absolute or relative thermal sensitivity. The former parameter reveals the absolute FIR change with temperature variation and is expressed as:

$$S_A = \frac{dFIR}{dT} = FIR\frac{\Delta E}{kT^2} \quad (2)$$

To compare the thermometers' quality, the relative sensitivity is usually used because this parameter determines the normalized change of FIR with temperature variation, and is defined as [44]:

$$S_R = \frac{1}{FIR}\frac{dFIR}{dT} \cdot 100\% = \frac{\Delta E}{kT^2} \cdot 100\% \quad (3)$$

For the samples under study, the luminescence spectra in the region 440–800 nm were recorded at different temperatures between 300 K and 800 K with steps of 25 K. Next, the experimental FIR values were evaluated by numerical integration of the recorded spectra. The best fit between the experimental temperature dependence of the FIR values and that predicted by Equation (1) provides the ΔE value involved. With these data, the S_A and S_R were determined as a function of the temperature from Equations (2) and (3), respectively. Figures 6–10 present the results obtained.

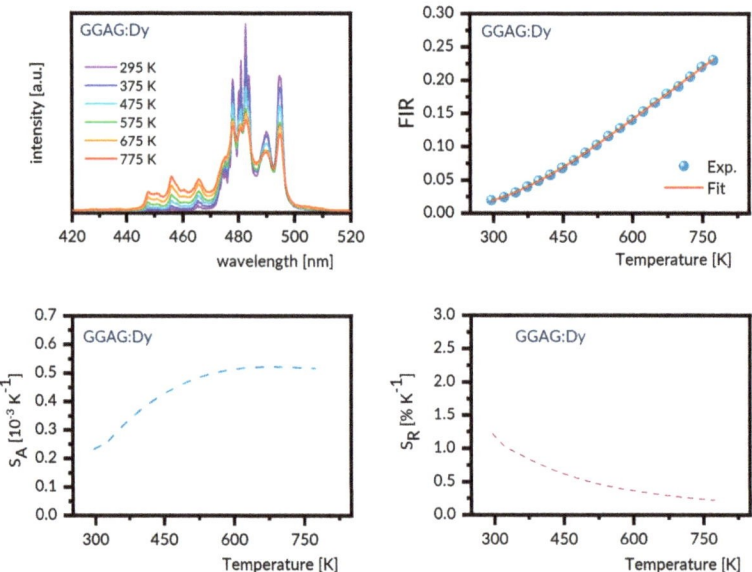

Figure 6. GGAG:Dy emission spectra recorded at several different temperatures (**upper left**), the plot of FIR versus temperature (**upper right**), the temperature dependence of S_A (**lower left**), and the temperature dependence of S_R (**lower right**).

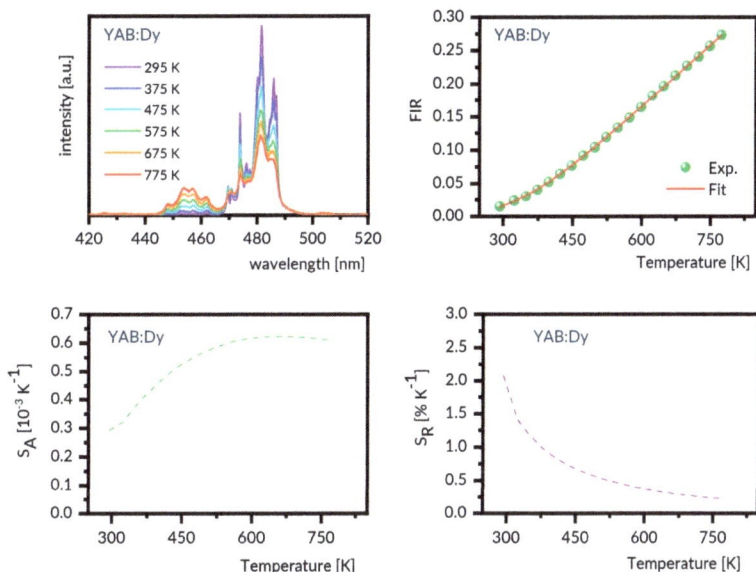

Figure 7. YAB:Dy emission spectra recorded at several different temperatures (**upper left**), the plot of FIR versus temperature (**upper right**), the temperature dependence of S_A (**lower left**), and the temperature dependence of S_R (**lower right**).

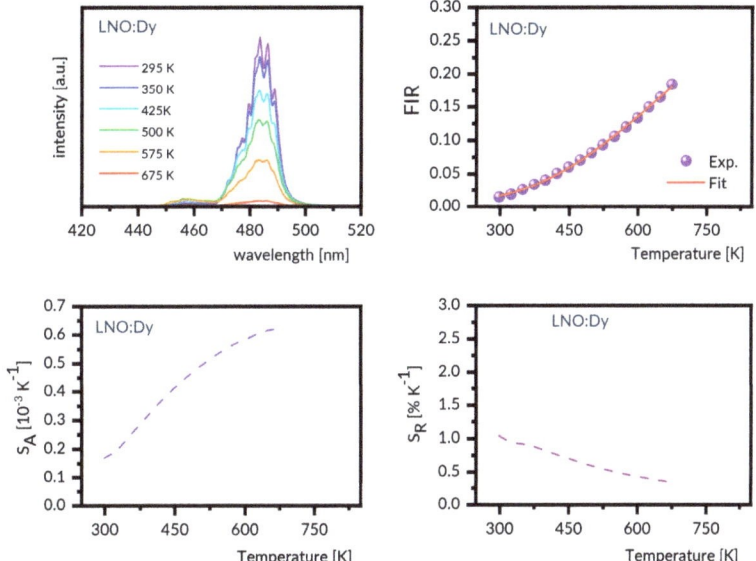

Figure 8. LNO:Dy emission spectra recorded at several different temperatures (**upper left**), the plot of FIR versus temperature (**upper right**), the temperature dependence of S_A (**lower left**), and the temperature dependence of S_R (**lower right**).

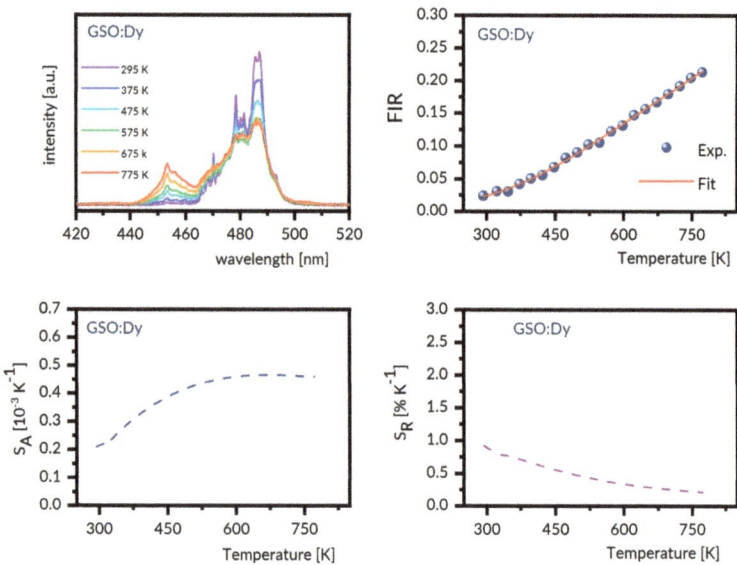

Figure 9. GSO:Dy emission spectra recorded at several different temperatures (**upper left**), the plot of FIR versus temperature (**upper right**), the temperature dependence of S_A (**lower left**), and the temperature dependence of S_R (**lower right**).

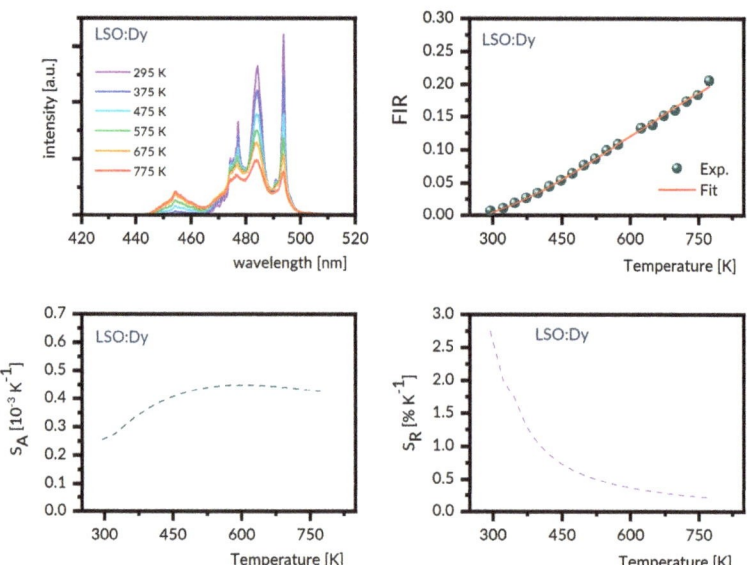

Figure 10. LSO:Dy emission spectra recorded at several different temperatures (**upper left**), the plot of FIR versus temperature (**upper right**), the temperature dependence of S_A (**lower left**), and the temperature dependence of S_R (**lower right**).

Each figure contains graphs showing spectra at several different temperatures (upper left), the plot of FIR versus temperature (upper right), the temperature dependence of S_A (lower left), and the temperature dependence of S_R (lower right). Our S_R values evaluated at 350 K and 750 K are compared in Table 3 to corresponding data reported for other Dy-doped crystals. It should be noted here that the luminescence intensity for LNO:Dy diminishes steeply with increasing temperature, restricting the reliability of the FIR and thermal sensitivity data at temperatures below 650 K. At this stage, the data in Table 3 deserve some comments. First, the S_R value at 350 K is greater than that at 750 K for all systems gathered, indicating that they are most suitable for near room temperature sensing. Second, the effect of temperature on S_R for different crystals is not the same. For instance, the change in temperature from 350 K to 750 K reduces the S_r by a factor of roughly seven for LSO:Dy and by a factor of three only for GSO:Dy. On the other hand, the change in crystal host is able to change the S_R values by no more than a factor of two, roughly. The ΔE values defined by Equation (1) and involved in plots of FIR versus temperature in Figures 6–10 are given in the second column of Table 4 as ΔE_{calc} values. When discussing our results, we refer to the luminescence spectra presented in Figures 6–10 and the energy level scheme in Figure 11 that was constructed based on the low temperature absorption spectra shown in References [31–35].

Table 3. Comparison of relative sensitivity S_r determined for Dy-doped crystals.

Dy-Doped Material	Relative Sensitivity S_r [% K^{-1}]	
	350 K	750 K
Gd$_3$Ga$_3$Al$_2$O$_5$ [this work]	0.93	0.23
YAl$_3$(BO$_3$)$_4$ [this work]	1.18	0.24
LiNbO$_3$ [this work]	0.92	0.37 (650 K)
Gd$_2$SiO$_5$ [this work]	0.77	0.23
Lu$_2$SiO$_5$ [this work]	1.73	0.23
Y$_2$SiO$_5$ [28]	-	0.41
GdVO$_4$ [22]	1.80	-
Na$_5$La$_{0.5}$Dy$_{0.5}$(WO$_4$)$_4$ [30]	1.80	0.30
NaDy(MoO$_4$)$_2$ [30]	0.75	0.38
YNbO$_4$ [24]	1.40	0.36
Y$_3$Al$_5$O$_{12}$ [28]	-	0.44
La$_3$Ga$_{5.5}$Ta$_{0.5}$O$_{14}$ [25]	1.47	0.34
BaYF$_5$ [26]	1.10	0.25
Gd$_2$Ti$_2$O$_7$ [27]	1.20	-
K$_3$Y(PO$_4$)$_2$ [45]	1.31	-
Ba$_3$Y$_4$O$_9$ [46]	1.34	-

Table 4. Calculated and experimental values of the energy gap ΔE between thermally coupled levels of Dy^{3+}.

Crystal	ΔE_{exp} (350 K) [cm^{-1}]	ΔE_{exp} (750 K) [cm^{-1}]	ΔE_{calc} [cm^{-1}]	ΔE_{exp} (350 K) − ΔE_{calc} [cm^{-1}]	ΔE_{exp} (750K) − ΔE_{calc} [cm^{-1}]
GAGG	1062	1158	945	−117	−213
YAB	1126	1116	916	−210	−200
LNO	1117	1161 *	1145	28	−16
GSO	1274	1212	934	−340	−278
LSO	1237	1261	839	−398	−422

* at 675 K for LNO:Dy.

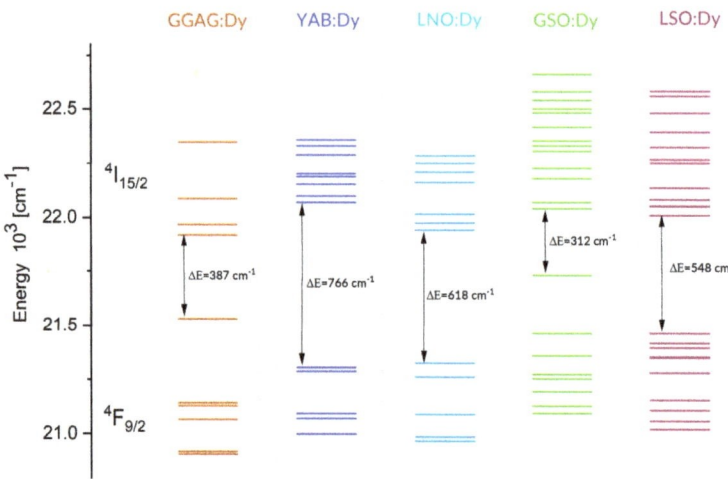

Figure 11. The crystal field splitting of the $^4I_{15/2}$ and $^4F_{9/2}$ multiplets of Dy^{3+}, determined based on low temperature absorption spectra.

The energy differences ΔE_{exp} between centroids of the $^4I_{15/2} \to {}^6H_{15/2}$ and $^4F_{9/2} \to {}^6H_{15/2}$ emission bands at 350 K and 750 K were determined numerically and given in the second and third columns, respectively. The resulting $\Delta E_{exp} - \Delta E_{calc}$ values appear in two end columns. It can be seen that, for all systems studied, the ΔE_{calc} and ΔE_{exp} values differ. Our results are consistent with those obtained recently by Perera and Rabufetti during their investigation of the thermosensitive properties of polycrystalline $NaLa_{1-x}Dy_x(MO_4)_2$ and $Na_5La_{1-x}Dy_x(MO_4)_4$ (M = Mo, W). It has been observed that the calculated energy gaps ΔE_{calc} are systematically smaller than the experimental values ΔE_{exp} at 350 K, and this dissimilarity ranged from 95 cm^{-1} to 350 cm^{-1} [30].

The reasons for these dissimilarities are not obvious, deserving, therefore, a closer investigation. When discussing our results, we refer to the luminescence spectra presented in Figures 6–10 and the energy level scheme in Figure 11 that was constructed based on low temperature absorption spectra. In principle, thermally induced changes of width and shape of the luminescence band related to the transition between multiplets of rare-earth ions in solids can be determined easily, provided the crystal field splitting of multiplets involved is known and the rates of transitions between individual crystal field levels are equal. Unfortunately, the latter condition is not always fulfilled in real systems. Hence, the former condition is not fulfilled frequently because a negligible intensity of some transitions prevents the location of levels involved.

Let us consider the LNO:Dy and LSO:Dy systems, which show the most significant disparity. It can be seen in Figure 11 that the $^4F_{9/2}$ metastable multiplet of Dy^{3+} in LNO is split by the crystal field into five components, all of them located from low temperature absorption spectra. The higher-energy thermally coupled multiplet $^4I_{15/2}$ is split by the crystal field into eight components, but only seven are located experimentally. The energy difference between the lowest component of the $^4I_{15/2}$ multiplet and the highest component of the $^4F_{9/2}$ multiplet is 618 cm^{-1}. When the temperature grows, the population of the higher energy components increases at the expense of the lower energy components for the $^4I_{15/2}$ and $^4F_{9/2}$ excited multiplets and for the ground $^6H_{15/2}$ multiplet. Anticipated changes of the luminescence bands consist of (a) a shift of the high energy wing towards shorter wavelengths and (b) an increase of intensity within the high energy wing due to a vanishing contribution of the self-absorption in this spectral region. It can be seen in Figure 8 that, in LNO:Dy, the anticipated changes are not corroborated by thermally induced changes of the experimental $^4F_{9/2} \to {}^6H_{15/2}$ luminescence band, which shows a nearly symmetric band-shape, weakly affected by the temperature. This may happen if thermally populated higher energy crystal field components of the initial $^4F_{9/2}$ multiplet have small transition rates.

Markedly different luminescent features were observed for LSO:Dy. For each of two Dy^{3+} sites in this host, the $^4F_{9/2}$ metastable multiplet is split by the crystal field into five components. In total, eight components were located from the low temperature absorption spectra. There are 16 components of the $^4I_{15/2}$ multiplet, but only 11 were located experimentally. The energy difference between the lowest component of the $^4I_{15/2}$ multiplet and the highest component of the $^4F_{9/2}$ multiplet is 548 cm^{-1}. It can be seen in Figure 10 that, unlike the LNO:Dy, the $^4F_{9/2} \to {^6H_{15/2}}$ luminescence band of LSO:Dy at 295 K shows a structure with well-defined peaks. The most intense and narrow one is located near the long wavelength edge of the band at about 494 nm, whereas the other, slightly less intense neighbor is located at about 484 nm. These positions coincide with those of the most prominent lines in the luminescence spectrum of LSO:Dy at 10 K [35], pointing at the distribution of transition rates, rather uncommon in that the highest rates have transitions bridging the lowest crystal field component of the $^4F_{9/2}$ with the highest energy crystal field components of the $^6H_{15/2}$. Increasing population of higher energy components of the initial $^4F_{9/2}$ multiplet counteracts this supremacy at higher temperatures, thereby changing the intensity distribution of the luminescence band components. The energy level schemes in Figure 11 are relevant to understanding the other peculiarities of the $^4F_{9/2} \to {^6H_{15/2}}$ luminescence bands shown in Figures 6–10. In particular, the overall spectral width of the bands complies with an obvious rule that it is a sum of the energy spreads of the two levels involved in a transition. As a result, the overall bandwidth is the smallest for YAB:Dy, slightly bigger for LNO:Dy, and markedly larger for the remaining systems. Additionally, disparities of the spectral positions of the $^4F_{9/2} \to {^6H_{15/2}}$ luminescence bands in the crystals can be well understood. The same concerns the $^4I_{15/2} \to {^6H_{15/2}}$ luminescence band, except for in the GGAG:Dy sample, where only four out of eight crystal field components of the initial multiplet were located experimentally. It is worth noticing that energy separation between the lowest crystal field component of the $^4I_{15/2}$ and the highest energy crystal field component of the $^4F_{9/2}$ is systematically smaller than the ΔE_{calc} values.

The comments expressed above indicate that the rates of radiative transitions between thermalized luminescent multiplets and the terminal ground state of Dy^{3+} ions are other important structure-sensitive factors relevant to the termographic features of the systems under study. To get more close insight, we followed the theoretical approach employed in the past to interpret the optical temperature sensing of Er^{3+}-doped calcium aluminate glass [47]. It was then proposed that the ratio of intensities I_{ik} and I_{jk} for a luminescence originated from a pair of thermally coupled levels can be calculated using the relation:

$$\frac{I_{ik}}{I_{jk}} = \frac{c_i(\nu) \, A_{ik} \, h\nu_{ik} \, g_i}{c_j(\nu) A_{jk} \, h\nu_{jk} \, g_j} \exp\left(-\frac{\Delta E}{kT}\right) \quad (4)$$

where $c(\nu)$ denotes coefficients related to the spectral response of the instrument at luminescence wavelengths, $h\nu$ denotes the energies of the emitted photons, A denotes the rates of radiative transitions related to the luminescence bands, g denotes the level of degeneracies, and ΔE denotes the energy separation between the two excited levels involved. The expression on the right-hand side of this general relation can be simplified when applied to Dy^{3+}-doped systems, assuming $c_i(\nu) \cong c_j(\nu)$ and $h\nu_{ik} \cong h\nu_{jk}$, since the energy separation between the $^4I_{15/2}$ and $^4F_{9/2}$ multiplets is small. Next, employing the Judd–Ofelt approach, the values $A_{ik} = A(^4I_{15/2} \to {^6H_{15/2}})$ and $A_{jk} = A(^4F_{9/2} \to {^6H_{15/2}})$ can be determined from the relation [48]:

$$A_{J'J} = \frac{64\pi^4 e^2}{3h(2J'+1)\overline{\lambda}^3} n \left(\frac{n^2+2}{3n}\right)^2 \sum_{t=2,4,6} \Omega_t \left|\left\langle \varphi_a \|U^{(t)}\| \varphi_b \right\rangle\right|^2 \quad (5)$$

where h is the Planck constant, $\bar{\lambda}$ is the mean wavelength of transition, n denotes the index of refraction, Ω_t are phenomenological intensity parameters, and $\left|\left\langle \varphi_a \| U^{(t)} \| \varphi_b \right\rangle\right|^2$ are doubly reduced matrix elements of unitary $U^{(t)}$ operators between the initial φ_a and terminal φ_b states. In this way, the rates $A(^4I_{15/2} \to {}^6H_{15/2})$ and $A(^4F_{9/2} \to {}^6H_{15/2})$ were calculated, inserting into Equation (5) the $\left|\left\langle \varphi_a \| U^{(t)} \| \varphi_b \right\rangle\right|^2$ values for Dy^{3+} taken from [49] and Ω_t parameters reported previously for systems under study. The calculated $A(^4I_{15/2} \to {}^6H_{15/2})/A(^4F_{9/2} \to {}^6H_{15/2})$ ratios are compared in the last column of Table 5.

Table 5. The Judd–Ofelt Ω_t parameters and calculated ratios R = $A(^4I_{15/2} \to {}^6H_{15/2})/A(^4F_{9/2} \to {}^6H_{15/2})$.

System	Ω_2 [10^{-20} cm^2]	Ω_4 [10^{-20} cm^2]	Ω_6 [10^{-20} cm^2]	R
GAGG:Dy	1.33	4.12	3.02	2.00
YAB:Dy	10.04	2.04	2.31	2.97
LNO:Dy	5.42	1.14	2.51	2.66
LNO:Dy [50]	9.75	2.63	2.52	2.79
GSO:Dy	3.22	2.16	3.76	2.32
LSO:Dy	9.06	1.88	3.12	2.77
LSO:Dy (without hypersensitive transitions)	4.31	1.28	3.49	2.48

The incertitude of the data presented in the above tables and graphs is worth commenting on at this stage to ascertain the meaningfulness of the generalizations proposed in the following. Obviously, the incertitude of data for LNO:Dy is regarded as the highest because a strong thermally induced increase of optical absorption in the UV-blue gradually reduces the luminescence intensity, adversely affecting the incertitude of S_A and S_R values at higher temperatures. Reliability of the data for the remaining four systems is believed to be reasonable, i.e., the incertitude of the S_A and S_R values is assessed to be below 10%, and that of radiative transition rates derived from the Judd–Ofelt treatment is within 20%. To be safe with interpreting the results, we focused our attention on the S_R values at 350, and notice that these values for the YAB:Dy and LSO:Dy systems are higher than those for the LNO:Dy, GGAG:Dy, and GSO:Dy systems. In view of the gathered data, this finding cannot be attributed to the dissimilarity of structural factors consisting of the geometry and symmetry of Dy^{3+} sites, the number of non-equivalent Dy^{3+} sites, and the host anisotropy. A straightforward attribution involving the peculiarities of the crystal field splitting of the $^6H_{15/2}$ ground state and the $^4I_{15/2}$ and $^4F_{9/2}$ excited multiplets or the nature and degree of spectral line broadening of Dy^{3+} transitions is not justified either. Instead, the S_R values mentioned above can be correlated meaningfully with R values gathered in the last column of Table 5. In fact, the R values for YAB:Dy and LSO:Dy are higher than those for GGAG:Dy and GSO:Dy, in agreement with the respective S_R values. For LNO:Dy, the R value is comparable to those for YAB:Dy and LSO:Dy. It disagrees with the rather low S_R value, likely because of the high incertitude of the luminescence data mentioned above.

At this stage, the correlation described above deserves some more detailed comments. The ratios R = $A(^4I_{15/2} \to {}^6H_{15/2})/A(^4F_{9/2} \to {}^6H_{15/2})$ involve radiative transition rates A that follow from Equation (5). With simplifications resulting from a small energy difference between the $^4I_{15/2}$ and $^4F_{9/2}$ multiplets, we obtain:

$$R = A\left(^4I_{15/2} \to {}^6H_{15/2}\right)/A\left(^4F_{9/2} \to {}^6H_{15/2}\right) \propto (0.0072\Omega_2 + 0.0003\Omega_4 + 0.0684\Omega_6)/(0.0047\Omega_4 + 0.0295\Omega_6) \quad (6)$$

Numbers preceding the Ω_t values in Equation (6) are values of matrix elements $\left|\left\langle \varphi_a \| U^{(t)} \| \varphi_b \right\rangle\right|^2$ of the unit tensor operators involved. It is worth noticing that the $\left|\left\langle \varphi_a \| U^{(2)} \| \varphi_b \right\rangle\right|^2 = 0$ for the $(^4F_{9/2} \to {}^6H_{15/2})$ transition. Therefore, Equation (6) predicts that the higher the Ω_2 value is, the higher the R value will be, as seen in Table 5. The examination of Table 2 corroborates this prediction, revealing the increase of β from the lowest value for YAB:Dy to the highest value for GAGG:Dy. The correspondence between S_R values in Table 3 and calculated R values gathered in Table 5 is not rigorous, but it can be regarded as a general trend. It follows from experimental data and comments presented above that the dissimilarity of the rates of radiative transitions from crystal field levels induces thermal changes of inter-multiplet luminescent transition, which are not predicted by the Ω_t parameters determined from spectra at 300 K. Nevertheless, in our opinion, the predictions following from Equation (6) and from the luminescence branching ratio analysis may be useful during a preliminary assessment of the thermosensitive properties of new phosphor materials.

4. Conclusions

Detailed spectroscopic investigation of single crystal samples of $Gd_2SiO_5:Dy^{3+}$, $Lu_2SiO_5:Dy^{3+}$, $LiNbO_3:Dy^{3+}$, and $Gd_3Ga_3Al_2O_{12}:Dy^{3+}$ fabricated by the Czochralski method and of $YAl_3(BO_3)_4:Dy^{3+}$ fabricated by the top-seeded high temperature solution method provided new and original information on their thermosensitive properties. Obtained results indicate that all of them are highly suitable for near room temperature sensing, with the relative thermal sensitivity S_R for $YAl_3(BO_3)_4:Dy^{3+}$ and $Lu_2SiO_5:Dy^{3+}$ undoubtedly higher than those for the remaining systems studied. A thermally induced increase of absorption intensity for $YAl_3(BO_3)_4:Dy^{3+}$ due to color centers was inferred from the optical absorption spectra in the UV-blue region, recorded as a function of temperature between 295 K and 725 K. For $LiNbO_3:Dy^{3+}$, the thermally induced increase of absorption intensity, which we interpret in terms of temperature-dependent charge transfer (CT) transitions, is particularly strong, and hides absorption bands of Dy^{3+} in the UV-blue region above about 500 K, restricting, thereby, the thermal sensitivity region. The difference between thermosensitive features cannot be attributed directly to the dissimilarity of structural factors consisting of the geometry and symmetry of Dy^{3+} sites, the number of non-equivalent Dy^{3+} sites, and the host anisotropy. Based on the crystal field splitting of Dy^{3+} multiplets inferred from low temperature spectra, we interpret observed disagreement of the energy difference ΔE_{calc} obtained from the intensity ratio (FIR), fitting with ΔE_{exp} inferred from the centers of gravity of luminescence bands in terms of dissimilarity of rates of radiative transitions between individual crystal field levels. It was found that a meaningful correlation between the values of relative thermal sensitivity S_R and rates of radiative transitions of Dy^{3+} inferred from the Judd–Ofelt treatment exists. It was also concluded that the resulting predictions based on the Judd–Ofelt parameters and the luminescence branching ratio analysis may be useful during a preliminary assessment of the thermosensitive properties of new phosphor materials.

Author Contributions: Conceptualization, R.L. and W.R.-R.; methodology, B.M., J.K., M.G. and M.B.; validation, R.L. and W.R-R.; formal analysis, R.L., W.R.-R., J.K. and B.M.; investigation, R.L., J.K. and B.M.; resources, W.R.-R. and M.B.; data curation, J.K. and B.M.; writing—original draft preparation, R.L. and W.R.-R.; writing—review and editing, R.L. and W.R.-R.; visualization, R.L., J.K. and B.M.; supervision, R.L. and W.R.-R.; funding acquisition, W.R.-R. and M.B. All authors have read and agreed to the published version of the manuscript.

Funding: The work was financially supported within the statutory funds of the Institute of Low Temperature and Structure Research, Polish Academy of Sciences in Wroclaw.

Institutional Review Board Statement: Not applicable.

Informed Consent Statement: Not applicable.

Data Availability Statement: The data presented in this study are available on request from the corresponding author.

Conflicts of Interest: The authors declare no conflict of interest.

References

1. Dramićanin, M.D. Trends in luminescence thermometry. *J. Appl. Phys.* **2020**, *128*, 040902. [CrossRef]
2. Bednarkiewicz, A.; Marciniak, L.; Carlos, L.D.; Jaque, D. Standardizing luminescence nanothermometry for biomedical applications. *Nanoscale* **2020**, *12*, 14405. [CrossRef] [PubMed]
3. Tingting, B.; Ning, G. Micro/Nanoscale Thermometry for Cellular Thermal Sensing. *Small* **2016**, *12*, 4590–4610. [CrossRef]
4. Wang, X.; Liu, Q.; Bu, Y.; Liu, C.S.; Liu, T.; Yan, X. Optical temperature sensing of rare-earth ion doped phosphors. *RSC Adv.* **2015**, *5*, 86219–86236. [CrossRef]
5. Jaque, D.; Ventrone, F. Luminescence nanothermometry. *Nanoscale* **2012**, *4*, 4301–4326. [CrossRef] [PubMed]
6. Kolesnikov, I.E.; Kurochkin, M.A.; Kalinichev, A.A.; Kolesnikov, E.Y.; Lähderanta, E. Optical temperature sensing in Tm^{3+}/Yb^{3+}-doped GeO_2–PbO–PbF_2 glass ceramics based on ratiometric and spectral line position approaches. *Sens. Actuators A* **2018**, *284*, 251–259. [CrossRef]
7. Liao, J.; Kong, L.; Wang, M.; Sun, Y.; Gong, G. Tunable upconversion luminescence and optical temperature sensing based on non-thermal coupled levels of Lu_3NbO_7:Yb^{3+}/Ho^{3+} phosphors. *Opt. Mater.* **2019**, *98*, 109452. [CrossRef]
8. Lisiecki, R.; Ryba-Romanowski, W. Silica-based oxyfluoride glass and glass-ceramic doped with Tm^{3+} and Yb^{3+} -VUV-VIS-NIR spectroscopy and optical thermometry. *J. Alloys Compd.* **2020**, *814*, 152304. [CrossRef]
9. Dai, W.; Hu, J.; Liu, G.; Xu, S.; Huang, K.; Zhou, J.; Xu, M. Thermometer of stable $SrAl_2Si_2O_8$: Ce^{3+}, Tb^{3+} based on synergistic luminescence. *J. Lumin.* **2020**, *217*, 116807. [CrossRef]
10. Zhao, Y.; Wang, X.; Zhang, Y.; Li, Y.; Yao, X. Optical temperature sensing of up-conversion luminescent materials: Fundamentals and progress. *J. Alloys Compd.* **2020**, *817*, 152691. [CrossRef]
11. Savchuk, O.A.; Carvajal, J.J.; Haro-Gonzalez, P.; Aguilo, M.; Díaz, F. Luminescent nanothermometry using short-wavelength infrared light. *J. Alloys Compd.* **2018**, *746*, 710–719. [CrossRef]
12. Kaczmarek, A.M.; Kaczmarek, M.K.; Van Deun, R. Er^{3+}-to-Yb^{3+} and Pr^{3+}-to-Yb^{3+} energy transfer for highly efficient near-infrared cryogenic optical temperature sensing. *Nanoscale* **2019**, *11*, 833. [CrossRef]
13. Zhang, J.; Chen, Y. $Y_{4.67}Si_3O_{13}$-based phosphors: Structure, morphology and upconversion luminescence for optical thermometry. *J. Am. Ceram. Soc.* **2019**, *102*, 5471–5483. [CrossRef]
14. Łukaszewicz, M.; Klimesz, B.; Szmalenberg, A.; Ptak, M.; Lisiecki, R. Neodymium-doped germanotellurite glasses for laser materials and temperature sensing. *J. Alloys Compd.* **2021**, *860*, 157923. [CrossRef]
15. Brites, C.D.S.; Fiaczyk, K.; Ramalho, J.F.C.B.; Sójka, M.; Carlos, L.D.; Zych, E. Widening the Temperature Range of Luminescent Thermometers through the Intra- and Interconfgurational Transitions of Pr^{3+}. *Adv. Opt. Mater.* **2018**, 1701318. [CrossRef]
16. Selvi, S.; Venkataiah, G.; Arunkumar, S.; Muralidharan, G.; Marimuthu, K. Structural and luminescence studies on Dy^{3+} doped lead boro-telluro-phosphate glasses. *Phys. B Condens. Matter* **2014**, *454*, 72–81. [CrossRef]
17. Rajendra, H.J.; Pandurangappa, C.; Monika, D.L. Luminescence properties of dysprosium doped YVO_4 phosphor. *J. Rare Earths* **2018**, *36*, 1245–1249. [CrossRef]
18. Karabulut, Y.; Canimoglu, A.; Kotan, Z.; Akyuz, O.; Ekdal, E. Luminescence of dysprosium doped strontium aluminate phosphors by codoping with manganese ion. *J. Alloys Compd.* **2014**, *583*, 91–95. [CrossRef]
19. Liu, Y.; Liu, G.; Wang, J.; Dong, X.; Yu, W. Multicolor photoluminescence and energy transfer properties of dysprosium and europium-doped Gd_2O_3 phosphors. *J. Alloys Compd.* **2015**, *649*, 96–103. [CrossRef]
20. Rimbach, A.C.; Steudel, F.; Ahrens, B.; Schweizer, S. Tb^{3+}, Eu^{3+}, and Dy^{3+} doped lithium borate and lithium aluminoborate glass: Glass properties and photoluminescence quantum efficiency. *J. Non-Cryst. Solids* **2018**, *499*, 380–386. [CrossRef]
21. Bowman, S.R.; O'Connor, S.; Condon, N.J. Diode pumped yellow dysprosium lasers. *Opt. Express* **2012**, *20*, 12906–12911. [CrossRef] [PubMed]
22. Antić, Ž.; Dramićanin, M.D.; Prashanthi, K.; Jovanović, D.; Kuzman, S.; Thundat, T. Pulsed Laser Deposited Dysprosium-Doped Gadolinium-Vanadate Thin Films for Noncontact, Self-Referencing Luminescence Thermometry. *Adv. Mater.* **2016**, *28*, 7745–7752. [CrossRef]
23. Nikolić, M.G.; Jovanović, D.J.; Dramićanin, M.D. Temperature dependence of emission and lifetime in Eu^{3+}- and Dy^{3+}-doped $GdVO_4$. *Appl. Opt.* **2013**, *52*, 1716–1724. [CrossRef] [PubMed]
24. Dačanin Far, L.; Lukić-Petrović, S.R.; Dordević, V.; Vuković, K.; Glais, E.; Viana, B.; Dramićanin, M.D. Luminescence temperature sensing in visible and NIR spectral range using Dy^{3+} and Nd^{3+} doped $YNbO_4$. *Sens. Actuators A* **2018**, *270*, 89–96. [CrossRef]
25. Komar, J.; Lisiecki, R.; Głowacki, M.; Berkowski, M.; Suszynska, M.; Ryba-Romanowski, W. Spectroscopic properties of Dy^{3+} ions in $La_3Ga_{5.5}Ta_{0.5}O_{14}$ single crystal. *J. Lumin.* **2020**, *220*, 116989. [CrossRef]
26. Cao, Z.; Zhou, S.; Jiang, G.; Chen, Y.; Duan, C.; Yin, M. Temperature dependent luminescence of Dy^{3+} doped $BaYF_5$ nanoparticles for optical thermometry. *Curr. Appl. Phys.* **2014**, *14*, 1067–1071. [CrossRef]
27. Ćulubrk, S.; Lojpur, V.; Ahrenkiel, S.P.; Nedeljković, J.M.; Dramićanin, M.D. Non-contact thermometry with Dy^{3+} doped $Gd_2Ti_2O_7$ nano-powders. *J. Lumin.* **2016**, *170*, 395–400. [CrossRef]
28. Hertle, E.; Chepyga, L.; Batentschuk, M.; Will, S.; Zigan, L. Temperature-dependent luminescence characteristics of Dy^{3+} doped in various crystalline hosts. *J. Lumin.* **2018**, *204*, 64–74. [CrossRef]
29. Hertle, E.; Chepyga, L.; Osvet, A.; Brabec, C.J.; Batentschuk, M.; Will, S.; Zigan, L. $(Gd,Lu)AlO_3$:Dy^{3+} and $(Gd,Lu)_3Al_5O_{12}$:Dy^{3+} as high-temperature thermographic phosphors. *Meas. Sci. Technol.* **2019**, *30*, 034001. [CrossRef]

30. Sameera Perera, S.; Rabuffetti, F.A. Dysprosium-activated scheelite-type oxides as thermosensitive phosphors. *J. Mater. Chem. C* **2019**, *7*, 7601. [CrossRef]
31. Lisiecki, R.; Solarz, P.; Niedźwiedzki, T.; Ryba-Romanowski, W.; Głowacki, M. $Gd_3Ga_3Al_2O_{12}$ single crystal doped with dysprosium: Spectroscopic properties and luminescence characteristics. *J. Alloys Compd.* **2016**, *689*, 733–739. [CrossRef]
32. Dominiak-Dzik, G.; Solarz, P.; Ryba-Romanowski, W.; Beregi, E.; Kovacs, L. Dysprosium-doped $YAl_3(BO_3)_4$ (YAB) crystals: An investigation of radiative and non-radiative processes. *J. Alloys Compd.* **2003**, *359*, 51–58. [CrossRef]
33. Dominiak-Dzik, G.; Ryba-Romanowski, W.; Palatnikov, M.N.; Sidorov, N.V.; Kalinnikov, V.T. Dysprosium-doped $LiNbO_3$ crystal. Optical properties and effect of temperature on fluorescence dynamics. *J. Mol. Struct.* **2004**, *704*, 139–144. [CrossRef]
34. Lisiecki, R.; Dominiak-Dzik, G.; Solarz, P.; Ryba-Romanowski, W.; Berkowski, M.; Głowacki, M. Optical spectra and luminescence dynamics of the Dy-doped Gd_2SiO_5 single crystal. *Appl. Phys. B* **2010**, *98*, 337–346. [CrossRef]
35. Dominiak-Dzik, G.; Ryba-Romanowski, W.; Lisiecki, R.; Solarz, P.; Berkowski, M. Dy-doped Lu_2SiO_5 single crystal: Spectroscopic characteristics and luminescence dynamics. *Appl. Phys. B* **2010**, *99*, 285–297. [CrossRef]
36. Megaw, H.E. A note on the structure of lithium niobate, $LiNbO_3$. *Acta Cryst. A* **1968**, *24*, 583. [CrossRef]
37. Dramicanin, M.D.; Jokanovic, V.; Viana, B.; Antic-Fidancev, E.; Matric, M.; Andric, Z. Luminescence and Structural Properties $Gd_2SiO_5:Eu^{3+}$ nanophosphors synthesized from the hydrothermal obtained silica sol. *J. Alloys Compd.* **2006**, *424*, 213. [CrossRef]
38. Gustafsson, T.; Klintenberg, M.; Derenzo, S.E.; Weber, M.J.; Thomas, J.O. Lu_2SiO_5 by single crystal X-ray and neutron diffraction. *Acta Cryst. C* **2001**, *57*, 668. [CrossRef] [PubMed]
39. Abrahams, S.C.; Reddy, J.M.; Bermstein, J.L. Ferroelectric lithium niobate. 3. Single crystal X-ray diffraction study at 24 °C. *J. Phys. Chem. Solids* **1966**, *27*, 997. [CrossRef]
40. Leonyuk, N.I. Crystal Growth of Multifunctional Borates and Related Materials. *Crystals* **2019**, *9*, 164. [CrossRef]
41. Sackville Hamilton, A.C.; Lampronti, G.I.; Rowley, S.E.; Dutton, S.E. Enhancement of the magnatocaloric effect driven by changes in the crystal structure of Al-doped GGG, $Gd_3Ga_{5-x}Al_xO_{12}$ ($0 \leq x \leq 5$). *J. Phys. Condens. Matter* **2014**, *26*, 116001. [CrossRef] [PubMed]
42. Lisiecki, R.; Macalik, B.; Kowalski, R.; Komar, J.; Ryba-Romanowski, W. Effect of Temperature on Luminescence of $LiNbO_3$ Crystals Single-Doped with Sm^{3+}, Tb^{3+}, or Dy^{3+} Ions. *Crystals* **2020**, *10*, 1034. [CrossRef]
43. Dramicanin, M.D. Sensing temperature via downshifting emissions of lanthanide-doped metal oxides and salts, A review. *Methods Appl. Fluoresc.* **2016**, *4*, 042001. [CrossRef]
44. Brites, C.D.S.; Balabhadra, S.; Carlos, L.D. Lanthanide-based thermometers: At the cutting-edge of luminescence thermometry. *Adv. Opt. Mater.* **2019**, 1801239. [CrossRef]
45. Zhang, J.; Zhang, Y.; Jiang, X. Investigations on upconversion luminescence of $K_3Y(PO_4)_2:Yb^{3+}\text{-}Er^{3+}/Ho^{3+}/Tm^{3+}$ phosphors for optical temperature sensing. *J. Alloys Comp.* **2018**, *748*, 438–445. [CrossRef]
46. Liu, S.; Ming, H.; Cui, J.; Liu, S.; You, W.; Ye, X.; Yang, Y.; Nie, H.; Wang, R. Color-Tunable Upconversion Luminescence and Multiple Temperature Sensing and Optical Heating Properties of $Ba_3Y_4O_9:Er^{3+}/Yb^{3+}$ Phosphors. *J. Phys. Chem. C* **2018**, *122*, 16289–16303. [CrossRef]
47. León-Luis, S.F.; Monteseguro, V.; Rodríguez-Mendoza, U.R.; Martín, I.R.; Alonso, D.; Cáceres, J.M.; Lavín, V. $2CaO\text{-}Al_2O_3:Er^{3+}$ glass: An efficient optical temperature sensor. *J. Lumin.* **2016**, *179*, 272–279. [CrossRef]
48. Hehlen, M.P.; Brik, M.G.; Kramer, K.W. 50th anniversary of the Judd–Ofelt theory: An experimentalist's view of the formalism and its application. *J. Lumin.* **2013**, *136*, 221–239. [CrossRef]
49. Carnall, W.T.; Fields, P.R.; Rajnak, K. Electronic Energy Levels in the Trivalent Lanthanide Aquo Ions. I. Pr^{3+}, Nd^{3+}, Pm^{3+}, Sm^{3+}, Dy^{3+}, Ho^{3+}, Er^{3+}, and Tm^{3+}. *J. Chem. Phys.* **1968**, *49*, 4424–4442. [CrossRef]
50. Malinowski, M.; Myziak, P.; Piramidowicz, R. Spectroscopic and Laser Properties of $LiNbO_3$:Dy Crystals. *Acta Phys. Pol.* **1996**, *90*, 181. [CrossRef]

Red Y_2O_3:Eu-Based Electroluminescent Device Prepared by Atomic Layer Deposition for Transparent Display Applications

José Rosa [1], Mikko J. Heikkilä [2], Mika Sirkiä [1] and Saoussen Merdes [1,*]

1. Beneq Oy, Olarinluoma 9, FI-02200 Espoo, Finland; jose.rosa@beneq.com (J.R.); mika.sirkia@beneq.com (M.S.)
2. Department of Chemistry, University of Helsinki, P.O. Box 55, FI-00014 Helsinki, Finland; mikko.j.heikkila@helsinki.fi
* Correspondence: saoussen.merdes@beneq.com

Abstract: Y_2O_3:Eu is a promising red-emitting phosphor owing to its high luminance efficiency, chemical stability, and non-toxicity. Although Y_2O_3:Eu thin films can be prepared by various deposition methods, most of them require high processing temperatures in order to obtain a crystalline structure. In this work, we report on the fabrication of red Y_2O_3:Eu thin film phosphors and multi-layer structure Y_2O_3:Eu-based electroluminescent devices by atomic layer deposition at 300 °C. The structural and optical properties of the phosphor films were investigated using X-ray diffraction and photoluminescence measurements, respectively, whereas the performance of the fabricated device was evaluated using electroluminescence measurements. X-ray diffraction measurements show a polycrystalline structure of the films whereas photoluminescence shows emission above 570 nm. Red electroluminescent devices with a luminance up to 40 cd/m² at a driving frequency of 1 kHz and an efficiency of 0.28 Lm/W were achieved.

Keywords: Y_2O_3:Eu; phosphor; photoluminescence; electroluminescence; atomic layer deposition

1. Introduction

Inorganic-based electroluminescent (EL) devices have been extensively studied for transparent flat panel display applications due to their distinct characteristics. Such technology allows for the creation of displays capable of withstanding harsh environments thanks to their exclusively solid structure, which leads to a high level of vibration and mechanical shock resistance [1]. Additionally, the electroluminescence phenomenon, which is not affected by temperature, allows EL devices to operate in a wide range of temperatures [2]. Furthermore, the ability to use alternating current to drive EL devices prevents charge accumulation, leading to long operating lifetimes [3].

Because the abovementioned characteristics are difficult to achieve with technologies such as organic-light emitting diodes (OLEDs), inorganic-based electroluminescent displays are very attractive from the commercial point of view. LUMINEQ thin film electroluminescent (TFEL) rugged displays and their transparent version TASEL displays are good examples of such commercial products which have been incorporated in industries such as automotive, industrial vehicles, and optical devices.

While yellow and green TFEL and TASEL displays are commercially available, demand for red EL devices has been increasing. Transparent red electroluminescent displays could, for example, be integrated to heavy vehicles, enabling them to display warning signs more effectively, thereby increasing the safety of operators. In the past, some attempts to develop red electroluminescent devices have been made by integrating phosphors such as CaS:Eu [4–6], CaY$_2$S$_4$:Eu [7], β-Ca$_3$(PO$_4$)$_2$:Eu [8], and ZnS:Sm,P [9] into the classic dielectric/semiconductor/dielectric (DSD) EL device structure. Red EL devices, with phosphors such as Eu$_2$O$_3$ [10], Ga$_2$O$_3$:Eu [11,12], and IGZO:Eu [13], were also developed using alternative device structures. However, only the use of a color filter with the yellow ZnS:Mn phosphor resulted in sufficiently high red luminescence to be used in commercial products [14]. This solution is unfortunately not suitable for transparent display applications as the use of filters reduces the overall transparency of the device.

Among the currently available red inorganic phosphors, Y_2O_3:Eu and Y_2O_2S:Eu are the most efficient [15,16]. Y_2O_3 and Y_2O_2S are known for their good chemical and photochemical stability. Furthermore, because Y^{3+} and Eu^{3+} have similar dimensions of the ionic radii, rare-earth materials such as Eu^{3+} can easily be integrated into Y_2O_3 and Y_2O_2S matrices [17]. However, Y_2O_3 exhibits a high electrical resistivity, with reported values in the 10^{11}–10^{12} Ωm range [18], which makes it incompatible with the classic DSD electroluminescent device structure. Nevertheless, several papers have demonstrated the successful use of Y_2O_3 and Y_2O_2S in red and green electroluminescent devices using multilayer structures where ZnS is used as a carrier accelerating layer [19,20].

Y_2O_3:Eu thin film phosphors can be grown by various methods such as wet chemistry [21], laser vaporization [22], hydrothermal [23], microwave hydrothermal [24,25], chemical precipitation with calcination [26], co-precipitation [27], Pechini [28], sol–gel [29,30], and pulse laser deposition [31] methods. Atomic layer deposition (ALD) is a well-known method that allows the growth of uniform and dense films with well-controlled stoichiometry and high chemical stability. Moreover, ALD, which is the method used for the fabrication of commercial electroluminescent displays, offers the advantage of an all-in-one growth step for the dielectric and phosphor layers in a DSD structure, thereby improving device resistance to moisture [1,32]. Years of advances in ALD technology have allowed the use of more elements and chemical precursors for the development of novel processes. As a result, opportunities for the fabrication of high-quality phosphors, and consequently more efficient electroluminescent devices, may arise in the future.

In a previous paper, we reported the growth of blue and red $Y_2O_{3-x}S_x$:Eu phosphors by ALD [33]. In this work, we focus on the fabrication and the performance evaluation of Y_2O_3:Eu-based multilayer structure electroluminescent devices that can potentially be used in red transparent display applications.

2. Materials and Methods

Atomic layer deposition processes for Y_2O_3, Eu_2O_3, Al_2O_3, and ZnS thin films were first developed on (100)-oriented Si substrates. All the films were grown at 300 °C in a Beneq TFS-200 ALD-reactor (Beneq Oy, Espoo, Finland) at a pressure of about 1.3 mbar. $(CH_3C_p)_3Y$ (98%, Intatrade, Anhalt-Bitterfeld Germany), Eu(thd)$_3$ (THD = 2,2,6,6-tetramethyl-3,5-heptanedionate) (99.5%, Intatrade, Anhalt-Bitterfeld Germany), Zn(OAc)$_2$ (99.9%, Alpha Aesar, Thermo Fisher GmbH, Germany), and trimethylaluminum (TMA, Al(CH3)3) (98%, Strem Chemicals UK Ltd., Cambridge, UK) were used as precursors for yttrium, europium, zinc and aluminum, respectively, while H_2O and/or O_3 were used as oxygen precursors for the Y_2O_3, Al_2O_3, and Eu_2O_3 processes. H_2S was used as sulfur precursor for the ZnS process. In all processes, N_2 was used as a carrier and purging gas. Details about the pulsing sequences and pulse and purge times are presented in Table 1. The doping level of the Y_2O_3 films with Eu was controlled by pulsing M number of Y_2O_3 cycles followed by N number of Eu_2O_3 cycles, resulting in an M:N doping ratio. To form the Y_2O_3:Eu layer, M:N cycles were repeated until achieving the expected thickness, always starting with a Y_2O_3 cycle and ending with a Eu_2O_3 cycle.

Table 1. Pulsing sequences and corresponding pulse time and purge time for the thin films prepared by atomic layer deposition (ALD). Growth per cycle (GPC) values displayed on the table were deduced from ellipsometry measurements.

Process	Pulsing Sequence	Pulsing Time (s)	GPC [nm]
Y_2O_3	Y(MeCp)$_3$/N$_2$/H$_2$O/N$_2$	2/6/0.2/7	0.16
Eu_2O_3	Eu(Thd)$_3$/N$_2$/O$_3$/N$_2$/H$_2$O/N$_2$	3/7/5/7/0.2/7	0.03
Al_2O_3	AlMe$_3$/N$_2$/H$_2$O/N$_2$	0.5/5/0.3/5	0.10
ZnS	Zn(OAc)$_2$/N$_2$/H$_2$S/N$_2$	2.5/6/0.3/3	0.24

The electroluminescent device was prepared using the structure proposed by T. Suyama et al. [19]. The multilayer structure was grown by ALD on a standard glass substrate coated with an ion-diffusion barrier and an ITO layer provided by LUMINEQ (Beneq Oy, Espoo, Finland). First, a 150 nm thick Al_2O_3 dielectric layer was grown by ALD. It was then followed by several ZnS (50 nm)/Y_2O_3:Eu (40 nm) multilayers. Finally, another 150 nm thick Al_2O_3 layer was deposited on the structure. The 1720 nm thick device was finalized by depositing a top contact. A schematic illustration of the device is presented in Figure 1. While it is possible to use a transparent top contact for a fully transparent device, for merely convenience purposes, top contact stripes of aluminum were sputtered here using a mechanical mask. The crossing of the ITO transparent contact and the aluminum stripes, which also comprises the sandwich multilayer Al_2O_3/ZnS/Y_2O_3:Eu/Al_2O_3 structure, creates a passive matrix with a pixel size of 3×5 mm^2. Note that prior to the deposition of the top Al_2O_3 layer, the multilayer sequence was always completed with a ZnS top layer.

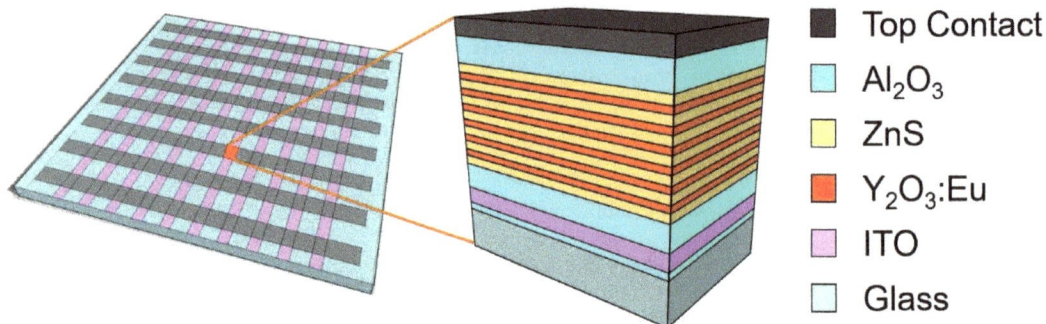

Figure 1. Schematic illustration of the passive matrix-like structure and the device cross section of the Y_2O_3:Eu/ZnS EL pixel prepared by ALD. The device is based on the multilayer electroluminescent structure proposed by T. Suyama et al. [19]. In this work, 6 layers of Y_2O_3:Eu and 7 layers of ZnS were used.

A SE400adv ellipsometer (SENTECH Instruments GmbH, Berlin, Germany) using a 633 nm wavelength at 70° angle of incidence, was used to determine the growth per cycle (GPC) for each material. GPC values were subsequently used to determine the thickness of the different layers. The crystallinity of Y_2O_3:Eu and ZnS thin films was investigated by X-ray diffraction (XRD) using the Cu Kα line in a Rigaku SmartLab (Rigaku Europe SE, Neu-Isenburg, Germany) high-resolution X-ray diffractometer equipped with in-plane arm. The XRD data were analyzed using the HighScore Plus 4.6 (PANalytical B.V., Almelo, The Netherlands). Photoluminescence (PL) emission was measured from Y_2O_3:Eu thin film phosphors with a Hitachi F-7100 Fluorescence Spectrophotometer (Hitachi High-Tech Analytical Science Ltd., Abingdon, UK) equipped with a 150 W xenon lamp. Measurements were performed at room temperature with an excitation slit of 5 nm, emission slit of 2.5 nm, and a photomultiplier tube voltage of 400 V. To determine the excitation wavelength, excitation spectra were recorded for maximum emission at 612 nm. Electroluminescent devices were powered by a Hewlett Packard 6811a source using AC mode at a frequency of 1 kHz. Electroluminescence spectra were recorded using a Konica Minolta CS-2000 spectrometer (Konica Minolta Sensing Europe B.V., Nieuwegein, The Netherlands) with a measurement angle of 1°.

For the calculation of the EL device efficiency, the Sawyer–Tower circuit was used to determine the charge density versus voltage (Q–V) characteristic. The used circuit is composed of a sense capacitor connected in series with the EL device. The total capacitance of the circuit was determined using a Fluke 76 digital multimeter. Data from the Q–V plot were acquired by measuring the voltage at each of the device terminals using a WaveSurfer 3104z (Teledyne Lecroy, Teledyne GmbH, Heidelberg, Germany) oscilloscope. The charge

(Q) of the device could be determined by multiplying the output voltage by the total capacitance of the circuit [32]. Simulations were performed using LTspice XVII.

3. Results

To optimize the emission of the Eu-doped Y_2O_3 thin film phosphors, films with three different Y_2O_3:Eu_2O_3 ratios were grown. Thus, three Eu doping concentrations (2:2, 3:2, and 4:2) were obtained by changing the number of Y_2O_3 and Eu_2O_3 sequences. As an example, a 4:2 doping configuration refers to a Y_2O_3:Eu thin film layer in which 4 layers of Y_2O_3 (Y(MeCp)$_3$/N$_2$/H$_2$O/N$_2$) were followed by 2 layers of Eu_2O_3 (Eu(Thd)$_3$/N$_2$/O$_3$/N$_2$/H$_2$O/N$_2$) during the ALD process. Taking into consideration Y_2O_3 and Eu_2O_3 densities, growth rates on Si substrate, and assuming that the Y_2O_3 and Eu_2O_3 films are stoichiometric, the 2:2, 3:2 and 4:2 doping configurations lead to calculated Eu concentrations of 16, 11, and 9 mol%, respectively.

3.1. Characterization of Y_2O_3:Eu and ZnS Thin Films

Figure 2a shows excitation spectra for a maximum emission at 612 nm, measured between 200 and 450 nm on a Y_2O_3:Eu sample grown on Si with a Y_2O_3:Eu_2O_3 layer ratio of 2:2. The excitation spectrum between 200 and 315 nm was measured using a Hitachi L-39 (UV-39) cutoff filter to remove a high intensity Rayleigh scattering peak located between 288 and 315 nm. The spectra show that the highest emission at 612 nm is obtained for an excitation of 238 nm. Figure 2b shows emission spectra for Y_2O_3:Eu_2O_3 layer ratios of 2:2, 3:2, and 4:2. The emission spectra were recorded for the excitation wavelength of 238 nm which was deduced from the excitation spectrum in Figure 2a. All PL spectra show an emission between 575 and 650 nm with a sharp line located at 612 nm. This red color emission is typical of the Eu^{3+} $^5D_0 \rightarrow {}^7F_J$ (J = 0, 1, 2, 3, and 4) transitions. Note that Y_2O_3:Eu samples grown with 2:2 and 4:2 doping configurations show much lower emission intensities compared to the sample grown with a doping concentration of 3:2.

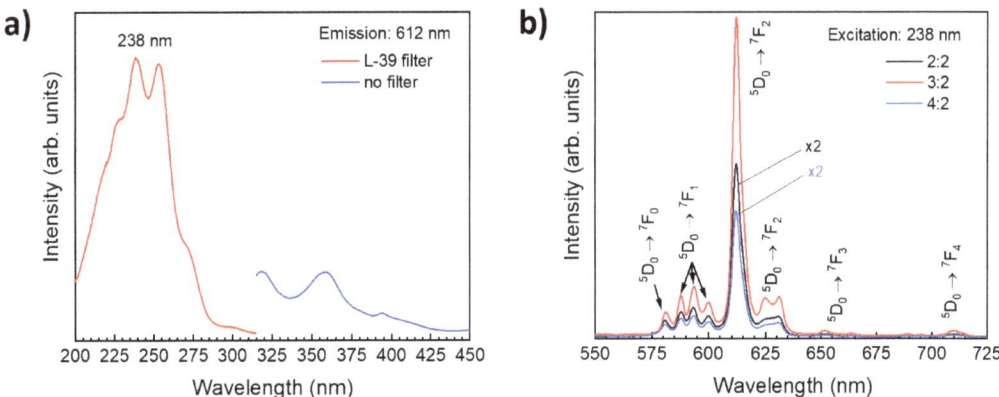

Figure 2. (**a**) Excitation spectra and (**b**) emission spectra of ALD Y_2O_3:Eu thin films prepared with different Eu concentrations. The measurements were performed at room temperature.

Figure 3a shows grazing incidence X-ray diffractograms for Y_2O_3:Eu and ZnS thin films measured between 15 and 65°. The Y_2O_3:Eu sample was prepared with a 3:2 (Y_2O_3:Eu_2O_3) cycle ratio. The Y_2O_3:Eu XRD diffractogram shows that the main phase of the film is polycrystalline (randomly orientated) cubic (pattern number 00-041-1105; Ia3) with some traces of monoclinic phase (marked with asterisk). The grazing incidence XRD data of the ZnS sample show clearly that the sample is highly orientated as only the (002) reflection is observed. The wide bump, between 45 and 60°, most likely originates from the substrate.

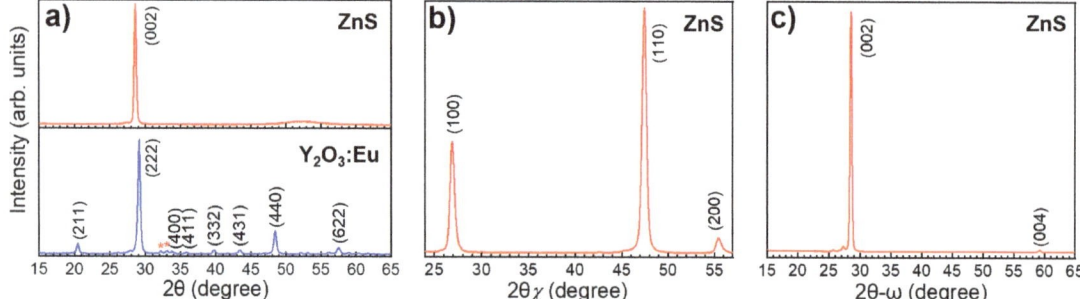

Figure 3. (a) Grazing incidence XRD for Y$_2$O$_3$:Eu and ZnS samples grown by ALD on Si substrates, (*) marks traces of monoclinic phase. The Y$_2$O$_3$:Eu sample was prepared with a 3:2 cycle ratio. (b) XRD spectrum for the ZnS sample measured in in-plane measurement mode. (c) XRD spectrum for the ZnS sample in the 2θ-ω measurement mode.

Further proof of the orientation was obtained by performing an in-plane measurement that probes the crystalline planes perpendicular to the surface normal as shown in Figure 3b. One can see only (hk0) family of planes meaning that (00l) planes are strongly orientated parallel to the surface. On Figure 3c, which shows the 2θ-ω measurement for the ZnS sample, the hump disappears supporting the idea that it was originated from the substrate. The peak at 59.1° reveals the (004) reflection related to the (002) intense reflection.

3.2. Red Electroluminescent Device

Figure 4a shows a photograph of a 3 × 5 mm^2 red Y$_2$O$_3$:Eu/ZnS-based EL pixel under a sinusoidal excitation of 1 kHz measured at 280 Vrms. The photograph was taken with a digital camera in automatic mode under normal room lighting. For this pixel, a brightness of 40 cd/m^2 was measured. Figure 4b shows the electroluminescence spectrum, at maximum luminance, of the Y$_2$O$_3$:Eu/ZnS EL device with a 3:2 (Y$_2$O$_3$:Eu$_2$O$_3$) cycle ratio. The EL spectrum, which was measured under an operating voltage of 280 Vmrs and a frequency of 1 kHz, clearly shows the typical $^5D_0 \rightarrow\ ^7F_J$ (J = 0, 1, 2, 3, and 4) transitions in Eu^{3+} emission centers. The sharp $^5D_0 \rightarrow\ ^7F_2$ line is located at 612 nm. Note the prominent $^5D_0 \rightarrow\ ^7F_4$ emission at 708 nm. The 1931 CIE color coordinates shown in Figure 4c were deduced from the EL spectrum in Figure 4b using OriginLab Chromaticity Diagram script (Origin Pro 2019, Northampton, MA, USA). Thus, the obtained red color emission corresponds to (x, y) values of (0.640, 0.348).

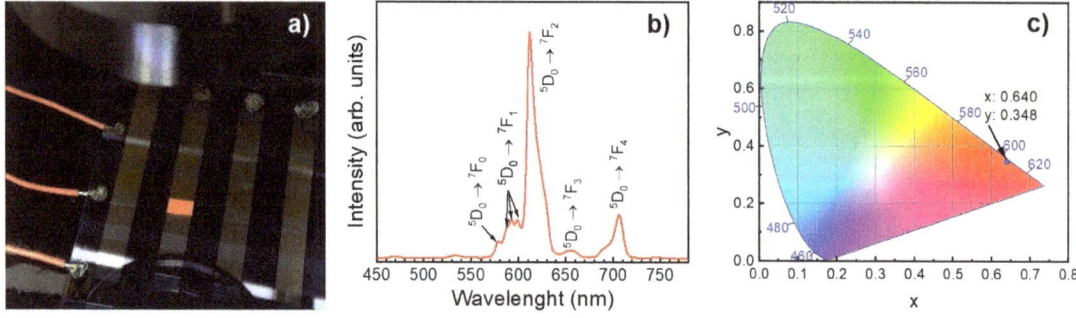

Figure 4. (a) Photograph of a 3 × 5 mm^2 red Y$_2$O$_3$:Eu/ZnS-based EL pixel emitting an intensity of 40 cd/m^2. The pixel was measured under normal room lighting. (b) Electroluminescence spectrum of the pixel shown in (a). (c) CIE 1931 chromatography diagram deduced from the spectrum in (b).

Figure 5a shows the luminance versus applied voltage characteristics of the Y_2O_3:Eu/ZnS electroluminescent device under a sinusoidal excitation of 1 kHz. The device shows a maximum brightness of 40 cd/m² at 280 Vrms. The threshold voltage of the device is not well-defined; it can, however, be considered as the voltage needed for the generation of 1 cd/m² [32]. Here, a luminance of 1 cd/m² is achieved for an excitation voltage of 180 Vrms. Figure 5b shows Q–V characteristics of a ZnS/Y_2O_3:Eu EL device, measured at 40 Vrms above the threshold voltage and 1 kHz sinusoidal wave. The measured sense capacitor and total capacitance of the circuit were 171 nF and 6.24 nF, respectively. The input power density, which was calculated by multiplying the area of the graphic in Figure 5b to the applied frequency, was determined to be 153 W/m². Based on these values, an efficiency of 0.28 Lm/W was calculated. Note the Y axis of the Q–V curve which is not centered in the position (0, 0) coordinates of the graphic.

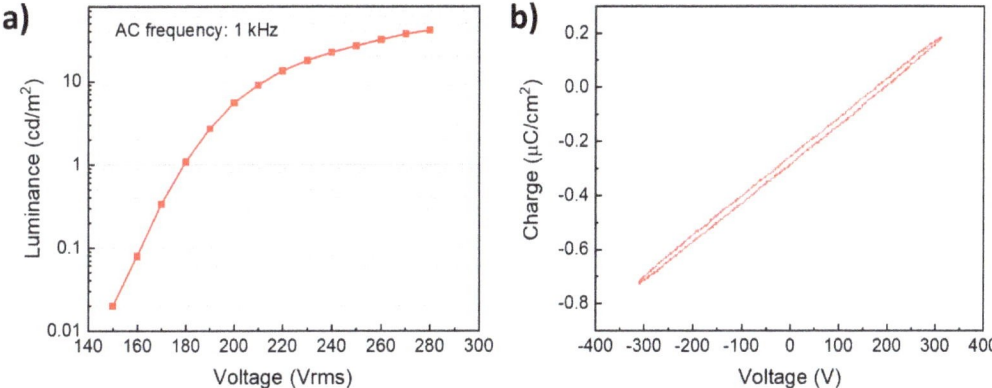

Figure 5. (**a**) Luminance versus the applied voltage and (**b**) charge–voltage (Q-V) characteristics for the Y_2O_3:Eu/ZnS electroluminescent device under a sinusoidal wave with a frequency of 1 kHz.

4. Discussion

Y_2O_3:Eu, ZnS, and Al_2O_3 thin films were successfully grown by ALD at 300 °C using commercial precursors. The processing temperature was limited to 300 °C because of the decomposition temperature of the metalorganic precursors and O_3. Y_2O_3:Eu thin film samples, grown with different Eu concentrations, show clearly red emission with a maximum intensity at 612 nm. This line is related to the $^5D_0 \rightarrow {}^7F_1$ magnetic dipole transition of Eu^{3+} [34]. With the process conditions described in this work, the optimum Eu concentration was found to be about 11 mol%. While a lower Eu concentration of 9 mol% led to lower PL intensities as expected, the well-known quenching that arises from energy transfer between the Eu^{3+} luminescent centers was observed for a Eu concentration of 16 mol%. These values are close to the ones reported by H. Huang et al. [35] in comparison with the optimum Eu concentration values of 20 and 5 mol% reported by J. Kaszewski et al. [25] and Y. Kumar et al. [27], respectively.

In a classic DSD electroluminescent device structure, an ideal phosphor should have a polycrystalline structure [32]. Therefore, the polycrystalline nature of our ALD Y_2O_3:Eu and ZnS thin film layers is advantageous to the multilayer Y_2O_3:Eu/ZnS electroluminescent device. Furthermore, in comparison with other reported Y_2O_3:Eu electroluminescent devices [36–38], our low processing temperature of 300 °C offers the possibility of building devices on some temperature-resistant polymer flexible substrates [39].

An all-in-one growth step for the Al_2O_3 dielectric, ZnS, and Y_2O_3:Eu phosphor layers was used for the fabrication of our EL device by ALD. In contrast to the photoluminescence spectrum, the electroluminescence spectrum shows a prominent $^5D_0 \rightarrow {}^7F_4$ emission at 708 nm. This could be due to the lower sensitivity of the PL equipment in comparison with the EL equipment, since most photomultiplier tubes have lower sensitivity in the $^5D_0 \rightarrow {}^7F_4$ transition region [34]. At 280 Vrms and under a sinusoidal excitation of 1 kHz, with the growth conditions reported in this paper, we achieved high-purity red color emission with an intensity up to 40 cd/m². This intensity could be significantly increased by further optimization of the different device layers, i.e., optimization of Y_2O_3:Eu and ZnS thicknesses and the dielectric layer (here a mere Al_2O_3 layer was used). Using multilayer structures, red and green Y_2O_3/Y_2O_2S-based electroluminescent devices with luminance up to 137 cd/m² (at 150 Vrms) and 124 cd/m² (at 300 Vrms), respectively, were reported by T. Suyama et al. [19] and K. Ohmi et al. [20]. While those values are higher than the ones we obtained for our devices, devices in [19,20] were measured under an excitation frequency of 5 kHz. Frequency has been reported to significantly influence the electroluminescence emission intensity. As an example, luminance values could be increased from 15 to 350 cd/m² in CaYS:Eu electroluminescent devices by increasing the frequency from 50 Hz to 1 kHz [7].

While it is difficult to compare the efficiency of our device with other red electroluminescent devices due to different measurement conditions, the calculated efficiency of 0.28 Lm/W for our ZnS/Y_2O_3:Eu multilayer EL device is lower than the 0.8 Lm/W value reported for the ZnS:Mn EL device with red filter and measured with a frequency of 60 Hz [16]. Q–V characteristics usually appear in a trapezoid shape where physical quantities such as threshold voltage, threshold voltage of the phosphor layer, threshold charge density, and transferred charge density are well-defined [32]. The elliptic shape of our Q–V characteristics is due to the multilayer structure of the ZnS/Y_2O_3:Eu EL device and possible presence of leakage current in the phosphor layer. Our Q–V curve appears negatively biased when the ITO layer of the EL device is connected to the power supply and the top contact is connected to the sense capacitor in the Sawyer–Tower circuit, as shown in Figure 6a. However, when the connections are inverted (the top contact is connected to the power supply and the ITO layer is connected to the sense capacitor), the Q–V curve appears positively biased. Therefore, one possible explanation for this behavior is the asymmetric structure of the device. During the growth process, each phosphor layer starts with the deposition of Y_2O_3 and finishes with Eu_2O_3 making ZnS surrounded on one side by Y_2O_3 and on the other by Eu_2O_3 as shown in Figure 6a. We believe this asymmetry might favor charge accumulation.

The Q–V characteristics could be reproduced by simulating the equivalent circuit (Figure 6b) of the EL device in the Sawyer–Tower circuit. Figure 6c shows the simulation results of two different scenarios: (i) in red, where the Sawyer–Tower circuit has the EL device with the ITO layer connected to the power supply and the top contact connected to the sense capacitor, as depicted in Figure 6a; and (ii) in blue, where the data were simulated with the top contact connected to the power supply and the ITO layer to the sense capacitor. This simulation requires high voltages and one Zener diode (related to the ZnS/Y_2O_3 or Eu_2O_3/ZnS interfaces) with higher threshold voltage than its counterpart. The simulation in Figure 6c matches Figure 5b when the Zener diode $D_{ZnS/Y2O3}$, which is related to the ZnS/Y_2O_3 interface, has a larger breakdown voltage than $D_{Eu2O3/ZnS}$.

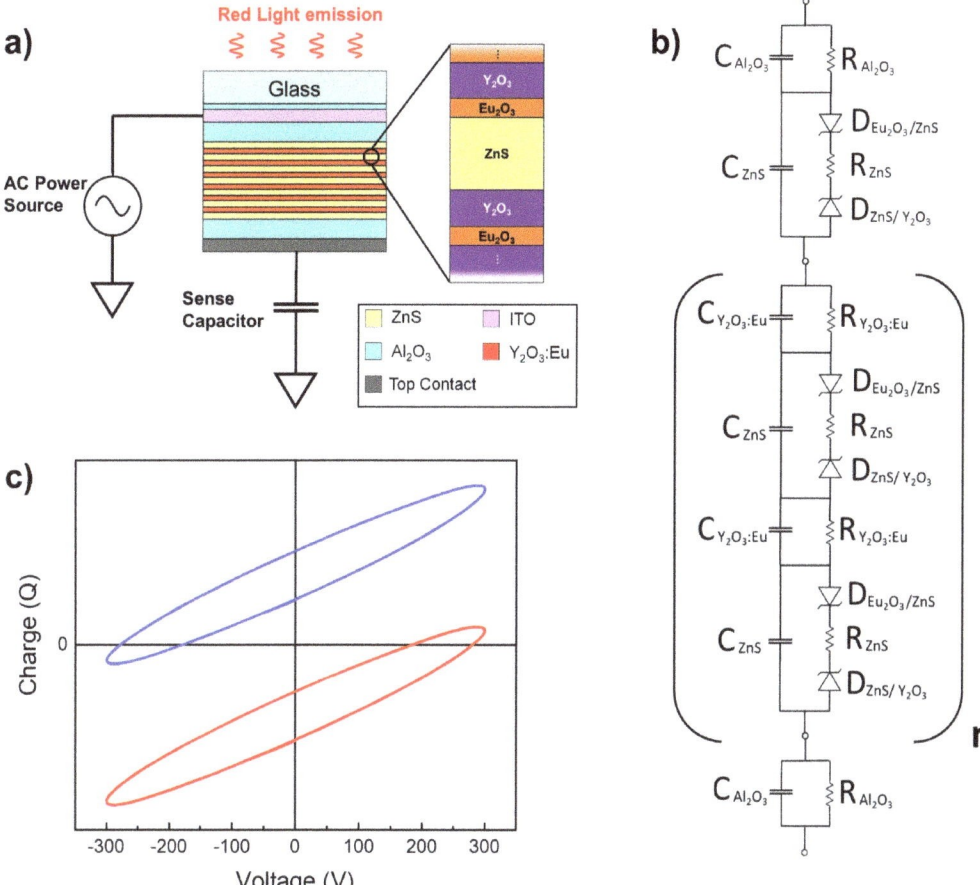

Figure 6. (a) Upside down representation of the 2D schematic of the Y_2O_3:Eu/ZnS EL device connected in a Sawyer-Tower circuit schematic with an amplification scheme of the ZnS layer and its surroundings. (b) Equivalent circuit of the Y_2O_3:Eu/ZnS EL device. (c) Simulated Q-V characteristics when (red) ITO is connected to the power supply and the top contact is connected to the sense capacitor; and (blue) ITO is connected to the sense capacitor and the top contact is connected to the power supply.

5. Conclusions

In this work, we demonstrate the feasibility of transparent red Y_2O_3:Eu-based electroluminescent devices by atomic layer deposition at relatively low temperature. Y_2O_3:Eu, ZnS, and Al_2O_3 thin films and related multilayer structure devices were prepared at 300 °C. XRD measurements showed high crystallinity of the Y_2O_3:Eu and ZnS films. Photoluminescence and electroluminescence measurements showed a bright red emission of the phosphors and electroluminescent devices, respectively. A luminance up to 40 cd/m² and an efficiency of 0.28 Lm/W were achieved. Further optimization of the phosphor and EL device is expected to lead to higher emission intensities.

Author Contributions: Conceptualization, S.M.; formal analysis, J.R., M.J.H., M.S. and S.M.; investigation, J.R., M.J.H., M.S. and S.M.; writing—original draft preparation, J.R. and S.M.; writing—review and editing, J.R. and S.M.; supervision, S.M. All authors have read and agreed to the published version of the manuscript.

Funding: The work in this paper was funded by the European Union's Horizon 2020 research and innovation program under the Marie Sklodowska-Curie grant agreement No 76495.

Institutional Review Board Statement: Not applicable.

Informed Consent Statement: Not applicable.

Data Availability Statement: Data sharing is not applicable to this article.

Acknowledgments: The authors would like to thank Pertti Malvaranta for the preparation of the ITO layer and Elina Haustola for sputtering the top contact.

Conflicts of Interest: The authors declare no conflict of interest. The funders had no role in the design of the study; in the collection, analyses, or interpretation of data; in the writing of the manuscript; or in the decision to publish the results.

References

1. Leskelä, M.; Mattinen, M.; Ritala, M. Review Article: Atomic layer deposition of optoelectronic materials. *J. Vac. Sci. Technol. B* **2019**, *37*, 030801. [CrossRef]
2. Smet, P.F.; Moreels, I.; Hens, Z.; Poelman, D. Luminescence in sulfides: A rich history and a bright future. *Materials* **2010**, *3*, 2834–2883. [CrossRef]
3. Wang, L.; Xiao, L.; Gu, H.; Sun, H. Advances in Alternating Current Electroluminescent Devices. *Adv. Opt. Mater.* **2019**, *7*, 1–30. [CrossRef]
4. Tanaka, K.; Mikami, A.; Ogura, T.; Taniguchi, K.; Yoshida, M.; Nakajima, S. High brightness red electroluminescence in CaS:Eu thin films. *Appl. Phys. Lett.* **1986**, *48*, 1730–1732. [CrossRef]
5. Poelman, D.; Vercaemst, R.; Van Meirhaeghe, R.L.; Laflère, W.H.; Cardon, F. Influence of the growth conditions on the properties of CaS:Eu electroluminescent thin films. *J. Lumin.* **1997**, *75*, 175–181. [CrossRef]
6. Abe, Y.; Onisawa, K.I.; Ono, Y.A.; Hanazono, M. Effects of oxygen in CaS:Eu active layers on emission properties of thin film electroluminescent cells. *Jpn. J. Appl. Phys.* **1990**, *29*, 1495–1498. [CrossRef]
7. Kawanishi, M.; Miura, N.; Matsumoto, H.; Nakano, R. New red-emitting CaY_2S_4:Eu thin-film electroluminescent devices. *Jpn. J. Appl. Phys. Part 2 Lett.* **2003**, *42*, 10–12. [CrossRef]
8. Koide, T.; Ito, M.; Kawai, T.; Matsushima, Y. An inorganic electroluminescent device using calcium phosphate doped with Eu^{3+} as the luminescent layer. *Mater. Sci. Eng. B Solid-State Mater. Adv. Technol.* **2013**, *178*, 306–310. [CrossRef]
9. Tohda, T.; Fujita, Y.; Matsuoka, T.; Abe, A. New efficient phosphor material ZnS:Sm,P for red electroluminescent devices. *Appl. Phys. Lett.* **1986**, *48*, 95–96. [CrossRef]
10. Yin, X.; Wang, S.; Mu, G.; Wan, G.; Huang, M.; Yi, L. Observation of red electroluminescence from an $Eu_2O_3/p + $-Si device and improved performance by introducing a Tb_2O_3 layer. *J. Phys. D Appl. Phys.* **2017**, *50*, 105103. [CrossRef]
11. Wellenius, P.; Suresh, A.; Muth, J.F. Bright, low voltage europium doped gallium oxide thin film electroluminescent devices. *Appl. Phys. Lett.* **2008**, *92*, 021111. [CrossRef]
12. Wellenius, P.; Suresh, A.; Foreman, J.V.; Everitt, H.O.; Muth, J.F. A visible transparent electroluminescent europium doped gallium oxide device. *Mater. Sci. Eng. B Solid-State Mater. Adv. Technol.* **2008**, *146*, 252–255. [CrossRef]
13. Wellenius, P.; Suresh, A.; Luo, H.; Lunardi, L.M.; Muth, J.F. An amorphous indium-gallium-zinc-oxide active matrix electroluminescent pixel. *IEEE/OSA J. Disp. Technol.* **2009**, *5*, 438–445. [CrossRef]
14. Wu, X.; Carkner, D. 8.1: Invited Paper: TDEL: Technology Evolution in Inorganic Electroluminescence. *SID Symp. Dig. Tech. Pap.* **2005**, *36*, 108. [CrossRef]
15. Ronda, C.R. Recent achievements in research on phosphors for lamps and displays. *J. Lumin.* **1997**, *72–74*, 49–54. [CrossRef]
16. Leskelä, M. Rare earths in electroluminescent and field emission display phosphors. *J. Alloys Compd.* **1998**, *275–277*, 702–708. [CrossRef]
17. Harazono, T.; Adachi, R.; Shimomura, Y.; Watanabe, T. Firing temperature dependence of Eu diffusion in $Eu-Y_2O_2S$ studied by ^{89}Y MAS NMR. *Phys. Chem. Chem. Phys.* **2001**, *3*, 2943–2948. [CrossRef]
18. Ivanĭ, R.; Novotn, I.; Vlastimil, K. Properties of Y_2O_3 thin films applicable in Micro–electrochemical Cells. *J. Electr. Eng.* **2003**, *54*, 83–87.
19. Suyama, T.; Okamoto, K.; Hamakawa, Y. New type of thin-film electroluminescent device having a multilayer structure. *Appl. Phys. Lett.* **1982**, *41*, 462–464. [CrossRef]
20. Ohmi, K.; Tanaka, S.; Kobayashi, H.; Nire, T. Electroluminescent Devices with (Y_2O_2S:Tb/ZnS) n Multilayered Phosphor Thin Films Prepared by Multisource Deposition. *Jpn. J. Appl. Phys.* **1992**, *31*, L1366–L1369. [CrossRef]
21. Adam, J.; Metzger, W.; Koch, M.; Rogin, P.; Coenen, T.; Atchison, J.S.; König, P. Light emission intensities of luminescent Y_2O_3:Eu and Gd_2O_3:Eu particles of various sizes. *Nanomaterials* **2017**, *7*, 26. [CrossRef] [PubMed]
22. Kostyukov, A.I.; Snytnikov, V.N.; Snytnikov, V.N.; Ishchenko, A.V.; Rakhmanova, M.I.; Molokeev, M.S.; Krylov, A.S.; Aleksandrovsky, A.S. Luminescence of monoclinic Y_2O_3:Eu nanophosphor produced via laser vaporization. *Opt. Mater.* **2020**, *104*, 109843. [CrossRef]

23. Zhu, P.; Wang, W.; Zhu, H.; Vargas, P.; Bont, A. Optical Properties of Eu^{3+}-Doped Y_2O_3 Nanotubes and Nanosheets Synthesized by Hydrothermal Method. *IEEE Photonics J.* **2018**, *10*, 1–10. [CrossRef]
24. Kaszewski, J.; Godlewski, M.M.; Witkowski, B.S.; Słońska, A.; Wolska-Kornio, E.; Wachnicki, Ł.; Przybylińska, H.; Kozankiewicz, B.; Szal, A.; Domino, M.A.; et al. Y_2O_3:Eu nanocrystals as biomarkers prepared by a microwave hydrothermal method. *Opt. Mater.* **2016**, *59*, 157–164. [CrossRef]
25. Kaszewski, J.; Rosowska, J.; Witkowski, B.S.; Wachnicki, Ł.; Wenelska, K.; Mijowska, E.; Bulyk, L.I.; Włodarczyk, D.; Suchocki, A.; Kozankiewicz, B.; et al. Shape control over microwave hydrothermally grown Y_2O_3:Eu by europium concentration adjustment. *J. Rare Earths* **2019**, *37*, 1206–1212. [CrossRef]
26. Li, H.; Kang, J.; Yang, J.; Wu, B. Fabrication of Aunanoparticle@$mSiO_2$@Y_2O_3:Eu nanocomposites with enhanced fluorescence. *J. Alloys Compd.* **2016**, *673*, 283–288. [CrossRef]
27. Kumar, Y.; Pal, M.; Herrera, M.; Mathew, X. Effect of Eu ion incorporation on the emission behavior of Y_2O_3 nanophosphors: A detailed study of structural and optical properties. *Opt. Mater.* **2016**, *60*, 159–168. [CrossRef]
28. De Oliveira Krauser, M.; de Souza Oliveira, H.H.; Cebim, M.A.; Davolos, M.R. Relationship between scintillation properties and crystallite sizes in Y_2O_3:Eu^{3+}. *J. Lumin.* **2018**, *203*, 100–104. [CrossRef]
29. García-Murillo, A.; Carrillo-Romo, F.; de, J.; Oliva-Uc, J.; Esquivel-Castro, T.A.; de la Torre, S.D. Effects of Eu content on the luminescent properties of Y_2O_3:Eu^{3+} aerogels and $Y(OH)_3/Y_2O_3$:Eu^{3+}@SiO_2 glassy aerogels. *Ceram. Int.* **2017**, *43*, 12196–12204. [CrossRef]
30. Unal, F.; Kaya, F.; Kazmanli, K. Effects of dopant rate and calcination parameters on photoluminescence emission of Y_2O_3:Eu^{3+} phosphors: A statistical approach. *Ceram. Int.* **2019**, *45*, 17818–17825. [CrossRef]
31. Ali, A.G.; Dejene, B.F.; Swart, H.C. The influence of different species of gases on the luminescent and structural properties of pulsed laser-ablated Y_2O_2S:Eu^{3+} thin films. *Appl. Phys. A Mater. Sci. Process.* **2016**, *122*, 1–9. [CrossRef]
32. Ono, Y.A. *Electroluminescent Displays*; Series On Information Display; reprint; World Scientific: Singapore, 1995; ISBN 9810219210.
33. Rosa, J.; Deuermeier, J.; Soininen, P.J.; Bosund, M.; Zhu, Z.; Fortunato, E.; Martins, R.; Sugiyama, M.; Merdes, S. Control of Eu Oxidation State in $Y_2O_{3-x}S_x$:Eu Thin-Film Phosphors Prepared by Atomic Layer Deposition: A Structural and Photoluminescence Study. *Materials* **2019**, *13*, 93. [CrossRef]
34. Binnemans, K. Interpretation of europium(III) spectra. *Coord. Chem. Rev.* **2015**, *295*, 1–45. [CrossRef]
35. Huang, H.; Xu, G.Q.; Chin, W.S.; Gan, L.M.; Chew, C.H. Synthesis and characterization of Eu:Y_2O_3 nanoparticles. *Nanotechnology* **2002**, *13*, 316. [CrossRef]
36. Sowa, K.; Tanabe, M.; Furukawa, Y.; Nakanishi, Y.; Hatanaka, Y. Characteristics of Y_2O_3:Eu/ZnS/Y_2O_3:Eu Red Light Emitting Elec Fluorescencetroluminescent Devices. *Jpn. J. Appl. Phys.* **1992**, *31*, 3598–3602. [CrossRef]
37. Gupta, A.; Brahme, N.; Bisen, D.P. Photoluminescence and Electroluminescence of Eu Doped Y_2O_3. *Phys. Procedia.* **2015**, *76*, 16–24. [CrossRef]
38. Rodionov, V.E.; Shmidko, I.N.; Zolotovsky, A.A.; Kruchinin, S.P. Electroluminescence of Y_2O_3:Eu and Y_2O_3:Sm films. *Mater. Sci. Pol.* **2013**, *31*, 232–239. [CrossRef]
39. Liu, J.; Ni, H.; Wang, Z.; Yang, S.; Zhou, W. Colorless and Transparent high—Temperature-Resistant Polymer Optical Films—Current Status and Potential Applications in Optoelectronic Fabrications. In *Optoelectronics-Materials and Devices*; InTech: London, UK, 2015.

Derivatives of Imidazole and Carbazole as Bifunctional Materials for Organic Light-Emitting Diodes

Oleksandr Bezvikonnyi [1,2], Ronit Sebastine Bernard [1], Viktorija Andruleviciene [1], Dmytro Volyniuk [1], Rasa Keruckiene [1], Kamile Vaiciulaityte [1], Linas Labanauskas [3] and Juozas Vidas Grazulevicius [1,*]

[1] Department of Polymer Chemistry and Technology, Faculty of Chemical Technology, Kaunas University of Technology, K. Baršausko g. 59, LT-51423 Kaunas, Lithuania
[2] Department of Physics, Faculty of Mathematics and Natural Science, Kaunas University of Technology, Studentų g. 50, LT-51369 Kaunas, Lithuania
[3] Center for Physical Sciences and Technology (FTMC), Department of Organic Chemistry, Sauletekio Ave. 3, LT-10257 Vilnius, Lithuania
* Correspondence: juozas.grazulevicius@ktu.lt

Abstract: New derivatives of carbazole and diphenyl imidazole for potential multiple applications were synthesized and investigated. Their properties were studied by thermal, optical, photophysical, electrochemical, and photoelectrical measurements. The compounds exhibited relatively narrow blue light-emission bands, which is favorable for deep-blue electroluminescent devices. The synthesized derivatives of imidazole and carbazole were tested as fluorescent emitters for OLEDs. The device showed deep-blue emissions with CIE color coordinates of (0.16, 0.08) and maximum quantum efficiency of 1.1%. The compounds demonstrated high triplet energy values above 3.0 eV and hole drift mobility exceeding 10^{-4} cm^2/V·s at high electric fields. One of the compounds having two diphenyl imidazole moieties and *tert*-butyl-substituted carbazolyl groups showed bipolar charge transport with electron drift mobility reaching 10^{-4} cm^2/V·s at electric field of 8×10^5 V/cm. The synthesized compounds were investigated as hosts for green, red and sky-blue phosphorescent OLEDs. The green-, red- and sky-blue-emitting devices demonstrated maximum quantum efficiencies of 8.3%, 6.4% and 7.6%, respectively.

Keywords: imidazole; carbazole; luminescence; emitter; host; organic light-emitting diode; upconversion

1. Introduction

Organic light-emitting diodes (OLEDs) are used in displays and lighting devices due to their low turn-on voltage, high power efficiency, mechanical flexibility and wide range of colors [1]. Generally, electrons and holes injected from the respective electrodes move through charge injection/transport layers into the emitting layer of a device. There, electrons and holes recombine and form singlet and triplet excitons [2]. Fluorescent OLEDs exploit only singlet excitons for light emission [3]. Organic fluorescent materials with high emission quantum yield, restricted intermolecular interactions, and appropriate charge injection/transporting properties are required for stable and efficient OLEDs [4,5]. Because of the high stability of blue singlet emission-based OLEDs, they are still used in commercial displays [6]. Therefore, development of stable, efficient and cost-effective fluorescent blue emitters remains an urgent task. Phosphorescent OLEDs (PHOLEDs) can utilize both singlet and triplet excitons for light emission. This property allows us to increase external quantum efficiency of PHOLEDs [7,8]. Phosphorescent emitters normally have to be doped in a suitable charge transporting host matrix due to their typically poor charge carrier mobility [9]. Host materials with appropriate energy levels and triplet energies, balanced charge carrier transport, as well as high thermal stability and morphological stability of their layers, can enable excellent PHOLED performance [10–12].

Many low-molar-mass electroactive compounds were synthesized for highly efficient OLEDs [13]. The advantages of low-molar-mass compounds are their precise chemical structure, and high purity [14]. Among the wide-bandgap materials, carbazole derivatives are among the most widely studied due to their good hole-transporting properties, high triplet energy and rigid molecular framework [15]. Carbazole ring can be easily modified by attaching various electron-accepting units, such as imidazole [16]. Derivatives of carbazole with benzimidazole [12,17] or -phenanthro [9,10-d]imidazole [18] substituents demonstrate suitable triplet energies, efficient bipolar charge transport and relatively high photoluminescence quantum efficiencies. Blue OLEDs, based on such emitters, demonstrate external quantum efficiency, reaching 3% [17,18]. Blue emitters can sometimes be utilized also as hosts to generate green or white light by energy transfer to an emissive compound [19–21]. Donor–acceptor derivatives of imidazole containing diphenylimidazole [22,23], pyrene-imidazole [24–26], phenanthroimidazole [26–34], benzimidazole [35,36], and theobromine [37] exhibit specific localization of molecular orbitals due to highly twisted molecular geometry and conformations. Consequently, these compounds can form a hybridized local charge transfer state (HLCT) [22–37] that can be exploited for practical electroluminescent applications, including blue-emitting OLEDs. Recently, there has been a great deal of interest in the use of HLCT compounds including imidazole derivatives [34,38] as hosts in systems utilizing prompt fluorescent [34], TADF [39], phosphorescent [38,39] and HLCT [40] emitters. Such an approach results in triplet harvesting via host similarly to the principle of hyperfluorescence involving TADF host matrices [41,42].

The simple and efficient synthesis of materials intended for OLEDs to a great extent contributes to the cost effectiveness of the devices [43]. Carbazole and imidazole derivatives correspond to the criterion of cost effectiveness. Carbazole derivatives are susceptible to various synthetic methods, including nucleophilic substitution, Suzuki, and Buchwald–Hartwig coupling [44]. Imidazole derivatives can be easily obtained by one-pot cyclization reaction from readily available starting compounds [45,46].

In this work, we present multifunctional derivatives of carbazole and imidazole obtained by simple one-pot condensation reaction. The appropriate properties of the obtained compounds enable them to be used as emitters for doping-free blue OLEDs and as hosts for green, red and sky-blue PHOLEDs, avoiding the photochemical oxidation of the imidazole ring by the device fabrication in an inert atmosphere.

2. Materials and Methods

The 9*H*-carbazole (95%, Reachem Slovakia, Bratislava, Slovakia), benzil (98%, Aldrich, St. Louis, MO, USA), 9-ethyl-9*H*-carbazole-3 carbaldehyde (98%, Aldrich), 9-ethyl-9*H*-carbazol-3-amine (98%, Aldrich), ammonium acetate (98%, Fluka, Dresden, Germany), acetic acid (99%, Reachem), anhydrous sodium sulfate (99%, Reachem), *tert*-butyl chloride (99%, Aldrich), aluminum(III) chloride (99%, Aldrich), 2-ethylhexyl bromide (95%, Aldrich), and tin(II) chloride (98%, Aldrich) were purchased as reagent grade chemicals and used as received.

3,3′-(4,5-Diphenyl-1*H*-imidazole-1,2-diyl)bis(9-ethyl-9*H*-carbazole) (**1**)

A mixture of benzil (1 g, 4.8 mmol), 9-ethyl-9*H*-carbazole-3-carbaldehyde (1.1 g, 4.8 mmol), 9-ethyl-9*H*-carbazol-3-amine (1.2 g, 5.7 mmol) and ammonium acetate (3.7 g, 47.6 mmol) in acetic acid (5 mL) was heated at the reflux temperature for 1 h. The mixture was poured into cold water and extracted with chloroform. The organic layer was dried with anhydrous sodium sulfate, filtered and distilled. The product was purified by column chromatography using a mixture of acetone and hexane at a volume ratio of 1:3 as an eluent. The compound **1** was crystallized from acetone. The yield of pale-yellow crystals was 2.38 g (82%), mp = 264–265 °C.

IR, cm^{-1}: 3055 (ar. C-H), 2978 (aliph. C-H), 1599, 1470, 1444 (ar. C=C), 1332, 1230 (C-N).

^{1}H NMR (400 MHz, CDCl$_3$-d, δ): 8.28 (d, J = 1.3 Hz, 1H, ar.), 7.84 (d, J = 7.8 Hz, 1H, ar.), 7.76–7.78 (m, 2H, ar.), 7.62 (d, J = 7.3 Hz, 2H, ar.), 7.37–7.44 (m, 2H, ar.), 7.30–7.33 (m 2H, ar.), 7.20–7.26 (m, 4H, ar.), 7.12–7.18 (m, 5H, ar.), 7.02–7.10 (m, 5H, ar.), 4.25 (q, J = 7.2 Hz, 2H, CH$_2$), 4.17 (q, J = 7.2 Hz, 2H, CH$_2$), 1.34 (t, J = 7.2 Hz, 3H, CH$_3$), 1.26 (t, J = 7.2 Hz, 3H, CH$_3$).

^{13}C NMR (101 MHz, CDCl$_3$-d, δ): 140.54, 140.26, 139.13, 131.24, 131.18, 128.29, 128.21, 127.74, 126.82, 126.36, 126.11, 125.74, 123.07, 122.70, 122.49, 120.81, 120.67, 120.60, 119.23, 119.03, 108.82, 108.55, 108.49, 107.95, 37.78, 37.58, 30.89, 13.79.

MS (APCl$^+$), m/z = 608 [M + H]$^+$.

6,6'-(4,5-Diphenyl-1H-imidazole-1,2-diyl)bis(3-(tert-butyl)-9-ethyl-9H-carbazole) (2)

A solution of compound 1 (1 g, 1.7 mmol) and aluminum(III) chloride (0.8 g, 6.6 mmol) in dichloromethane (10 mL) was cooled down using an ice bath. The solution of tert-butyl chloride (0.4 mL, 3.6 mmol) in dichloromethane (5 mL) was added dropwise to the reaction mixture. The mixture was stirred at room temperature for 1 h, then poured into water and extracted with chloroform. The organic layer was dried with anhydrous sodium sulfate, filtered and distilled. The product was purified by column chromatography using a mixture of acetone and hexane at a volume ratio of 1:6 as an eluent. The yield of pale-yellow solid was 0.82 g (69%).

IR, cm^{-1}: 3053 (ar. C-H), 2963 (aliph. C-H), 1604, 1481 (ar. C=C), 1301, 1232 (C-N).

^{1}H NMR (400 MHz, DMSO-d6, δ): 8.20–8.34 (m, 2H, ar.), 8.10–8.13 (m, 1H, ar.), 7.94–8.05 (m, 1H, ar.), 7.68–7.76 (m, 1H, ar.), 7.57–7.60 (m, 2H, ar.), 7.50–7.56 (m, 3H, ar.), 7.45–7.49 (m 2H, ar.), 7.33–7.42 (m, 4H, ar.), 7.16–7.31 (m, 6H, ar.), 4.25–4.43 (m, 4H, CH$_2$), 1.18–1.38 (m, 24H, CH$_3$).

^{13}C NMR (101 MHz, DMSO-d6, δ): 149.73, 148.06, 142.30, 141.90, 140.55, 139.59, 138.90, 138.74, 138.51, 136.92, 136.83, 135.43, 132.06, 132.03, 131.64, 131.48, 128.84, 128.59, 126.96, 126.70, 126.31, 125.89, 124.90, 124.83, 124.13, 123.04, 122.95, 122.17, 122.13, 121.88, 121.85, 121.80, 120.63, 120.19, 119.96, 119.79, 117.33, 116.04, 109.58, 109.43, 109.33, 109.03, 108.88, 106.02, 37.64, 37.48, 37.21, 35.48, 35.42, 34.86, 34.53, 32.17, 32.02, 14.27, 14.20, 14.15.

MS (APCl$^+$), m/z = 720 [M + H]$^+$.

3,3'-(1,1'-(9-(2-Ethylhexyl)-9H-carbazole-3,6-diyl)bis(4,5-diphenyl-1H-imidazole-2,1-diyl))bis(9-ethyl-9H-carbazole) (3)

Compound 3 was synthesized from benzil (1.35 g, 6.4 mmol), 9-ethyl-9H-carbazole-3-carbaldehyde (1.59 g, 7.1 mmol), diaminocarbazole 2 (1 g, 3.2 mmol) and ammonium acetate (4.98 g, 64.6 mmol) by the same procedure as compound 1. The product was purified by column chromatography using the mixture of acetone and hexane at a volume ratio of 1:4 as an eluent. The yield of white solid was 2.63 g (74%).

IR, cm^{-1}: 3054 (ar. C-H), 2932 (aliph. C-H), 1602, 1475 (ar. C=C), 1333, 1231 (C-N).

^{1}H NMR (400 MHz, CDCl$_3$-d, δ): 8.16 (s, 2H, ar.), 7.69 (d, J = 7.7 Hz, 2H, ar.), 7.60–7.50 (m, 6H, ar.), 7.43 (d, J = 8.5 Hz, 2H, ar.), 7.33 (t, J = 7.6 Hz, 2H, ar.), 7.26 (d, J = 8.2 Hz, 2H, ar.), 7.18 (t, J = 7.5 Hz, 6H, ar.), 7.14–7.07 (m, 7H, ar.), 7.07–7.00 (m, 7H, ar.), 6.89–7.00 (m, 4H, ar.), 4.10–4.24 (m, 4H, CH$_2$), 3.94–4.02 (m, 2H, CH$_2$), 1.84–1.91 (m, 1H, CH), 1.35–0.98 (m, 14H, CH$_2$, CH$_3$), 0.73 (t, J = 7.4 Hz, 3H, CH$_3$), 0.68 (t, J = 6.9 Hz, 3H, CH$_3$).

^{13}C NMR (101 MHz, CDCl$_3$-d, δ): 148.33, 140.50, 140.24, 139.62, 137.94, 134.87, 131.25, 131.16, 131.05, 129.32, 128.14, 127.50, 126.88, 126.77, 126.41, 125.73, 123.06, 122.64, 122.31, 121.63, 121.44, 120.66, 120.44, 119.01, 109.46, 108.51, 107.87, 57.27, 47.85, 39.16, 37.69, 30.91, 28.51, 24.37, 22.87, 13.90, 10.89.

MS (APCl$^+$), m/z = 1103 [M + H]$^+$.

6,6'-(1,1'-(9-(2-Ethylhexyl)-9H-carbazole-3,6-diyl)bis(4,5-diphenyl-1H-imidazole-2,1-diyl))bis(3-(tert-butyl)-9-ethyl-9H-carbazole) (4)

Compound 4 was synthesized from compound 3 (1 g, 0.9 mmol), AlCl$_3$ (0.48 g, 3.6 mmol) and tert-butyl chloride (0.22 mL, 2 mmol) by the same procedure as compound 2. The product was purified by column chromatography using the mixture of acetone and hexane at a volume ratio of 1:5 as an eluent. The yield of pale-yellow solid was 0.62 g (57%).

IR, cm^{-1}: 3052 (ar. C-H), 2958 (aliph. C-H), 1605, 1480 (ar. C=C), 1300, 1234 (C-N).

¹H NMR (400 MHz, CDCl₃-d, δ): 8.36 (s, 2H, ar.), 7.85 (s, 2H, ar.), 7.56 (d, J = 7.6 Hz, 4H, ar.), 7.50 (s, 2H, ar.), 7.42 (d, J = 8.5 Hz, 2H, ar.), 7.23–7.15 (m, 8H), 7.07–7.15 (m, 8H, ar.), 7.02–7.07 (m, 4H, ar.), 6.89–7.01 (m, 6H, ar.), 4.07–4.17 (m, 4H, CH2), 3.95–4.02 (m, 2H, CH₂), 1.83–1.92 (m, 1H, CH), 1.30 (s, 18H, CH₃), 1.27–1.11 (m, 11H, CH₂, CH₃), 0.99–1.09 (m, 3H, CH₃), 0.77 (t, J = 7.4 Hz, 3H, CH₃), 0.67 (t, J = 6.5 Hz, 3H, CH₃).

¹³C NMR (101 MHz, CDCl₃-d, δ): 148.53, 146.00, 142.10, 140.54, 139.93, 138.43, 138.01, 134.94, 133.67, 131.24, 131.14, 130.85, 129.39, 129.36, 128.14, 128.09, 127.50, 127.48, 126.40, 126.27, 123.98, 123.66, 123.17, 123.04, 122.99, 122.75, 122.68, 122.40, 121.68, 120.64, 116.79, 109.30, 107.99, 107.39, 38.97, 37.67, 34.38, 31.95, 24.58, 22.64, 13.90, 10.91.

MS (APCI⁺), m/z = 1216 [M + H]⁺.

3. Results and Discussions

3.1. Synthesis

The synthetic routes for preparation of derivatives of carbazole and diphenyl imidazole are presented in Scheme 1. The target compound **1** was synthesized using the commercial reagents. During the condensation reaction, a *tetra*-substituted imidazole ring was obtained. *Tert*-butyl groups were attached to carbazole moieties to obtain compound **2**. Compounds **3** and **4** were prepared by the same methods as **1** and **2**, respectively.

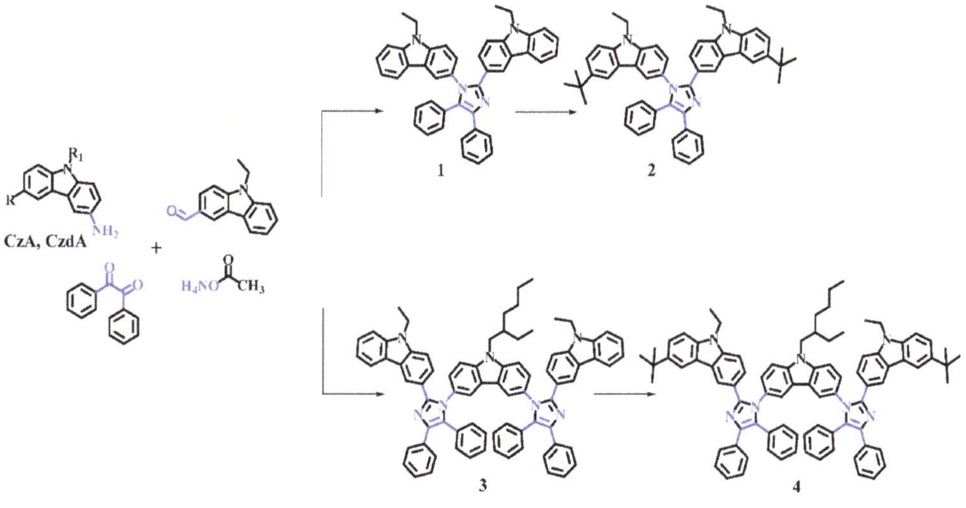

	CzA	CzdA
R	H	NH₂
R₁	ethyl	ethylhexyl

Scheme 1. Synthesis of derivatives of carbazole and diphenyl imidazole.

The yields of derivatives **1** and **3** were of 82% and 74%, respectively, whereas the Friedel–Crafts alkylations gave compounds **2** and **4** in 69% and 57% yields, respectively. All the compounds were found to be soluble in common organic solvents. Their structures were confirmed by ¹H NMR, ATR-IR, and mass spectrometry.

3.2. Thermal Properties

The behavior of compounds **1–4** under heating was studied by TGA and DSC. The obtained results are summarized in Table 1.

Table 1. Thermal and photophysical characteristics of compounds 1–4.

Compound	T_D, °C	T_m, °C	T_g, °C	T_{cr}, °C	λ_{ab}, nm	λ_{fl}, nm	E_T, eV
1	399	270	134	236	299	398	3.18
2	400	-	147	-	303	398	3.16
3	477	-	151	-	303	402	3.08
4	487	-	172	-	304	403	3.05

T_D—5% weight-loss temperature; T_m—melting point; T_g—glass transition temperature; T_{cr}—crystallization temperature, λ_{ab}—lowest energy band of absorption; λ_{fl}—wavelength of maximum intensity of fluorescence of the solutions in THF; E_T—triplet energy determined from phosphorescence spectrum.

The synthesized compounds showed high thermal stability confirmed by TGA. The 5% weight-loss temperatures (T_D) of the compounds with one imidazole ring (**1** and **2**) were found to be ca. 400 °C, while compounds with two imidazole units (**3** and **4**) showed TD higher than 477 °C (Figures 1a and S1a). Compound **1** was obtained as a crystalline substance. In the first DSC heating scan of **1** (Figure 1b), an endothermic melting signal was observed at 270 °C. The second heating scan revealed glass transition temperature (T_g) at 134 °C and further heating showed crystallization and melting signals. Compounds **2**, **3** and **4** were isolated after the synthesis and purification as amorphous substances with T_g of 146 °C, 151 °C and 172 °C, respectively, observed in the heating and cooling scans (Figure S1b–d). DSC results demonstrate that increased molecular weight by attachment additional carbazole–imidazole moiety or *tert*-butyl groups results in an increase of T_g.

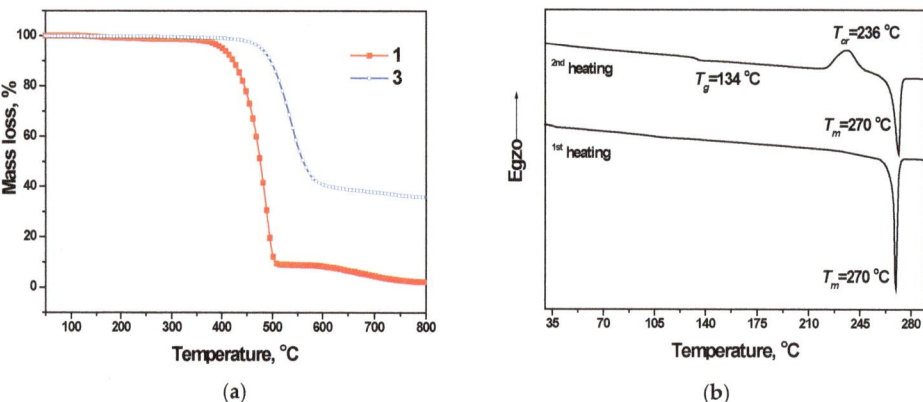

Figure 1. (a) TGA curves of compounds **1** and **3**; (b) DSC curves of compound **1**.

3.3. Photophysical Properties

Absorption spectra of dilute solutions of compounds **1–4** are presented in Figures 2a and S2a–S5a. Optical characteristics of the compounds are collected in Table 1. The dilute THF solutions of **1–4** absorb UV radiation up to 380 nm. The attachment of additional fragments of carbazole and diphenyl imidazole did not contribute to the extension of the systems of conjugated π-electrons (cf. UV spectra of THF solutions of **1** and **2** with those of the solutions of **3** and **4**). The investigation of absorption and emission spectra of **1–4** dissolved in solvents of different polarity was carried out (Figures S2–S5). No dependence of the wavelengths of absorption peaks on the polarity of solvents was observed (Figures S2a–S5a). THF solutions of **1–4** emitted blue light with the photoluminescence (PL) intensity maxima in the range of 398–403 nm (Figure 2a, Table 1).

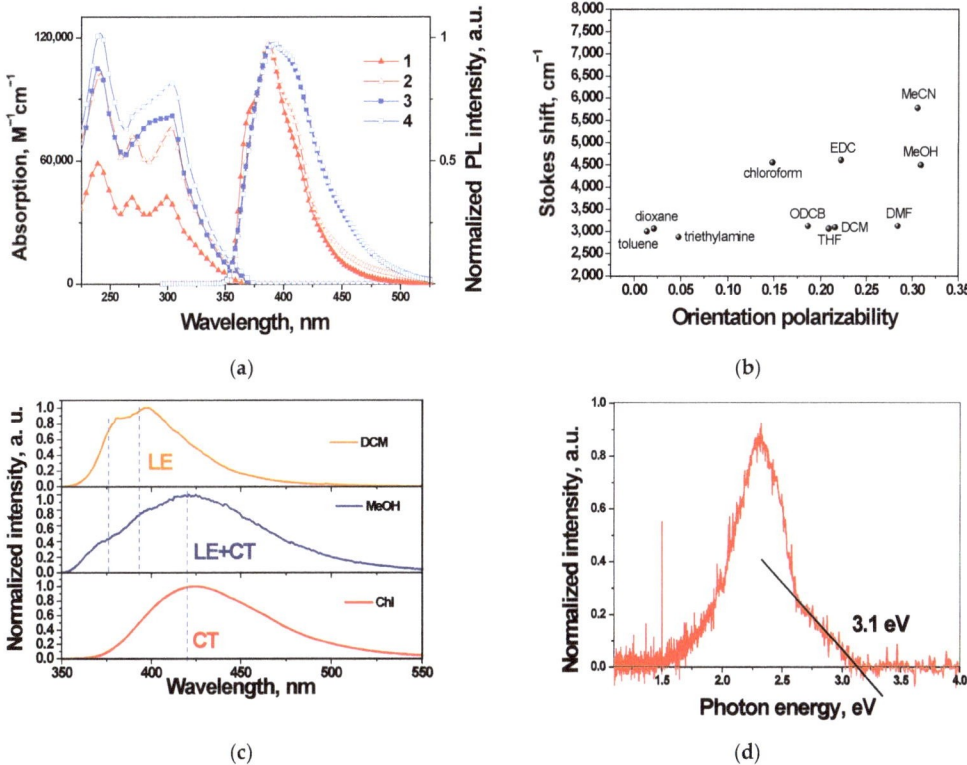

Figure 2. (**a**) Absorption and PL spectra of the dilute THF solutions of **1–4**; (**b**) the Lippert–Mataga plots for **2**; (**c**) PL spectra of the dilute DCM, MeOH and chloroform solutions of **2**; (**d**) phosphorescence spectrum of the dilute THF solution of compound **4** recorded at 77 K.

The aspecific electrostatic interactions between molecules of the solvent and the materials in the context of Onsager interpretation [47] is in the core of the Lippert–Mataga approach [48–50] used for the description of the solvatochromic effect. The Lippert–Mataga plots were built for solutions of **1–4**, which demonstrated the correlation of the Stokes shift and orientation polarizability (Figures 2b and S6a–c). It was established that the Lippert–Mataga plots cannot be sufficiently fitted linearly for the solutions of **1–4** with appropriate value of errors. For example, the toluene and THF solutions of **1–4** exhibited local excited (LE) state emission solely despite of the different polarity of the solvents represented by the different values of orientation polarizability. Additionally, the bathochromic shift of charge transfer (CT) bands of **1–4** appeared with the increase in polarity of the solvents without a strict order. These observations point to the fact that the Lippert–Mataga approach is not applicable for the description of compounds **1–4** in terms of dipole moment change [51]. The interaction of the solvent and the solute is ongoing in an individual manner for the particular solvent, resulting in the different distribution of LE and CT components. Apparently, the emission of **1–4** in solutions is affected by HLCT, which is characterized as combined transition of the LE and CT states. The observed LE, CT and mixed LE and CT states of the solutions of compounds **1–4** in different solvents may originate from the twisting of the angles between the carbazole and imidazole moieties (Figure 2c) [45].

The wavelengths of intensity maxima of phosphorescence spectra of THF solutions of the compounds recorded at 77 K ranged from 533 nm to 537 nm (Figures 2d and S7a–c). The triplet energies (E_T) of the compounds were estimated from the onset of the phosphorescence spectra (Figure 2d). Compound **1** exhibited the highest E_T of 3.18 eV, while compounds **2**, **3** and **4** exhibited slightly smaller triplet energy levels 3.16 eV, 3.08 eV and 3.10 eV, respectively (Table 1).

The emission spectra of the layers of compounds **1–4** had two peaks at ca. 400 nm and a wide red-shifted emission peak at ~550 nm (Figure 3a). It is known that an imidazole ring in the presence of oxygen under UV-irradiation is subjected to the ring-opening reaction, which results in the formation of new materials [52,53]. The PL spectra of 1 wt.% solid solution of compound **1** in ZEONEX 480 were recorded after different periods of UV irradiation. They showed consistent increase in intensity of emission band, peaking at 550 nm (Figure 3b). The small concentration of compound **1** in ZEONEX 480 ensures absence of intermolecular interactions as well as formation of intermolecular excimers. PL decay curves of the layers of **1–4** recorded at the different wavelengths of ca. 410 nm and ca. 550 nm exhibited prompt fluorescence (Figure S8). The observed difference between PL lifetimes demonstrates the different origin of emission. Thus, PL decay curves recorded at 550 nm have a biexponential character. The slightly longer-lived component of the prompt fluorescence is attributed to the oxidized product. Thus, these experiments prove the photochemical process occurring in the layers of compounds **1–4** [54]. The photochemical process was also detected for the THF solutions of the compounds (Figure S9a–c). It was manifested by substantial decrease of intensity and the redshift of the peak after continuous UV irradiation.

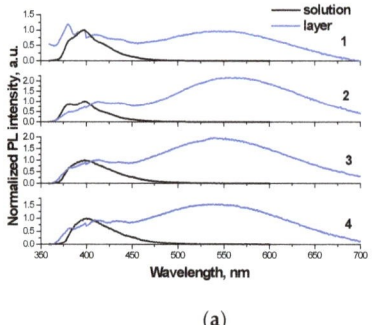
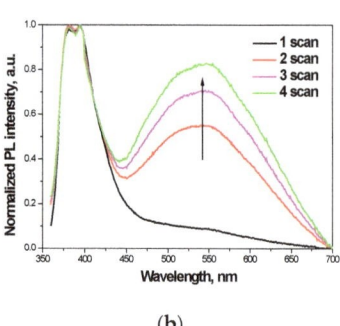

(a) (b)

Figure 3. (a) PL of the dilute THF solutions and the layers of **1–4**; (b) PL spectra of the solid solution of **1** in ZEONEX 480 after UV-irradiation for 10 min after every measurement.

3.4. Electrochemical and Photoelectrical Properties

The values of ionization potential (IP) estimated by cyclic voltammetry (CV) and UV photoelectron (PE) spectroscopy are presented in Table 2. The IP_{CV} values of the compounds were estimated from the half-wave potential of the first oxidation relative to ferrocene (Figure S10a–d). The electron affinity (EA_{CV}) values were obtained from the IP_{CV} values and the optical bandgaps, which were deduced from the edges of the absorption spectra of the dilute THF solutions of **1–4**. The IP_{CV} and EA_{CV} values of **1–4** are in the small ranges of 5.21–5.26 eV and 1.81–1.94 eV, respectively. The comparable values of the IP_{CV} and EA_{CV} observed for all the studied compounds demonstrate the same oxidation and reduction sites.

Table 2. Electrochemical and photoelectrical characteristics of compounds **1–4**.

Compound	IP_{CV} [1], eV	EA_{CV}, eV	IP_{PE} [2], eV	μ_h, cm^2/V·s [3]	μ_e, cm^2/V·s [3]
1	5.25	1.81	5.35	2.9×10^{-4}	-
2	5.21	1.87	5.18	3.8×10^{-5}	-
3	5.26	1.92	5.29	1.1×10^{-4}	-
4	5.25	1.94	5.22	6.3×10^{-5}	1.2×10^{-4}

[1] Determined by cyclic voltammetry; [2] measured by UV photoelectron spectrometry in air; [3] charge mobility at 6.4×10^5 V/cm.

The photoelectron emission spectra of the solid samples of **1–4** are presented in Figure 4a. The IP_{PE} values of the compounds are in the slightly larger range of 5.18–5.35 eV comparing with that of IP_{CV}. This observation can be explained by the different environments in solutions and solid layers of the compounds. IP_{CV} and IP_{PE} values of compounds **2** and **4**, bearing *tert*-butyl groups, are slightly lower than those of their counterparts **1** and **3**, respectively. This trend can be explained by the σ donor effect of *tert*-butyl groups [55] which reduce the ionization potential values.

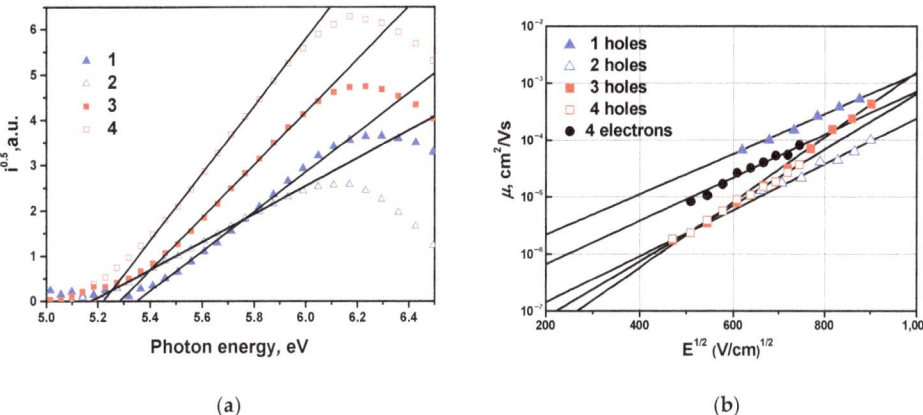

Figure 4. (a) Photoelectron emission spectra of the layers of compounds **1–4**; (b) electric field dependence of hole and electron mobilities for the layers of compounds **1–4**.

The TOF technique was used to study charge transporting properties of the compounds. TOF current transients for holes and electrons for vacuum deposited films of the compounds **1–4** were recorded at different electric fields (Figure S11). When the transit times were well recognized from the TOF current transients in log–log scales, charge mobilities were calculated. The electric field dependence values of hole and electron mobilities of the layers of compounds **1–4** are shown in Figure 4b. The layers of compounds demonstrated hole drift mobilities ranging from 10^{-5} cm^2/V·s to 10^{-4} cm^2/V·s at electric field of 6.4×10^5 V/cm (Table 2). Electron mobility was detected only in the layer of compound **4**, which reached 10^{-4} cm^2/V·s at high electric fields (>8.1×10^5 V/cm). Thus, the difference between hole and electron mobilities in the layer of compound **4** is only one order of magnitude. To achieve high efficiency of PHOLED, host materials with a balanced charge transport are required [56]. This allows generation of a broad charge recombination zone in the emissive layer.

3.5. Performance in Fluorescent OLEDs

To investigate electroluminescent properties of compounds **1–4** as materials for doping-free light-emitting layers (EML), OLEDs A–D were fabricated using device structure ITO/MoO$_3$ (1 nm)/NPB (55 nm)/EML (40 nm)/TSPO1 (5 nm)/TPBi (65 nm)/LiF (0.5 nm)/Al. A diagram of energy levels of devices A–D is shown in Figure 5a. The electroluminescent characteristics are summarized in Table 3.

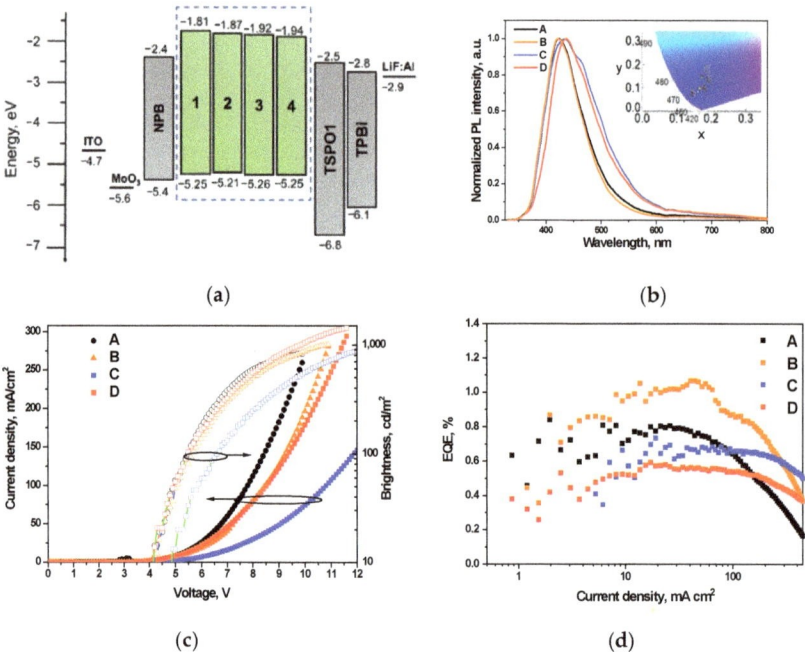

Figure 5. Diagram of energy levels of OLEDs A–D (**a**); electroluminescence spectra of OLEDs A–D recorded at applied voltage of 10 V; inset shows CIE 1931 color diagram with the corresponding CIE color coordinates (**b**); current density and luminance versus voltage (**c**); external quantum efficiency versus current density (**d**). Arrow denotes the correspondence of data to axis.

Table 3. Electroluminescent characteristics of devices A–M.

OLED	Emitting Layer (EML)	λ_{em}, nm	CIE, xy	V_{on}, V	Max. Brightness, cd/m^2	Max. Current Efficiency, cd/A	Max. Power Efficiency, lm/W	Max. EQE, %
\multicolumn{9}{c}{Doping-free blue fluorescent devices ITO/MoO$_3$/NPB/EML/TSPO1/TPBi/LiF/Al}								
A	**1**	420	(0.17, 0.09)	4.0	840	0.7	0.38	0.9
B	**2**	418	(0.16, 0.08)	4.1	1030	0.8	0.42	1.1
C	**3**	431	(0.17, 0.15)	4.8	1350	0.9	0.43	0.8
D	**4**	437	(0.17, 0.14)	4.0	1430	0.8	0.41	0.6
\multicolumn{9}{c}{Green phosphorescent devices ITO/m-MTDATA/EML/Bphen/LiF/Al}								
E	Ir(ppy)$_3$:**1**	422	(0.21, 0.17)	5.2	5580	2.7	1.1	2.5
F	Ir(ppy)$_3$:**2**	510	(0.27, 0.58)	3.7	4630	3.5	1.6	0.97
G	Ir(ppy)$_3$:**3**	510	(0.29, 0.60)	3.1	24,600	31.1	13.7	8.3
H	Ir(ppy)$_3$:**4**	510	(0.25, 0.39)	3.9	5800	1.9	1.1	0.74
\multicolumn{9}{c}{Red phosphorescent devices ITO/MoO$_3$/NPB/EML/TSPO1/TPBi/LiF/Al}								
I	(piq)$_2$Ir(acac):**1**	625	(0.56, 0.29)	4.4	5700	2.7	0.9	3.0
J	(piq)$_2$Ir(acac):**2**	625	(0.59, 0.29)	4.0	4800	2.6	1.4	2.7
K	(piq)$_2$Ir(acac):**3**	632	(0.67, 0.32)	3.2	8400	4.4	3.3	6.4
L	(piq)$_2$Ir(acac):**4**	629	(0.63, 0.31)	3.3	4700	3.9	1.7	4.5
\multicolumn{9}{c}{Sky-blue phosphorescent device ITO/HAT-CN/NPB/EML/TSPO1/TPBi/LiF/Al}								
M	FIrpic:**3**	475	(0.15, 0.36)	3.2	3300	17.9	17.6	7.6

In OLEDs, the layers of MoO$_3$ and LiF were used as the charge injections layers. The layer of N,N'-di(1-naphthyl)-N,N'-diphenyl-(1,1'-biphenyl)-4,4'-diamine acted as the hole-transporting layer. The layers of diphenyl [4-(triphenylsilyl)phenyl]phosphine oxide (TSPO1) and 2,2',2''-(1,3,5-benzinetriyl)-tris(1-phenyl-1-H-benzimidazole) (TPBi) were utilized as the hole/exciton blocking and electron transporting layers, respectively.

Electroluminescence (EL) spectra of OLEDS A–D are shown in Figures 5b and S12. Electroluminescence intensity maxima of devices A–D were observed in the range of 418–437 nm. According to the CIE color coordinates, A, B and C, D emitted deep-blue and blue light, respectively (Figure 5b inset, Table 3). The shapes and positions of the peaks of electroluminescence spectra of the devices were very similar to those of the fluorescence spectra of the solutions of compounds **1–4** in THF (Figures 2a and 5b). The shapes of EL spectra were practically the same at different applied voltages (Figure S12). The emission peaks of devices C and D were wider and slightly red-shifted compared to those of devices A and B. It was also observed that during operation of the devices at various electrical voltages, no additional emission peak appeared in the longer wavelength region.

All OLEDs showed relatively low turn-on voltages (V_{on}) of 4.0–4.8 V (Figure 5c). This observation confirms that injection and transport of holes and electrons towards the emission layer was efficient. Device D showed maximum brightness of 1430 cd/m^2 and minimum turn-on voltage of 4.0 V. Device B, with emissive layer of emitter **2**, demonstrated the best electroluminescence properties (Figures 5b,c and S13). Maximum current efficiency (0.8 cd/A) and maximum energy efficiency (0.42 lm/W) of device B were similar to those of devices A, C and D. However, the maximum external quantum efficiency (EQE) of device B was the highest one (1.1%) (Figure 5d). This observation can be attributed to the narrowest EL spectrum of device B (Figure 5b).

The obtained results show that the synthesized derivatives of carbazole and diphenyl imidazole (**1–4**) are suitable for the formation of functional layers of OLED. The photooxidation process of the imidazole ring can be successfully suppressed by fabrication of the devices in an inert atmosphere. This is confirmed by the absence of the additional low-energy peaks in the electroluminescence spectra.

3.6. Performance in Phosphorescent OLEDs

To study the performance of compounds **1–4** as hosts, the simple PHOLEDs E-H and I-L were fabricated using green or red phosphorescent emitters, i.e., tris [2-phenylpyridinato-C2,N]iridium(III) (Ir(ppy)$_3$) or bis [2-(1-isoquinolinyl-N)phenyl-C](2,4-pentanedionato-O2,O4)iridium(III) ((piq)$_2$Ir(acac)), respectively. The concentration of emitters in the hosts was of 10 wt.%. The device structures and energy levels of the materials used in the devices are schematically shown in Figures 6a and S14. The electroluminescence and efficiency characteristics of the devices E–H are given in Table 3 and Figures 6 and S15.

Despite the exploitation of the green phosphorescent emitter Ir(ppy)$_3$, most of the PHOLEDs of the series E–H were not characterized by green emission (Figure S16). Low-intensity violet emission bands were observed in EL spectra of PHOLEDs E–H. The wavelengths of these emission bands were close to those of fluorescence bands of pure **1–4**. Thus, the recombination of excitons occurred not only in the emitter but also in hosts. This observation can be attributed to inefficient host–guest energy transfer. Device G containing compound **3** as a host was characterized by green emission with the intensity maximum at ca. 510 nm, confirming the radiative recombination of excitons mainly on Ir(ppy)$_3$. Thus, device G was characterized by the most efficient host–guest energy transfer. PHOLEDs E–H were characterized by the relatively low values of V_{on} (3.1–5.2 V), confirming the very efficient injection from the electrodes and transport of holes and electrons to the emitting layer. The highest brightness of 24,600 cd/m^2 and the lowest V_{on} of 3.1 V observed for device G compared to those recorded for the other devices shows more effective exciton recombination and radiative transition in the emitting layer with host **3**. Thus, compound **3** can be regarded as the effective host for PHOLEDs. Device G with host **3** exhibited the maximum current, power, and external quantum efficiencies of 31 cd/A, 13.7 lm/W,

8.3%, respectively, in the absence of light outcoupling enhancement. It has to be noted that the simple and unoptimized PHOLED based on host **3** was fabricated. Therefore, the maximum efficiency of device was lower compared to those reported for other PHOLEDs containing Ir(ppy)$_3$ [57].

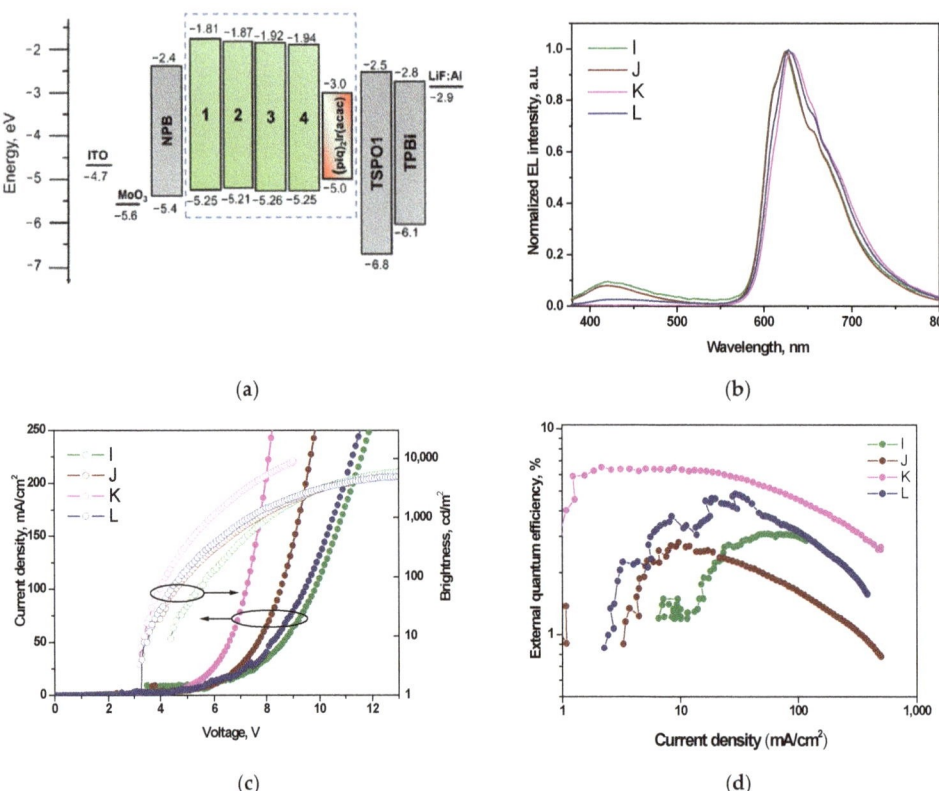

Figure 6. Diagram of energy levels of OLEDs I–L (**a**); electroluminescence spectra of OLEDs I–L recorded at applied voltage of 10 V (**b**); current density and luminance versus voltage plots (**c**); external quantum efficiency versus brightness plots (**d**). Arrow denotes the correspondence of data to axis.

Similarly, the highest maximum EQE of 6.4% was obtained for the red PhOLED (device K) based on host **3**. In contrast to other red PhOLEDs fabricated in this work, emission from the host was not observed in EL spectra of device K recorded at the different voltages. This observation demonstrates perfect host–guest energy transfer from **3** to (piq)$_2$Ir(acac) (Figures 6a and S17). Red PHOLEDs I–K also showed low turn-on voltages of 3.2–4.4 V (Figure 6c, Table 3). Maximum brightness of 8400 cd/m^2 was observed for device K. This device exhibited maximum quantum efficiency of 6.4 % (Figure 6d).

Taking into account that **3** showed the best performance among the compounds of the series, PHOLED M was fabricated with the same structure as that of the red PHOLEDs except the layer of hexaazatriphenylenehexacarbonitrile (HAT-CN) and the layer of FIrpic doped in **3** as an emitting layer (Figure 7). The V_{on} of 3.2 V (Figure 7c) and appropriate spectral properties show complete electronic excitation energy transfer from **3** to FIrpic. Device M with a sky-blue EL reached of maximum EQE of 7.6% (Figure 7d).

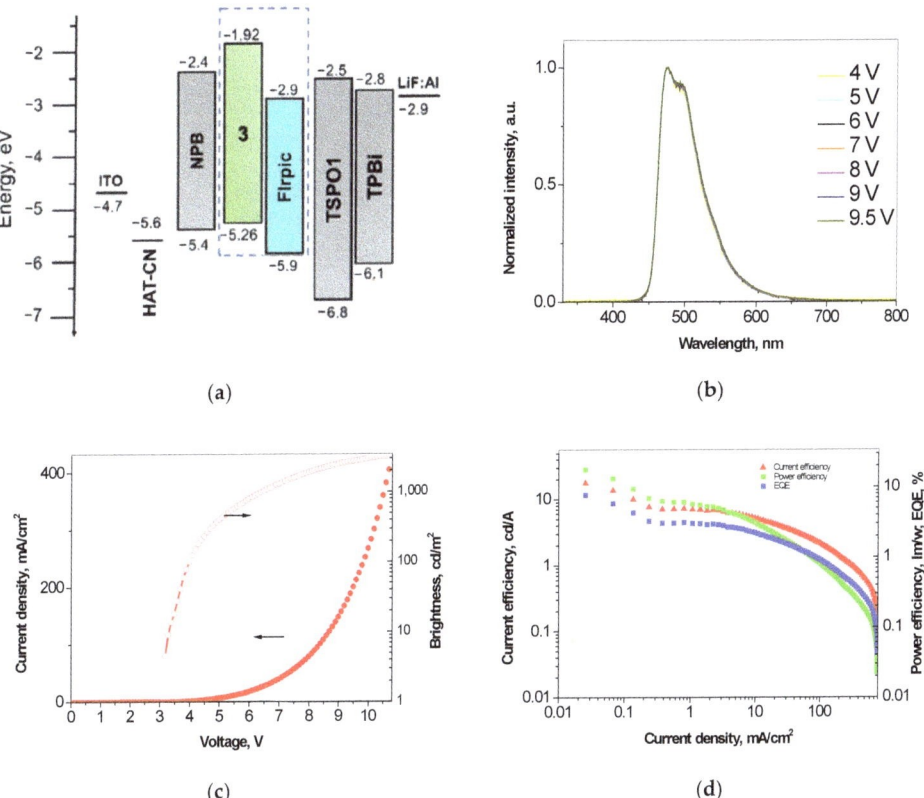

Figure 7. Diagram of energy levels (**a**); electroluminescence spectra of OLEDs recorded at different applied voltage (**b**); current density and luminance versus voltage plots (**c**); current efficiency, power efficiency, external quantum efficiency versus brightness plots (**d**) of PHOLED M. Arrow denotes the correspondence of data to axis.

4. Conclusions

Four new derivatives of carbazole and diphenyl imidazole were synthesized and characterized. Optical, photophysical, electrochemical, and photoelectric properties of the compounds were found to be similar, indicating that attachment of the additional carbazole–diphenyl imidazole fragment has no significant impact on these properties. Compounds with two imidazole rings demonstrated higher thermal stability and glass transition temperatures compared with those of the derivatives having one imidazole ring. OLEDs fabricated using the synthesized compounds as emitters showed deep-blue and blue emissions with maximum quantum efficiency ranging from 0.6% to 1.1%. The suitable values of energy levels, triplet energies and charge carrier mobilities allowed the use of the synthesized compounds as hosts for green and red phosphorescent OLEDs. The devices showed sky-blue, green or red emissions, confirming the radiative recombination of excitons in phosphorescent emitters. Maximum quantum efficiencies of 7.6%, 8.3%, and 6.4%, respectively, were observed without optimization of the structures.

Supplementary Materials: The following supporting information can be downloaded at: https://www.mdpi.com/article/10.3390/ma15238495/s1. Figure S1. (a) TGA curves of compounds **2** and **4**; (b–d) DSC curves of compounds **2**–**4**. Figure S2. Absorption (a) and PL (b) spectra of **1** solutions. Figure S3. Absorption (a) and PL (b) spectra of **2** solutions. Figure S4. Absorption (a) and

PL (b) spectra of **3** solutions. Figure S5. Absorption (a) and PL (b) spectra of **4** solutions. Figure S6. Lippert–Mataga plots for **1** (a), **3** (b) and **4** (c). Figure S7. (a–c) Phosphorescence spectrum of dilute solutions of **1**–**3** in THF at 77 K. Figure S8. PL decay curves of solutions of **1**–**4** in THF. Figure S9. PL spectra of THF solutions of **1** (a) and **3** (b), normalized PL spectra of **3** (c) before and after 40 min photoexcitation by the UV light. Figure S10. (a–d) Cyclic voltammograms of compounds **1**–**4**. Figure S11. TOF signals for holes/electrons in vacuum-deposited layers of compounds **1**–**4** at different electric fields. Figure S12. EL spectra of non-doped devices A–D at different voltages. Figure S13. Electroluminescence characteristics of OLEDs A–D: (a) current efficiency versus current density, (b) power efficiency versus current density. Figure S14. Diagram of energy levels of green PHOLEDs. 4,4′,4′′-tris[phenyl(m-tolyl)amino]triphenylamine (m-MTDATA) and bathophenanthroline (Bphen) were used as hole and electron transporting layers, respectively. Figure S15. Electroluminescence characteristics of PHOLEDs E–H: (a) current density and luminance versus voltage, (b) current efficiency versus current density, (c) power efficiency versus current density, (d) external quantum efficiency versus current density. Figure S16. Electroluminescence spectra of PHOLEDs E–H recorded at applied voltage of 10 V. Figure S17. EL spectra of devices I–L at different voltages. Refs. [58,59] are cited in supplementary materials.

Author Contributions: Conceptualization, O.B., D.V. and V.A.; methodology, O.B., D.V. and V.A.; validation, R.S.B., O.B., R.K. and J.V.G.; formal analysis, D.V., V.A., R.K. and J.V.G.; investigation, O.B., R.S.B., D.V., R.K., K.V. and L.L.; resources, O.B., D.V., L.L. and J.V.G.; data curation, D.V. and V.A.; writing—original draft preparation, O.B., D.V., V.A. and J.V.G.; visualization, O.B. and D.V.; supervision, D.V., V.A. and J.V.G. All authors have read and agreed to the published version of the manuscript.

Funding: This research was funded by the European Social Fund under the 09.3.3-LMT-K-712 "Development of Competences of Scientists, other Researchers and Students through Practical Research Activities" measure (project 09.3.3-LMT-K-712-19-0136).

Institutional Review Board Statement: Not applicable.

Informed Consent Statement: Not applicable.

Data Availability Statement: All experimental data to support the findings of this study are available upon request by contacting the corresponding authors.

Acknowledgments: This research was supported by Lietuvos mokslo taryba (LMT, Research Council of Lithuania) and the European Social Fund. Ervinas Urbonas is thanked for the help in measurements of the photoluminescence spectra of solutions.

Conflicts of Interest: The authors declare no conflict of interest.

References

1. Ye, Z.; Ling, Z.; Chen, M.; Yang, J.; Wang, S.; Zheng, Y.; Wei, B.; Li, C.; Chen, G.; Shi, Y. Low Energy Consumption Phosphorescent Organic Light-Emitting Diodes Using Phenyl Anthracenone Derivatives as the Host Featuring Bipolar and Thermally Activated Delayed Fluorescence. *RSC Adv.* **2019**, *9*, 6881–6889. [CrossRef]
2. Shirota, Y.; Kageyama, H. Charge Carrier Transporting Molecular Materials and Their Applications in Devices. *Chem. Rev.* **2007**, *107*, 953–1010. [CrossRef] [PubMed]
3. Volz, D.; Wallesch, M.; Fléchon, C.; Danz, M.; Verma, A.; Navarro, J.M.; Zink, D.M.; Bräse, S.; Baumann, T. From Iridium and Platinum to Copper and Carbon: New Avenues for More Sustainability in Organic Light-Emitting Diodes. *Green Chem.* **2015**, *17*, 1988–2011. [CrossRef]
4. Zhu, M.; Yang, C. Blue Fluorescent Emitters: Design Tactics and Applications in Organic Light-Emitting Diodes. *Chem. Soc. Rev.* **2013**, *42*, 4963. [CrossRef] [PubMed]
5. Jou, J.H.; Kumar, S.; Agrawal, A.; Li, T.H.; Sahoo, S. Approaches for Fabricating High Efficiency Organic Light Emitting Diodes. *J. Mater. Chem. C* **2015**, *3*, 2974–3002. [CrossRef]
6. Xu, Z.; Tang, B.Z.; Wang, Y.; Ma, D. Recent Advances in High Performance Blue Organic Light-Emitting Diodes Based on Fluorescence Emitters. *J. Mater. Chem. C* **2020**, *8*, 2614–2642. [CrossRef]
7. Yook, K.S.; Lee, J.Y. Organic Materials for Deep Blue Phosphorescent Organic Light-Emitting Diodes. *Adv. Mater.* **2012**, *24*, 3169–3190. [CrossRef]
8. Kappaun, S.; Slugovc, C.; List, E.J.W. Phosphorescent Organic Light-Emitting Devices: Working Principle and Iridium Based Emitter Materials. *Int. J. Mol. Sci.* **2008**, *9*, 1527–1547. [CrossRef]

9. Tao, Y.; Yang, C.; Qin, J. Organic Host Materials for Phosphorescent Organic Light-Emitting Diodes. *Chem. Soc. Rev.* **2011**, *40*, 2943–2970. [CrossRef]
10. Li, Q.; Cui, L.S.; Zhong, C.; Jiang, Z.Q.; Liao, L.S. Asymmetric Design of Bipolar Host Materials with Novel 1,2,4-Oxadiazole Unit in Blue Phosphorescent Device. *Org. Lett.* **2014**, *16*, 1622–1625. [CrossRef] [PubMed]
11. Liu, X.K.; Zheng, C.J.; Xiao, J.; Ye, J.; Liu, C.L.; Wang, S.D.; Zhao, W.M.; Zhang, X.H. Novel Bipolar Host Materials Based on 1,3,5-Triazine Derivatives for Highly Efficient Phosphorescent OLEDs with Extremely Low Efficiency Roll-Off. *Phys. Chem. Chem. Phys.* **2012**, *14*, 14255–14261. [CrossRef] [PubMed]
12. Ban, X.; Jiang, W.; Sun, K.; Xie, X.; Peng, L.; Dong, H.; Sun, Y.; Huang, B.; Duan, L.; Qiu, Y. Bipolar Host with Multielectron Transport Benzimidazole Units for Low Operating Voltage and High Power Efficiency Solution-Processed Phosphorescent OLEDs. *ACS Appl. Mater. Interfaces* **2015**, *7*, 7303–7314. [CrossRef]
13. Krucaite, G.; Grigalevicius, S. A Review on Low-Molar-Mass Carbazole- Based Derivatives for Organic Light Emitting Diodes. *Synth. Met.* **2019**, *247*, 105422. [CrossRef]
14. Yang, X.; Xu, X.; Zhou, G. Recent Advances of the Emitters for High Performance Deep-Blue Organic Light-Emitting Diodes. *J. Mater. Chem. C* **2015**, *3*, 913–944. [CrossRef]
15. Nishimoto, T.; Yasuda, T.; Lee, S.Y.; Kondo, R.; Adachi, C. A Six-Carbazole-Decorated Cyclophosphazene as a Host with High Triplet Energy to Realize Efficient Delayed-Fluorescence OLEDs. *Mater. Horizons* **2014**, *1*, 264–269. [CrossRef]
16. Kulkarni, A.P.; Tonzola, C.J.; Babel, A.; Jenekhe, S.A. Electron Transport Materials for Organic Light-Emitting Diodes. *Chem. Mater.* **2004**, *16*, 4556–4573. [CrossRef]
17. Hung, W.Y.; Chi, L.C.; Chen, W.J.; Chen, Y.M.; Chou, S.H.; Wong, K.T. A New Benzimidazole/Carbazole Hybrid Bipolar Material for Highly Efficient Deep-Blue Electrofluorescence, Yellow-Green Electrophosphorescence, and Two-Color-Based White OLEDs. *J. Mater. Chem.* **2010**, *20*, 10113–10119. [CrossRef]
18. Gao, Z.; Wang, Z.; Shan, T.; Liu, Y.; Shen, F.; Pan, Y.; Zhang, H.; He, X.; Lu, P.; Yang, B.; et al. High-Efficiency Deep Blue Fluorescent Emitters Based on Phenanthro [9,10-d]Imidazole Substituted Carbazole and Their Applications in Organic Light Emitting Diodes. *Org. Electron.* **2014**, *15*, 2667–2676. [CrossRef]
19. Ban, X.; Jiang, W.; Sun, K.; Yang, H.; Miao, Y.; Yang, F.; Sun, Y.; Huang, B.; Duan, L. Systematically Tuning the ΔeST and Charge Balance Property of Bipolar Hosts for Low Operating Voltage and High Power Efficiency Solution-Processed Electrophosphorescent Devices. *J. Mater. Chem. C* **2015**, *3*, 5004–5016. [CrossRef]
20. Chen, Y.M.; Hung, W.Y.; You, H.W.; Chaskar, A.; Ting, H.C.; Chen, H.F.; Wong, K.T.; Liu, Y.H. Carbazole-Benzimidazole Hybrid Bipolar Host Materials for Highly Efficient Green and Blue Phosphorescent OLEDs. *J. Mater. Chem.* **2011**, *21*, 14971–14978. [CrossRef]
21. Hung, W.Y.; Chi, L.C.; Chen, W.J.; Mondal, E.; Chou, S.H.; Wong, K.T.; Chi, Y. A Carbazole-Phenylbenzimidazole Hybrid Bipolar Universal Host for High Efficiency RGB and White PhOLEDs with High Chromatic Stability. *J. Mater. Chem.* **2011**, *21*, 19249–19256. [CrossRef]
22. Wang, P.; Fan, S.; Liang, J.; Ying, L.; You, J.; Wang, S.; Li, X. Carbazole-Diphenylimidazole Based Bipolar Material and Its Application in Blue, Green and Red Single Layer OLEDs by Solution Processing. *Dye. Pigment.* **2017**, *142*, 175–182. [CrossRef]
23. Jayabharathi, J.; Goperundevi, G.; Thanikachalam, V.; Panimozhi, S. Regulation of Singlet and Triplet Excitons in a Single Emission Layer: Efficient Fluorescent/Phosphorescent Hybrid White Organic Light-Emitting Diodes. *ACS Omega* **2019**, *4*, 15030–15042. [CrossRef]
24. Liu, Y.; Yang, L.; Bai, Q.; Li, W.; Zhang, Y.; Fu, Y.; Ye, F. Highly Efficient Nondoped Blue Electroluminescence Based on Hybridized Local and Charge-Transfer Emitter Bearing Pyrene-Imidazole and Pyrene. *Chem. Eng. J.* **2021**, *420*, 129939. [CrossRef]
25. Liu, Y.; Liu, H.; Bai, Q.; Du, C.; Shang, A.; Jiang, D.; Tang, X.; Lu, P. Pyrene [4,5-d]Imidazole-Based Derivatives with Hybridized Local and Charge-Transfer State for Highly Efficient Blue and White Organic Light-Emitting Diodes with Low Efficiency Roll-Off. *ACS Appl. Mater. Interfaces* **2020**, *12*, 16715–16725. [CrossRef]
26. Chen, W.C.; Zhu, Z.L.; Lee, C.S. Organic Light-Emitting Diodes Based on Imidazole Semiconductors. *Adv. Opt. Mater.* **2018**, *6*, 1800258. [CrossRef]
27. Tian, X.; Sheng, J.; Zhang, S.; Xiao, S.; Gao, Y.; Liu, H.; Yang, B. A Novel Deep Blue Le-Dominated Hlct Excited State Design Strategy and Material for Oled. *Molecules* **2021**, *26*, 4560. [CrossRef]
28. Jayabharathi, J.; Anudeebhana, J.; Thanikachalam, V.; Sivaraj, S. Efficient Fluorescent OLEDS Based on Assistant Acceptor Modulated HLCT Emissive State for Enhancing Singlet Exciton Utilization. *RSC Adv.* **2020**, *10*, 8866–8879. [CrossRef]
29. Tagare, J.; Vaidyanathan, S. Recent Development of Phenanthroimidazole-Based Fluorophores for Blue Organic Light-Emitting Diodes (OLEDs): An Overview. *J. Mater. Chem. C* **2018**, *6*, 10138–10173. [CrossRef]
30. Yu, Y.; Zhao, R.; Liu, H.; Zhang, S.; Zhou, C.; Gao, Y.; Li, W.; Yang, B. Highly Efficient Deep-Blue Light-Emitting Material Based on V-Shaped Donor-Acceptor Triphenylamine-Phenanthro [9,10-d]Imidazole Molecule. *Dye. Pigment.* **2020**, *180*, 108511. [CrossRef]
31. Li, Z.; Xie, N.; Xu, Y.; Li, C.; Mu, X.; Wang, Y. Fluorine-Substituted Phenanthro [9,10-d]Imidazole Derivatives with Optimized Charge-Transfer Characteristics for Efficient Deep-Blue Emitters. *Org. Mater.* **2020**, *2*, 11–19. [CrossRef]
32. Zhang, S.; Li, W.; Yao, L.; Pan, Y.; Shen, F.; Xiao, R.; Yang, B.; Ma, Y. Enhanced Proportion of Radiative Excitons in Non-Doped Electro-Fluorescence Generated from an Imidazole Derivative with an Orthogonal Donor-Acceptor Structure. *Chem. Commun.* **2013**, *49*, 11302–11304. [CrossRef]

33. Cao, C.; Chen, W.C.; Tian, S.; Chen, J.X.; Wang, Z.Y.; Zheng, X.H.; Ding, C.W.; Li, J.H.; Zhu, J.J.; Zhu, Z.L.; et al. A Novel D-π-A Blue Fluorophore Based on [1,2,4]Triazolo [1,5-a] Pyridine as an Electron Acceptor and Its Application in Organic Light-Emitting Diodes. *Mater. Chem. Front.* **2019**, *3*, 1071–1079. [CrossRef]
34. Chen, L.; Zhang, S.; Li, H.; Chen, R.; Jin, L.; Yuan, K.; Li, H.; Lu, P.; Yang, B.; Huang, W. Breaking the Efficiency Limit of Fluorescent OLEDs by Hybridized Local and Charge-Transfer Host Materials. *J. Phys. Chem. Lett.* **2018**, *9*, 5240–5245. [CrossRef]
35. Gao, Y.H.; Chen, C.; Tang, Q.; Su, B.; Zhang, G.; Bo, B.X.; Jiang, W.L. Comparison Study of Two Isomers of Benzimidazole for Effective Blue OLEDs. *J. Mater. Sci. Mater. Electron.* **2017**, *28*, 7204–7211. [CrossRef]
36. Liu, Y.; Tao, T.; Hu, H.C.; Li, H.; Ouyang, X. Fine Regulation of Linker and Donor Moieties to Construct Benzimidazole-Based Blue Emitters for High-Efficient Organic Light-Emitting Diodes. *Dye. Pigment.* **2021**, *188*, 109191. [CrossRef]
37. Huang, Y.; Liu, Y.; Sommerville, P.J.W.; Kaminsky, W.; Ginger, D.S.; Luscombe, C.K. Theobromine and Direct Arylation: A Sustainable and Scalable Solution to Minimize Aggregation Caused Quenching. *Green Chem.* **2019**, *21*, 6600–6605. [CrossRef]
38. Ouyang, X.; Li, X.L.; Ai, L.; Mi, D.; Ge, Z.; Su, S.J. Novel "Hot Exciton" Blue Fluorophores for High Performance Fluorescent/Phosphorescent Hybrid White Organic Light-Emitting Diodes with Superhigh Phosphorescent Dopant Concentration and Improved Efficiency Roll-Off. *ACS Appl. Mater. Interfaces* **2015**, *7*, 7869–7877. [CrossRef]
39. Bucinskas, A.; Bezvikonnyi, O.; Gudeika, D.; Volyniuk, D.; Grazulevicius, J.V. Methoxycarbazolyl-Disubstituted Dibenzofurans as Holes- and Electrons-Transporting Hosts for Phosphorescent and TADF-Based OLEDs. *Dye. Pigment.* **2020**, *172*, 107781. [CrossRef]
40. Chen, X.; Ma, D.; Liu, T.; Chen, Z.; Yang, Z.; Zhao, J.; Yang, Z.; Zhang, Y.; Chi, Z. Hybridized Local and Charge-Transfer Excited-State Fluorophores through the Regulation of the Donor–Acceptor Torsional Angle for Highly Efficient Organic Light-Emitting Diodes. *CCS Chem.* **2021**, *4*, 1285–1295. [CrossRef]
41. Nakanotani, H.; Higuchi, T.; Furukawa, T.; Masui, K.; Morimoto, K.; Numata, M.; Tanaka, H.; Sagara, Y.; Yasuda, T.; Adachi, C. High-Efficiency Organic Light-Emitting Diodes with Fluorescent Emitters. *Nat. Commun.* **2014**, *5*, 4016. [CrossRef]
42. Chan, C.Y.; Tanaka, M.; Lee, Y.T.; Wong, Y.W.; Nakanotani, H.; Hatakeyama, T.; Adachi, C. Stable Pure-Blue Hyperfluorescence Organic Light-Emitting Diodes with High-Efficiency and Narrow Emission. *Nat. Photonics* **2021**, *15*, 203–207. [CrossRef]
43. Forrest, S.R. The Path to Ubiquitous and Low-Cost Organic Electronic Appliances on Plastic. *Nature* **2004**, *428*, 911–918. [CrossRef]
44. Mayer, L.; Kohlbecher, R.; Müller, T.J.J. Concatenating Suzuki Arylation and Buchwald–Hartwig Amination by A Sequentially Pd-Catalyzed One-Pot Process—Consecutive Three-Component Synthesis of C,N-Diarylated Heterocycles. *Chem.—A Eur. J.* **2020**, *26*, 15130–15134. [CrossRef]
45. Li, W.; Liu, D.; Shen, F.; Ma, D.; Wang, Z.; Feng, T.; Xu, Y.; Yang, B.; Ma, Y. A Twisting Donor-Acceptor Molecule with an Intercrossed Excited State for Highly Efficient, Deep-Blue Electroluminescence. *Adv. Funct. Mater.* **2012**, *22*, 2797–2803. [CrossRef]
46. Butkute, R.; Lygaitis, R.; Mimaite, V.; Gudeika, D.; Volyniuk, D.; Sini, G.; Grazulevicius, J.V. Bipolar Highly Solid-State Luminescent Phenanthroimidazole Derivatives as Materials for Blue and White Organic Light Emitting Diodes Exploiting Either Monomer, Exciplex or Electroplex Emission. *Dye. Pigment.* **2017**, *146*, 425–437. [CrossRef]
47. Onsager, L. Electric Moments of Molecules in Liquids. *J. Am. Chem. Soc.* **1936**, *58*, 1486–1493. [CrossRef]
48. Mataga, N.; Kaifu, Y.; Koizumi, M. Solvent Effects upon Fluorescence Spectra and the Dipolemoments of Excited Molecules. *Bull. Chem. Soc. Jpn.* **1956**, *29*, 465–470. [CrossRef]
49. Sumalekshmy, S.; Gopidas, K.R. Photoinduced Intramolecular Charge Transfer in Donor-Acceptor Substituted Tetrahydropyrenes. *J. Phys. Chem. B* **2004**, *108*, 3705–3712. [CrossRef]
50. Lippert, E. Dipolmoment Und Elektronenstruktur von Angeregten Molekülen. *Z. Naturforsch.—Sect. A J. Phys. Sci.* **1955**, *10*, 541–545. [CrossRef]
51. Copp, S.M.; Faris, A.; Swasey, S.M.; Gwinn, E.G. Heterogeneous Solvatochromism of Fluorescent DNA-Stabilized Silver Clusters Precludes Use of Simple Onsager-Based Stokes Shift Models. *J. Phys. Chem. Lett.* **2016**, *7*, 698–703. [CrossRef]
52. Yu, Y.; Zhao, R.; Zhou, C.; Sun, X.; Wang, S.; Gao, Y.; Li, W.; Lu, P.; Yang, B.; Zhang, C. Highly Efficient Luminescent Benzoylimino Derivative and Fluorescent Probe from a Photochemical Reaction of Imidazole as an Oxygen Sensor. *Chem. Commun.* **2019**, *55*, 977–980. [CrossRef]
53. Liu, J.; Chen, J.; Dong, Y.; Yu, Y.; Zhang, S.; Wang, J.; Song, Q.; Li, W.; Zhang, C. The Origin of the Unusual Red-Shifted Aggregation-State Emission of Triphenylamine-Imidazole Molecules: Excimers or a Photochemical Reaction? *Mater. Chem. Front.* **2020**, *4*, 1411–1420. [CrossRef]
54. Mailhot, B.; Morlat-Thérias, S.; Bussière, P.O.; Le Pluart, L.; Duchet, J.; Sautereau, H.; Gérard, J.F.; Gardette, J.L. Photoageing Behaviour of Epoxy Nanocomposites: Comparison between Spherical and Lamellar Nanofillers. *Polym. Degrad. Stab.* **2008**, *93*, 1786–1792. [CrossRef]
55. Tan, J.; Kuang, Y.; Wang, Y.; Huang, Q.; Zhu, J.; Wang, Y. Axial Tri-Tert-Butylphosphane Coordination to Rh2(OAc)4: Synthesis, Structure, and Catalytic Studies. *Organometallics* **2016**, *35*, 3139–3147. [CrossRef]
56. Ko, S.B.; Kang, S.; Kim, T. A Silane-Based Bipolar Host with High Triplet Energy for High Efficiency Deep-Blue Phosphorescent OLEDs with Improved Device Lifetime. *Chem.—A Eur. J.* **2020**, *26*, 7767–7773. [CrossRef] [PubMed]
57. Kuo, H.H.; Zhu, Z.L.; Lee, C.S.; Chen, Y.K.; Liu, S.H.; Chou, P.T.; Jen, A.K.Y.; Chi, Y. Bis-Tridentate Iridium(III) Phosphors with Very High Photostability and Fabrication of Blue-Emitting OLEDs. *Adv. Sci.* **2018**, *5*, 1800846. [CrossRef] [PubMed]

58. Miyamoto, E.; Yamaguchi, Y.; Yokoyama, M. Ionization Potential of Organic Pigment Film by Atmospheric Photoelectron Emission Analysis. *Electrophotography* **1989**, *28*, 364–370. [CrossRef]
59. Okamoto, S.; Tanaka, K.; Izumi, Y.; Adachi, H.; Yamaji, T.; Suzuki, T. Simple measurement of quantum efficiency in organic electroluminescent devices. *Jpn. J. Appl. Phy. Part 2 Lett.* **2001**, *40*, 783–784. [CrossRef]

MDPI AG
Grosspeteranlage 5
4052 Basel
Switzerland
Tel.: +41 61 683 77 34

Materials Editorial Office
E-mail: materials@mdpi.com
www.mdpi.com/journal/materials

Disclaimer/Publisher's Note: The title and front matter of this reprint are at the discretion of the . The publisher is not responsible for their content or any associated concerns. The statements, opinions and data contained in all individual articles are solely those of the individual Editor and contributors and not of MDPI. MDPI disclaims responsibility for any injury to people or property resulting from any ideas, methods, instructions or products referred to in the content.